Pharmacogenomics of Alcohol and Drugs of Abuse

Pharmacogenomics of Alcohol and Drugs of Abuse

Amitava Dasgupta

Loralie J. Langman

CRC Press
Taylor & Francis Group
Boca Raton London New York

CRC Press is an imprint of the
Taylor & Francis Group, an **informa** business

CRC Press
Taylor & Francis Group
6000 Broken Sound Parkway NW, Suite 300
Boca Raton, FL 33487-2742

First issued in paperback 2019

© 2012 by Taylor & Francis Group, LLC
CRC Press is an imprint of Taylor & Francis Group, an Informa business

No claim to original U.S. Government works

ISBN-13: 978-1-4398-5611-6 (hbk)
ISBN-13: 978-0-367-38149-3 (pbk)

Visit the Taylor & Francis Web site at
http://www.taylorandfrancis.com

and the CRC Press Web site at
http://www.crcpress.com

Contents

Preface..vii
The Editors...ix
Contributors xi

Chapter 1 Pharmacogenomics Principles of Alcohol and DOA............................ 1
            Christine L.H. Snozek and Loralie J. Langman

Chapter 2 Alcohol: Use and Abuse... 11
            Amitava Dasgupta

Chapter 3 Slate and Trait Markers of Alcohol Abuse... 47
            Joshua Bornhorst, Annjanette Stone, John Nelson, and Kim Light

Chapter 4 Introduction to Drugs of Abuse... 93
            Larry Broussard and Catherine Hammett-Stabler

Chapter 5 Pharmacogenomics of Amphetamine and Related Drugs 129
            Steven C. Kazmierczak

Chapter 6 Pharmacogenomics of Cocaine.. 137
            Loralie J. Langman and Christine L.H. Snozek

Chapter 7 Genetic Aspect of Marijuana Metabolism and Abuse 155
            Pradip Datta

Chapter 8 Genetic Aspect of Opiate Metabolism and Addiction 165
            Jorge L. Sepulveda

Chapter 9 Pharmacogenomics Aspects of Addiction Treatment 195
            F. Gerard Moeller

Chapter 10 Methodologies in Pharmacogenetics Testing................................... 211
            Jorge L. Sepulveda

Index.. 229

Preface

For centuries the typical practice of medicine has been and continues to be one where the clinician is concerned with an individual patient's symptoms, medical and family history, and, more recently, data from the laboratory, and, if applicable, imaging information for diagnosis and treatment. This, however, is a reactive approach rather than a preventive approach. After complete characterization of the human genome, scientists hoped to translate these findings for better treatment of diseases and possible cures for genetically inherited diseases. In the 1970s, expansion of therapeutic drug monitoring (TDM) services for drugs with narrow therapeutic indices certainly improved patient management by reducing incidences of drug toxicity as well as achieving better drug response by personalizing the dosage of a drug for a particular person. Pharmacogenomics can supplement traditional therapeutic drug monitoring, potentially predicting correct dosage before initiation of drug therapy. This may also be superior because the approach is proactive rather than reactive. A good example is warfarin, where dosage based on polymorphism of CYP2C9 and VCKOR1 has been stated in the package insert.

The same pharmacogenomics principles that are applicable for therapeutic drugs are also applicable for drugs of abuse. Metabolisms of certain narcotic analgesics such as codeine, hydrocodone, and oxymorphone are influenced by polymorphism of CYP2D6. Polymorphism of CYP2D6 also affects metabolism of methamphetamine, a drug used for treating attention deficit disorder and one that is frequently abused. Polymorphism of opiate receptor genes greatly influences response to opioids. Currently, abuse of drugs is considered a psychiatric illness, and both genes and environment play an important role in determining which individuals are going to be more prone to addiction. The majority of clinical trials examining the role of OPRM1 Asp 40 alleles in naltrexone treatment response for alcoholism find that patients lacking Asp40 allele do more poorly with the treatment.

There are excellent books on pharmacogenomics aspects of therapeutics, but current research has also looked at the role of pharmacogenomics in alcohol and drugs of abuse. The goal of this book is to provide a platform for readers to become familiar with the current state of knowledge in this area. Each chapter is written by experts in their field, covering all aspects of pharmacogenomics in alcohol and drugs of abuse. Chapter 1 covers the basic aspects of pharmacogenomics applicable to alcohol and drugs of abuse. In Chapter 3, both slate and trait markers of alcohol are addressed so that readers not only become familiarized with this timely topic, but also consider setting appropriate alcohol biomarker tests in their clinical laboratory. In Chapter 10, techniques used in pharmacogenomics testing are discussed in detail so that readers may consider setting up pharmacogenomics testing in their laboratory if desired. Each chapter contains a brief review of the drugs discussed; therefore, readers do not need a background of pharmacogenomics to follow the information presented.

We hope you will enjoy this book.

The Editors

Amitava Dasgupta received his Ph.D. in 1986 in chemistry from Stanford University and completed his postdoctoral training in clinical chemistry from the Department of Laboratory Medicine at the University of Washington at Seattle. He is board certified in both toxicology and clinical chemistry by the American Board of Clinical Chemistry and has been a member of AACC since 1987. Currently, he is a tenured full professor of pathology and laboratory medicine at The University of Texas Health Science Center at Houston and the director of the Clinical Chemistry and Toxicology Laboratory of Memorial-Hermann Laboratory Services, the teaching hospital of The University of Texas Medical School at Houston. Dasgupta has published 190 scientific papers, written many invited review articles and abstracts as well as five books, and edited another seven books. He is on the editorial boards of five major journals including the *American Journal of Clinical Pathology, Archives of Pathology and Laboratory Medicine, Therapeutic Drug Monitoring, Clinica Chimica Acta*, and the *Journal of Clinical Laboratory Analysis*. He is the recipient of the 2009 Irving Sunshine Award from the International Association for Therapeutic Drug Monitoring and Clinical Toxicology (IATDMCT) for outstanding contributions in clinical toxicology. In 2010 he received the AACC (American Association for Clinical Chemistry) Outstanding Contributions in Education award.

Loralie J. Langman completed her Ph.D. in laboratory medicine and pathology at the University of Alberta. She completed her clinical chemistry fellowship at the University of Toronto, specializing in forensic toxicology and molecular genetics. She is certified with the Canadian Academy of Clinical Biochemistry (CACB) and is the first individual to have achieved Diplomate status with the American Board of Clinical Chemistry (ABCC) in all three disciplines (clinical chemistry, molecular diagnostics, and toxicological chemistry). She is also a diplomate with the American Board of Forensic Toxicology.

Langman is a member of and serves on committees for several professional organizations including the Society of Forensic Toxicologists (SOFT), the American Academy of Forensic Sciences (AAFS), the National Academy of Clinical Biochemistry (NACB), the Clinical and Laboratory Standards Institute (CLSI), the Canadian Society of Clinical Chemists (CSCC), the International Association of Forensic Toxicologists (TIAFT), the American Association for Clinical Chemistry (AACC), the International Federation of Clinical Chemistry and Laboratory Medicine (IFCC), and the International Association of Therapeutic Drug Monitoring and Clinical Toxicology (IATDMCT). She is currently director, Toxicology and Drug Monitoring Laboratory, Mayo Clinic Rochester, Minnesota, and associate professor of laboratory med/pathology, Mayo Clinic, College of Medicine. She has authored over 40 peer-reviewed publications and 7 book chapters, and has presented more than 100 abstracts/presentations at national and international meetings.

Contributors

Joshua Bornhorst
Department of Pathology
University of Arkansas for Medical
  Sciences
Little Rock, Arkansas

Larry Broussard
Department of Clinical Laboratory
  Sciences
Louisiana State University Health
  Sciences Center
New Orleans, Louisiana

Amitava Dasgupta
Department of Pathology and
  Laboratory Medicine
The University of Texas Medical School
  at Houston
Houston, Texas

Pradip Datta
Siemens Diagnostics
Tarrytown, New York

Catherine Hammett-Stabler
Department of Pathology and
  Laboratory Medicine
University of North Carolina at
  Chapel Hill
Chapel Hill, North Carolina

Steven C. Kazmierczak
Department of Pathology
Oregon Health & Science University
Portland, Oregon

Loralie J. Langman
Department of Laboratory Medicine
  and Pathology
Mayo Clinic
Rochester, Minnesota

Kim Light
Department of Pharmaceutical Sciences
University of Arkansas for Medical
  Sciences
Little Rock, Arkansas

F. Gerard Moeller
Department of Psychiatry
The University of Texas Medical School
  at Houston
Houston, Texas

John Nelson
University of Arkansas for Medical
  Sciences
Little Rock, Arkansas

Jorge L. Sepulveda
Pathology and Laboratory Medicine
  Service
Philadelphia VA Medical Center
and
University of Pennsylvania
Philadelphia, Pennsylvania

Christine L.H. Snozek
Department of Laboratory Medicine
  and Pathology
Mayo Clinic
Rochester, Minnesota

Annjanette Stone
Pharmacogenomics Analysis
  Laboratory
University of Arkansas for Medical
  Sciences
Little Rock, Arkansas

1 Pharmacogenomics Principles of Alcohol and DOA

Christine L.H. Snozek and Loralie J. Langman

CONTENTS

1.1 Introduction 1
1.2 Pharmacokinetics 3
1.3 Pharmacodynamics.. 6
1.4 The Genetic Gap.. 7
1.5 Conclusions.. 8
References... 9

1.1 INTRODUCTION

A great deal of interest and research has been dedicated toward pharmacogenomics—that is, the study of genetic variation as it pertains to the use of drugs. Most commonly, the term is used to refer to heritable variability in the processes of pharmacokinetics (PK)—what the body does to the drug—and pharmacodynamics (PD)—what the drug does to the body. However, particularly in the context of drugs of abuse (DOA), the field can be expanded to include concepts that go beyond pharmacology into the realms of psychology and neurology [1]. Pharmacogenetic studies have addressed the importance of genetic variation in areas such as individual susceptibility to addiction, preference for a given DOA, or ability to cease drug use.

Substance dependence is in many ways a more complicated area than clinical pharmacological therapy (Box 1.1), despite the fact that there are relatively few drugs that are commonly abused. Development of addiction requires initial exposure to a DOA (frequently illicitly obtained) followed by continued use of the agent, often despite detrimental effects on the user's financial, emotional, societal, and psychological situation [1,2]. Environmental factors are therefore inextricably intertwined with pharmacogenetics—individuals who are never exposed to a DOA cannot become dependent upon it, regardless of the strength of their genetic predisposition toward addiction.

In addition to the difficulty presented by the complex gene–environment relationship, the typical concerns related to genetic studies must be considered, including differences in experimental strategies, study populations, and gene variants. The prevalence of any given variant can be vastly different between ethnic groups; there are numerous instances of alleles that are rare in one population being quite common

---

BOX 1.1   CHALLENGES OF DRUGS OF ABUSE (DOA) PHARMACOGENETICS

Genetic gap: Although most estimates of the heritability of addiction are high (>50%), there are very few confirmed associations between genetic variants and aspects of substance abuse such as response to a drug or risk of addiction. Genome-wide association studies (GWAS) in particular have been criticized for their failure to identify the "missing" polymorphisms responsible for susceptibility or resistance to drug use and abuse.

Environmental contribution: Substance abuse requires that certain environmental conditions be met, most notably that the individual has been exposed to the drug of interest. This has significant implications for selection of study control groups. Currently, few studies account for additional environmental factors such as traumatic life events or mental health issues as contributors to drug use.

Genetics of behavior: Addiction-related genes may affect phenomena that are not directly related to drug pharmacokinetics or pharmacodynamics, but rather control behaviors such as risk seeking and impulsivity. The resulting behavior patterns may encompass a wide range of phenotypes that include, but are not limited to, drug use.

Control groups: Depending on the substance and endpoints being studied, it can be extremely difficult to define and acquire a good control population. Never-users are essentially an unscreened group, as it is uncertain how a naïve individual would respond if exposed to a drug. Former or occasional users might be more appropriate as controls, but, especially for highly addictive drugs, can be difficult to recruit in sufficient numbers for high-power studies.

Affected individuals: The definition of an affected study participant varies greatly in the literature. There are several tools for assessing whether an individual is addicted to a drug, and if so, the severity of the addiction. It is quite common for these tools to show very poor correlation; some may measure the degree of physical dependence while others incorporate the individual's perception of his or her drug use and desire for change.

---

in a different group (3,4). Although substance abuse is a worldwide problem, most pharmacogenetic studies have focused on genetic variants in a limited number of ethnicities and have not addressed the applicability of their findings to other ethnic groups.

Approximately half of the risk for addiction is thought to be due to genetic variation, yet few genes have been solidly associated with substance abuse (5,6). It appears increasingly likely that the heritable component of substance abuse is composed of small contributions from multiple genes, rather than large effects from fewer genes. Candidate gene studies, that is, where specific genes are chosen for examination based on known mechanisms, have provided the majority of pharmacogenetic information to date. The advantage of this strategy is that logical targets (e.g., gene products known to be involved in drug metabolism or response) can be examined closely for association with substance abuse. However, a major drawback to such studies is that they may miss potential influences from genes that lack known mechanistic links to the drug(s) of interest (5).

In contrast to these targeted studies, genome-wide association studies (GWAS) are able to interrogate a large number of polymorphisms on all chromosomes, thus permitting detection of important variants in genes without known associations to DOAs (1,5,7). One major caveat to GWAS, though, is that most arrays include only single nucleotide polymorphisms (SNPs) with a relatively high prevalence (typically >5%), leading to underrepresentation of rare or ethnic-specific variants. Although some promising genetic data have come from GWAS reports (8), thus far this technology has not contributed as greatly to the study of addiction as was anticipated. Various strategies to improve this have been proposed: for example, small sample size is a frequent issue given the large number of polymorphisms being interrogated simultaneously. It may also be that GWAS arrays composed of lower-prevalence variants could improve the ability of these studies to detect smaller contributions from individual genes.

The chapters of this book will detail the current state of pharmacogenetic knowledge relevant to specific commonly abused drugs. Although some genes only affect addiction to a particular DOA, many variants have been associated with multiple drugs or with addiction as a general phenomenon. This is not surprising, given the degree of overlap between the PK and PD of various DOAs. The remainder of this chapter provides background into some of the major PK and PD pathways relevant to substance abuse, and into the debate surrounding the "genetic gap" between heritability predictions and identified associations.

1.2 PHARMACOKINETICS

The majority of pharmacogenetic studies have focused on genes involved in PK—that is, the processes of absorption, distribution, metabolism, and elimination. Together, these steps describe what the body does to a drug, from initial entry of the compound until it is removed in waste. Each process in PK is subject to both environmental and genetic variability, which creates a great deal of complexity for clinicians attempting to address individual differences in drug response or treatment. Not surprisingly, many clinicians feel unprepared to incorporate genetic information into their practices (9).

The enormous amount of information available regarding drug–drug interactions, genetic alterations, and other sources of PK variability have only recently begun to be summarized into readily accessible formats, such as online tools for predicting interactions between therapeutic drugs or for determining genetic influences on initial drug dosing. However, whereas these resources are now emerging for therapeutic agents, there are notable problems related to DOA. One issue of particular concern is that these tools, and the clinicians who employ them, often neglect the influences of DOAs. Just as certain therapeutic agents can alter PK parameters of other drugs,

TABLE 1.1
Genes of Pharmacokinetic Relevance to Drugs of Abuse (DOA) Pharmacogenetics

Gene	Activity	Relevant Drugs of Abuse	Key Variants
ADH	Metabolism	Alcohol	ADH1B*2, ADH1C*1, ADH4 C-136A
ALDH	Metabolism	Alcohol	ALDH2*2
BCHE	Metabolism	Cocaine	A, F, K, J, and S forms
CYP2A6	Metabolism	Nicotine	Several
CYP2B6	Metabolism	Methadone, nicotine	CYP2B6*5, CYP2B6*6
CYP2D6	Metabolism	Opioids, nicotine	Several
ABCB1	Transport	Opioids	ABCB1 C3435T

there are DOAs with similar capabilities; for example, cocaine is a potent inhibitor of the metabolic enzyme cytochrome P450 (CYP) 2D6 (10). Yet online tools and scientific publications rarely explore this clinically relevant area of abused drugs.

Within PK, the greatest pharmacogenetic knowledge exists regarding polymorphisms in metabolic enzymes and drug transporters (Table 1.1). The most commonly studied metabolic enzymes are the *CYP*s, but clinically important genetic variations have also been characterized in uridine diphosphate glucuronyl transferases (UGTs), esterases, and other enzyme families (11). By comparison, drug transporters are less well characterized, but there is great interest in the role of P-glycoprotein (P-gp, encoded by *ABCB1/MDR1*) and related proteins, which mediate drug entry into and exit from target sites.

CYP2D6 is one of the most polymorphic genes of pharmacogenetic relevance, with over 80 different alleles described to date (www.cypalleles.ki.se). The enzyme is involved in metabolism of roughly one-fourth of drugs, including nicotine and certain opioids; various genotypes result in an extremely wide range of phenotypes ranging from null alleles to expression many-fold higher than average (3,6,12).

The CYP3A4/5 enzyme family also metabolizes a wide array of therapeutic and abused drugs. There are fewer relevant polymorphisms in these genes, although a substantial fraction of the population (up to 90% in Caucasians) carries the *CYP3A5*3* null allele and therefore does not express CYP3A5 (4). However, environmental regulation of this system appears to trump genetic variation, and a wide array of enzyme inducers and inhibitors have been identified for CYP3A4/5 (13).

Other CYP enzymes are important in the metabolism of individual DOAs, including CYP2A6, which metabolizes nicotine, and CYP2B6, which metabolizes methadone (6, 13). The chemical structure of methadone is given in Figure 1.2. Most *CYP* genes have polymorphisms described in the literature, although the clinical significance varies greatly depending on the prevalence of alleles with altered function, as well as the contribution of that particular enzyme to the drug in question. An example of this is *CYP2B6* polymorphism in relation to methadone metabolism: mutant alleles are relatively common and typically decrease enzyme activity (3). However, although CYP2B6 is generally the dominant enzyme in methadone metabolism, there are alternate biotransformation pathways that may mask any effect of *CYP2B6* reduced-function alleles (14).

Metabolic enzymes and drug transporters affect virtually all aspects of PK, from absorption to elimination. Most DOAs do not require active transport for oral absorption; however, several are subject to first-pass metabolism, thus reducing their oral bioavailability. Morphine, for example, is only 20% to 25% bioavailable after an oral dose, due in large part to extensive first-pass metabolism converting the drug to inactive metabolites (15). The desire to avoid first-pass metabolism and the delay inherent in intestinal absorption are major factors behind the rationale for drug abusers finding alternate routes of administration for oral formulations (e.g., crushing pills and insufflating the powder, or inserting the drug into the rectum).

Distribution is often heavily dependent on drug transporters, particularly as concerns the blood-brain barrier (BBB). P-gp and related proteins serve to enforce the distribution barrier between the central nervous system (CNS) and the systemic circulation, thus altered function of these drug transporters can easily result in unexpected consequences due to atypical distribution. An excellent example of this is seen with loperamide, an opioid antidiarrheal agent that normally does not efficiently pass the BBB, and therefore does not have the CNS effects (e.g., respiratory depression) associated with opioids. Inhibition of P-gp function during administration of loperamide can induce respiratory depression due to the inability to exclude the drug from the CNS (16). As with *CYP3A4*, environmental effects (17) on *ABCB1* transcription and enzyme activity typically predominate over the influence of polymorphisms, yet some studies have associated atypical drug reactions with variant alleles of this gene (7).

Although there are examples of pharmacogenetic influences on DOAs and addiction in all areas of PK, by far the greatest knowledge pertains to the implications of genetic variants in drug metabolism. Two of the earliest and best-characterized examples of metabolic enzymes affecting drug abuse are the genes encoding alcohol dehydrogenase (*ADH*) and aldehyde dehydrogenase (*ALDH*), both essential in the conversion of ethanol to acetic acid (8,18). ADH converts ethanol to acetaldehyde, a toxic compound that is thought to be responsible for facial flushing and other negative effects associated with ethanol consumption. For this reason, variants resulting in accumulation of acetaldehyde (e.g., increased ADH activity or decreased ALDH activity) tend to be protective against alcoholism.

There are a number of other target enzymes, both confirmed and theoretical, whose genetic variability could play a role in addiction to various DOAs. For example, butyrylcholinesterase (also called pseudocholinesterase) is present in the metabolic cascades of both cocaine (to benzoylecgonine) and heroin (to 6-monoacetylmorphine) (19). Partial or complete deficiency of this enzyme could prolong the CNS effects of both these drugs, potentially affecting an individual's risk of addiction or negative response. Although just a few genes encoding metabolic enzymes have been conclusively associated with substance abuse, given the rich complexity of drug metabolism it is likely that current studies have only begun to address the contributions of this area to DOA pharmacogenetics.

1.3 PHARMACODYNAMICS

The flip side to PK is PD—the effects a drug has on the body. This includes not only the receptor responsible for binding a given drug, but also the downstream signaling, regulatory processes, gene transcription, and interacting signaling pathways. Most DOAs affect one or more of the neural pathways involved in mediating reward, particularly those relating to monoamine neurotransmitter and opioid receptor signaling (20). Select genes involved in the PD of abused drugs are shown in Table 1.2. PD responses to different DOAs display a great deal of overlap and crosstalk, as will be evident from the detailed discussions in later chapters of this book.

Some of the most important neurological pathways implicated in substance abuse are those regulated by monoamine neurotransmitters such as dopamine and serotonin (18,20). Many DOAs affect the release or regulation of monoamines at the synapse; most commonly abused stimulants exert a great deal of their pharmacological activity through this mechanism. Dopamine in particular has been associated strongly with the reinforcing properties of cocaine, amphetamines, and other stimulants, with lesser (though still significant) roles ascribed to other monoamines (20). Monoaminergic signaling is thus essential in the development of addiction to cocaine and other stimulants that directly affect one or more components of the system, but intriguingly, monoamines also play an indirect role in addiction to non-stimulant DOAs including alcohol (5).

Regulation of monoaminergic signaling is quite complex and can be altered at a variety of levels by genetic or environmental influences. The presynaptic neuron synthesizes and stores monoamine neurotransmitters until stimulated to release them into the synaptic cleft; from there they may be reclaimed by reuptake transporters on the presynaptic neuron, or bound by receptors on the postsynaptic neuron (21). Studies have tentatively linked aspects of substance abuse to gene variants in all stages from the beginning to the end of this process. Examples include enzymes involved in monoamine synthesis (tryptophan hydroxylase, *TPH1/2*) and catalysis (dopamine hydroxylase, *DβH*, and catechol-*o*-methyl transferase, *COMT*); monoamine transporters (dopamine and serotonin transporters, *SLC6A3* and *SLC6A4*, respectively); and receptors (dopamine receptors, *DRD2*, *DRD3*, and *DRD4*), among others (18,21). In general, variants leading to increased concentration or duration of

TABLE 1.2
Genes of Pharmacodynamic Relevance to Drugs of Abuse (DOA) Pharmacogenetics

Gene Product	Genes	Relevant Drugs of Abuse
Dopamine receptors	*DRD1–DRD5*	Stimulants, nicotine, opioids
Monoamine transporters	*SLC6A3, SLC6A4*	Cocaine, nicotine, opioids
Monoamine metabolism	*DBH, MAO, COMT*	Stimulants, nicotine, opioids
Opioid receptors	*OPRM1, OPRK1, OPRD1*	Opioids, alcohol, cocaine
GABA receptors	*GABR A2* and *G2, GABBR2*	Alcohol, nicotine, stimulants
Nicotinic acetylcholine receptor	*CHRN A3-A5, B4*	Nicotine, cocaine

monoamine neurotransmitters at the synapse will tend to reinforce the rewarding effects of a drug stimulating that pathway.

Another important neurological network in the pharmacogenetics of DOAs is that related to opioid receptors. The mu (μ), delta (δ), and kappa (κ) opioid receptors interact with endogenous opioid peptides (e.g., endorphins and enkephalins) as well as exogenous opioid drugs, to regulate pain sensation, stress response, respiration, gastrointestinal motility, and numerous other activities (20,22). Although the opioidergic system has a variety of physiological effects, the most important considerations for substance abuse are its role in nociception (i.e., the perception of pain) and crosstalk with monoaminergic reward pathways. Much of the analgesic capacity of endogenous and exogenous opioids is thought to be mediated by the mu receptor, encoded by *OPRM1*. The delta and kappa receptors (encoded by *OPRD1* and *OPRK1*, respectively) also function to regulate nociception, although their contributions are less well understood as compared to the mu receptor (22).

All opioid receptors are of interest as genetic influences in the development of addiction. In addition to obvious potential implications in abuse of heroin and prescription opioids, the opioidergic system has also been associated with dependence on alcohol and cocaine, neither of which is thought to directly interact with opioid receptors (5,21). The kappa opioid receptor in particular is capable of modulating the effects of dopamine surges, such as those that occur during repeated administration of cocaine (20). Further support for an integral role of opioid receptor signaling in nonopioid addiction is the utility of naloxone and other opioid receptor antagonists in the treatment of alcoholism (18,20).

A common missense mutation in *OPRM1*, A118G, results in altered binding of the receptor to opioid ligands and has been associated (although not definitively) with differences in opiate addiction and treatment of alcoholism (18,21). Other polymorphisms in the opioidergic system are less well characterized but may still be important in addiction. Current studies are examining the influence of additional *OPRM1* alleles, as well as variants in *OPRK1*, *OPRD1*, genes encoding endogenous ligands such as dynorphin and enkephalin, and other genes relevant to the opioid system.

Although less well characterized to date, several additional PD targets show promise as being relevant to DOA pharmacogenetics. Examples include the hypothalamic stress-response axis, cannabinoid receptors, nicotinic acetylcholine receptors, and gamma-aminobutyric acid (GABA)-responsive signaling (5,6,8,20). Understanding of the genes related to these pathways and their specific influences in substance abuse lags behind the depth of knowledge of monoaminergic and opioidergic pathways, but what is known will be discussed in later chapters of this book.

1.4   THE GENETIC GAP

Despite fairly high estimates (often upwards of 50%) of the degree to which addiction is a product of genetics (5), very few concrete associations have been made linking substance abuse with heritable variation. The reasons for this apparent gap are a subject of much discussion in the literature (2,5,23), including suggestions that the previous estimates of heritability are overinflated. Most of the proposed explanations for the genetic gap relate to the design of studies to date (e.g., the technique used, the number of individuals included, or the control population chosen). Another likely suggestion is that, rather than just a few genes having large effects on determining DOA addiction, there may be smaller contributions from a larger number of genes, making the influence of any given variant much more difficult to detect (8). This has been a major criticism of GWAS, because most chips for these studies include only high-prevalence alleles. A third possibility is that the "missing" heritability of addiction does not reside wholly within the DNA code: for example, microRNAs have become a hot area of study for understanding gene regulation, and likely play a role in substance abuse as well (5). There has so far been comparatively little examination of the contribution of epigenetics, which encompasses heritable variation stemming from sources other than differences in DNA sequence (e.g., DNA methylation and histone modification) (7,23).

Other reasons for the genetic gap might include the array of behavioral phenotypes that may be related to a given genotype. For example, a genetic predisposition toward risk-taking behavior may manifest as experimentation with addictive substances, or as non-drug-related activities such as skydiving or extreme sports (2). The former would provide the opportunity to develop into drug abuse, yet the genetic variant(s) involved may not be directly related to DOAs per se. Likewise, even genes that play a direct role in the addictive process may be expressed through different phenotypes: For example, in ultra-marathoners, the compulsion to exercise and the physical damage caused by running 50+ miles certainly seem to parallel the cravings and detrimental effects felt by substance abusers, but the former "addiction" is a far healthier and more socially acceptable manifestation than the latter.

Given these complexities, study design is an extremely important consideration in identifying genetic associations with DOAs. Specifically, selection of an appropriate control population can be quite difficult compared to other clinical studies. Controls that are "never-users" of a given DOA may actually have relevant genetic variants that predispose toward addiction, but the individuals simply have not been exposed to the drug. Ideally, the control group should consist of persons who have used the substance being studied but did not develop addiction to it. For alcohol, nicotine, prescription opioids, and marijuana, such individuals are in relatively high abundance; in contrast, for many other DOAs this type of control is substantially more difficult to find given the issues of social unacceptability, illegality, and higher addictive potential. Regardless, closing the genetic gap can only occur through careful selection of study populations and controlled design of experimental strategies.

1.5 CONCLUSIONS

As will be discussed in detail throughout the remaining chapters of this book, many of the same pharmacogenetic targets relevant to therapeutic drugs are also pertinent to DOAs and addiction. Current work is expanding the knowledge base into genes involved in PD and the addictive process, as well as deepening the comparatively richer body of literature surrounding PK-related genes. Careful experimental design that incorporates the lessons learned from previous studies may make progress toward closing the genetic gap between heritability estimates and specific variants linked definitively to substance abuse.

There remain notable challenges facing the application of pharmacogenetics to clinical treatment and prevention of drug abuse. As is also the case for the pharmacogenetics of therapeutic drugs, few variants relevant to DOAs and addiction show consistent association with a particular outcome in multiple studies. The most promising polymorphism for clinical implementation in the near future is likely the *OPRM1* A118G allele, as it relates to naloxone therapy for alcoholism. Although genotyping to optimize substance abuse treatment is a very promising application of pharmacogenetics, ideally the information could be used to target more intense prevention strategies to those at greater risk of addiction, rather than waiting until dependence develops. However, there are some serious concerns regarding such an approach, including the question of whether third-party payers would reimburse for this application of molecular testing, and the risk of stigmatizing individuals as potential addicts based solely on genetic predisposition. Yet despite the challenges facing clinical implementation of DOA pharmacogenetics, the potential benefits justify further study beyond a shadow of a doubt.

REFERENCES

1. Wong, C.C., and Schumann, G. (2008) Review. Genetics of addictions: strategies for addressing heterogeneity and polygenicity of substance use disorders, *Philos Trans R Soc Lond B Biol Sci 363*, 3213–3222.
2. Buckland, P.R. (2008) Will we ever find the genes for addiction?, *Addiction 103*, 1768–1776.
3. Johansson, I., and Ingelman-Sundberg, M. (2011) Genetic polymorphism and toxicology—with emphasis on cytochrome p450, *Toxicol Sci 120*, 1–13.
4. Lamba, J.K., Lin, Y.S., Schuetz, E.G., and Thummel, K.E. (2002) Genetic contribution to variable human CYP3A-mediated metabolism, *Adv Drug Deliv Rev 54*, 1271–1294.
5. Ho, M.K., Goldman, D., Heinz, A., Kaprio, J., Kreek, M.J., Li, M.D., Munafo, M.R., and Tyndale, R.F. (2010) Breaking barriers in the genomics and pharmacogenetics of drug addiction, *Clin Pharmacol Ther 88*, 779–791.
6. Rutter, J.L. (2006) Symbiotic relationship of pharmacogenetics and drugs of abuse, *AAPS J 8*, E174–E184.
7. Yuferov, V., Levran, O., Proudnikov, D., Nielsen, D.A., and Kreek, M.J. (2010) Search for genetic markers and functional variants involved in the development of opiate and cocaine addiction and treatment, *Ann N Y Acad Sci 1187*, 184–207.
8. Bierut, L.J. (2011) Genetic vulnerability and susceptibility to substance dependence, *Neuron 69*, 618–627.
9. Shields, A.E., and Lerman, C. (2008) Anticipating clinical integration of pharmacogenetic treatment strategies for addiction: are primary care physicians ready? *Clin Pharmacol Ther 83*, 635–639.
10. Shen, H., He, M.M., Liu, H., Wrighton, S.A., Wang, L., Guo, B., and Li, C. (2007) Comparative metabolic capabilities and inhibitory profiles of CYP2D6.1, CYP2D6.10, and CYP2D6.17, *Drug Metab Dispos 35*, 1292–1300.
11. Meyer, M.R., and Maurer, H.H. (2011) Absorption, distribution, metabolism and excretion pharmacogenomics of drugs of abuse, *Pharmacogenomics 12*, 215–233.

12. Shimada, T., Yamazaki, H., Mimura, M., Inui, Y., and Guengerich, F.P. (1994) Interindividual variations in human liver cytochrome P-450 enzymes involved in the oxidation of drugs, carcinogens and toxic chemicals: studies with liver microsomes of 30 Japanese and 30 Caucasians, *J Pharmacol Exp Ther 270*, 414–423.

13. Musshoff, F., Stamer, U.M., and Madea, B. (2010) Pharmacogenetics and forensic toxicology, *Forensic Sci Int 203*, 53–62.

14. Somogyi, A.A., Barratt, D.T., and Coller, J.K. (2007) Pharmacogenetics of opioids, *Clin Pharmacol Ther 81*, 429–444.

15. Brunton, L., Lazo, J., and Parker, K. (Eds.) (2006) *Goodman & Gilman's the Pharmacological Basis of Therapeutics*, 11th ed., McGraw-Hill, New York.

16. Linnet, K., and Ejsing, T.B. (2008) A review on the impact of P-glycoprotein on the penetration of drugs into the brain. Focus on psychotropic drugs, *Eur Neuropsychopharmacol 18*, 157–169.

17. Miller, D.S. (2010) Regulation of P-glycoprotein and other ABC drug transporters at the blood-brain barrier, *Trends Pharmacol Sci 31*, 246–254.

18. Khokhar, J.Y., Ferguson, C.S., Zhu, A.Z., and Tyndale, R.F. (2010) Pharmacogenetics of drug dependence: role of gene variations in susceptibility and treatment, *Annu Rev Pharmacol Toxicol 50*, 39–61.

19. Kamendulis, L.M., Brzezinski, M.R., Pindel, E.V., Bosron, W.F., and Dean, R.A. (1996) Metabolism of cocaine and heroin is catalyzed by the same human liver carboxylesterases, *J Pharmacol Exp Ther 279*, 713–717.

20. Kreek, M.J., LaForge, K.S., and Butelman, E. (2002) Pharmacotherapy of addictions, *Nat Rev Drug Discov 1*, 710–726.

21. Kreek, M.J., Bart, G., Lilly, C., LaForge, K.S., and Nielsen, D.A. (2005) Pharmacogenetics and human molecular genetics of opiate and cocaine addictions and their treatments, *Pharmacol Rev 57*, 1–26.

22. Vanderah, T.W. (2010) Delta and kappa opioid receptors as suitable drug targets for pain, *Clin J Pain 26 Suppl 10*, S10–S15.

23. Wong, C.C., Mill, J., and Fernandes, C. (2011) Drugs and addiction: an introduction to epigenetics, *Addiction 106*, 480–489.

2 Alcohol: Use and Abuse

Amitava Dasgupta

CONTENTS

2.1 Introduction 12
2.2 Alcohol: Historical Perspective 13
2.3 Alcohol: Definition of a Standard Drink and Calorie 14
2.4 Alcohol Metabolism 15
2.5 Use and Abuse of Alcohol: U.S. Statistics 17
2.6 Moderate, Heavy, and Binge Drinking 18
2.7 Blood Alcohol Concentrations and Number of Drinks 19
    2.7.1 Widmark Formula 20
    2.7.2 Alcoholic Odor in Breath and Endogenous Alcohol Production 21
2.8 Benefits of Moderate Drinking 22
    2.8.1 Moderate Alcohol Consumption and Reduced Risk of Heart
        Disease 22
    2.8.2 Wine versus Other Alcoholic Beverages on Preventing Heart
        Disease 25
    2.8.3 Consumption of Alcohol and Reduced Risk of Stroke 25
    2.8.4 Moderate Consumption of Alcohol and Type 2 Diabetes 25
    2.8.5 Moderate Alcohol Consumption and Lower Risk of Dementia
        and Alzheimer's Disease 26
    2.8.6 Moderate Alcohol Consumption and Cancer 27
    2.8.7 Moderate Alcohol Consumption and Prolonging Life? 27
    2.8.8 Moderate Alcohol Consumption and Arthritis 28
    2.8.9 Can Moderate Drinking Prevent Common Cold? 28
2.9 Hazards of Heavy Drinking and Alcohol Abuse 29
    2.9.1 Chronic Alcohol Abuse and Reduced Life Span 29
    2.9.2 Alcohol Abuse and Violent Behavior 30
    2.9.3 Alcohol Abuse and Liver Disease 31
    2.9.4 Heavy Alcohol Consumption and Brain Damage 33
    2.9.5 Heavy Alcohol Consumption and Risk of Heart Disease and
        Stroke 36
    2.9.6 Heavy Alcohol Consumption and Immune System 36
    2.9.7 Alcohol Abuse and Damage to Endocrine System Including
        Reproductive System and Bone 37

     2.9.8   Alcohol Abuse Related to Higher Risk of Cancer............................37
     2.9.9   Fetal Alcohol Syndrome...38
2.10 Conclusions...39
References...40

2.1   INTRODUCTION

Alcohol use can be traced back to 10,000 BC, and today drinking alcoholic beverages is common in many civilizations. Moderate alcohol consumption is beneficial to health, and drinking one (or less) drink a day may actually help an individual to prolong his or her life compared to a nondrinker while consumption of five or more drinks per day for a prolonged period results in alcohol dependency and related health problems.

Originally proposed by Professor Robert Dudley of the University of California, Berkeley, the Drunken Monkey Hypothesis speculates that the human attraction to alcohol may have a genetic basis due to the high dependence of primate ancestors of *Homo sapiens* on fruit as a major source of food. For 40 million years, primate diets were rich in fruits. In the humid tropical climate where the early evolution of humans took place, yeasts on the fruit skin and within fruit converted fruit sugars into ethanol. The small alcohol molecules diffused out of the fruit and the alcoholic smell could help a primate to identify a food as ripe and ready to consume. In tropical forests where monkeys lived, competition for ripe fruits was intense and hungry monkeys capable of following the smell of alcohol to identify ripe foods and eating them rapidly survived better than others. Eventually "natural selection" favored monkeys with a keen appreciation for the smell and taste of alcohol. Fossilized teeth show that fruit was a major component of the primate diet between 45 million and 34 million years ago, and some of the closest ancestors of the human species (gorillas, chimpanzees, and orangutans) ate diets based on fruit. Primates are known to have a higher olfactory sensitivity to alcohol than other mammals. By the time humans evolved from apes approximately 1 to 2 million years ago, fruits were mostly replaced by consuming roots, tubers, and meat. Although our ancestors stopped relying mainly on fruit as diet, it is possible that the taste for alcohol arose during our long shared ancestry with primates. For example, unripe palm contains no alcohol, but ripened palm has about 0.6% alcohol content and overripe palm falling on the ground has approximately 4% alcohol. Monkeys usually prefer ripe fruits with approximately 1% alcohol content but avoid overripe fruits with 4% alcohol but lower sugar content. Anecdotally, humans often consume alcohol with food, suggesting that drinking with food is a natural instinct. For millions of years, the amount of alcohol consumed by our ancestors was strictly limited. The situation did not change even 10,000 years ago when humans had knowledge of agriculture and could produce plenty of barley and malt, the raw material for fermentation; thus, the human history of consuming alcohol was initiated. However, yeasts stop producing alcohol when alcohol reaches between 10% and 15% because yeasts start dying at this alcohol concentration. The ancient beers and wines probably contained only 5% alcohol until alcohol distillation was invented in Central Asia around AD 700. Then drinks with higher alcoholic content became available and the history of alcohol abuse by humans began. Alcohol abuse can also be considered a disease of nutritional excess (1).

2.2   ALCOHOL: HISTORICAL PERSPECTIVE

The first historical evidence of alcoholic beverages came from the archeological discovery of Stone Age beer jugs approximately 10,000 years ago. The first palm date wine was most likely brewed in Mesopotamia, while wine was probably consumed by Egyptians approximately 6,000 years ago. Osiris was worshipped as a wine god throughout the nation. The first beer was probably brewed in ancient Egypt and both wine and beer were offered to the gods. Egyptians used alcoholic beverages for pleasure, rituals, as well as both medical and nutritional purposes, although ancient Egyptians were also aware of the harmful effect of alcohol abuse. The earliest evidence of alcohol use in China dates back to 5000 BC when alcohol was mainly produced from rice, honey, and fruits. A Chinese imperial edict of about 1116 BC made it clear that the use of alcohol in moderation was the key and was prescribed from the heavens. In ancient India, alcohol beverages were known as "sura," a favorite drink of Indra, the king of all gods and goddesses. Use of such drinks was known in 3000 to 2000 BC, and Ancient Ayurvedic texts concluded that alcohol is a medicine if consumed in moderation but a poison if consumed in excess. Beer was known to Babylonians as early as 2700 BC. In ancient Greece wine making was common in 1700 BC. Hippocrates identified numerous medicinal properties of wine but was critical of drunkenness (2).

In ancient civilizations, alcohol was used primarily to quench thirst because water was contaminated with bacteria. Hippocrates specifically cited that water from only springs, deep wells, and rainfall were safe for human consumption. In the event that water was contaminated, alcohol was preferred for drinking because alcoholic beverages are free from bacteria and other pathogens due to the antiseptic property of alcohol. Beer was a drink for common people while wine was reserved for elites. Around 30 BC, wine became available to common Romans due to expansion of the vineyards. During ancient times, beer and wine produced from fermentation of cereals, grapes, or fruits had much lower alcohol content than today's beer and wine and were safer for human consumption in larger quantities. In ancient Eastern civilization, drinking alcoholic beverages for thirst quenching was less common than in Western civilization because drinking tea was very popular in Eastern countries. During boiling to prepare tea, all pathogens died, thus making tea drinking very safe (3).

Yeast can produce alcoholic beverages with up to 15% alcohol content. In order to produce higher alcohol content, distillation is needed. Distilled spirits originated in China and India about 800 BC, but the distillation process became common in Europe only during the eleventh century and later. Alcohol consumption was common during the Middle Ages, and monasteries produced alcoholic beverages to nourish their monks as well as to sell to the public for generating revenue. During early American history, colonials showed little concern over drunkenness, and production of alcoholic beverages was a major source of commerce. In 1791, however, a tax was introduced known popularly as the "Whiskey tax" on both privately and publicly brewed distilled whiskey. The Whiskey tax was repealed by President Thomas Jefferson in 1802, but a new alcohol tax was temporally imposed between 1814 and 1817 to pay for the War of 1812. In 1862 President Abraham Lincoln introduced an alcohol tax to pay for Civil War costs. The act also created the office of Internal Revenue. In 1906 the Pure Food and Drug Act was passed regulating the labeling of products containing alcohol, opiates, cocaine, and cannabis (marijuana), among others. The law became effective in January 1907. In 1920 alcohol was prohibited in the United States but congress repealed the law in 1933. In 1978 President Jimmy Carter signed a bill to legalize home brewing of beer for personal use for the first time since Prohibition (4).

2.3 ALCOHOL: DEFINITION OF A STANDARD DRINK AND CALORIE

Alcohol content of various alcoholic beverages varies widely. The average alcohol content of beer is 5%, while the average alcohol content of wine is 10%, and the average alcohol content of whiskey is 40%. However, serving sizes also vary according to the type of beverage. For example, beer usually comes in a 10- or 12-ounce bottle, while a shot of tequila in a mixed drink is only 1.5 ounces. Therefore, regardless of the alcoholic beverage, a standard drink contains roughly the same amount of alcohol. In the United States, a standard drink is defined as a bottle of beer (12 ounces) containing 5% alcohol, 8.5 ounces of malt liquor containing 7% alcohol, a 5-ounce glass of wine containing 12% alcohol, 3.5 ounces of fortified wine like sherry or port containing about 17% alcohol, 2.5 ounces of cordial or liqueur containing 24% alcohol, or one shot of distilled spirits such as gin, rum, vodka, or whiskey (1.5 ounces). Each standard drink contains approximately 0.6 ounce of alcohol. For example, beer contains an average of 5% alcohol, and the total alcohol content of 12 ounces of beer is $12 \times 0.05 = 0.6$ ounce of alcohol. In general, the average bottle of beer contains an average of 0.56 ounce of alcohol, but a standard wine drink may contain 0.66 ounce of alcohol, while distilled spirits may contain up to 0.89 ounce of alcohol (5).

Historically, the alcoholic contents of various drinks were expressed as "proof." The term originated in the eighteenth century when British sailors were paid with rum as well as money. In order to ensure that the rum was not diluted with water, it was "proofed" by dousing gunpowder with it and setting it on fire. If the gunpowder failed to ignite, it indicated that the rum had too much water and was considered "under proof." A sample of rum that was 100 proof contained approximately 57% alcohol by volume. In the United States, proof to alcohol by volume is defined as a ratio of 1:2. Therefore, a beer that has 4% alcohol by volume is defined as 8 proof. In the United Kingdom, alcohol by volume to proof is a ratio of 4:7. Therefore, multiplying alcohol by volume content with a factor of 1.75 would provide the "proof" of the drink.

Currently, in the United States, the alcohol content of a drink is measured by the percentage of alcohol by the volume. The Code of Federal Regulations requires that the label of alcoholic beverages must state the alcohol content by volume. The

TABLE 2.1

Alcohol Content of Various Drinks

Beverage	One Standard Drink	Alcohol Content
Standard American beer	12 ounce (355 mL)	4%–7%
Table wine	5 ounce (148 mL)	7%–14%
Sparkling wine	5 ounce (148 mL)	8%–14%
Fortified wine	2–5 ounce (59–148 mL)	14%–24%
Whiskey	0.6 ounce (18 mL)	40%–75%
Vodka	0.6 ounce (18 mL)	40%–50%
Gin	0.6 ounce (18 mL)	40%–49%
Rum	0.6 ounce (18 mL)	40%–80%
Tequila	0.6 ounce (18 mL)	45%–50%
Brandies	0.6 ounce (18 mL)	40%–44%

regulation permits but does not require the "proof" of the drink to be printed. In the United Kingdom and in European countries, alcohol content of a beverage is expressed also as the percentage of alcohol in the drink. Alcohol content of various popular beverages is given in Table 2.1.

Alcoholic drinks primarily consist of water, alcohol, and variable amounts of sugars and carbohydrates (residual sugar and starch left after fermentation). Sometimes sugars are also added before fermentation to enhance the alcohol content of the beverage. Alcoholic drinks contain negligible amounts of other nutrients such as proteins, vitamins, or minerals. Therefore any calories derived from drinking alcoholic beverages come mostly from alcohol content and very little come from carbohydrate and sugar. However, distilled liquors such as cognac, vodka, whiskey, and rum contain no sugars. Red wine and dry white wines contain 2 to 10 gm of sugar per liter, while sweet wines and port wines may contain up to 120 gm of sugar per liter of wine. Beer and dry sherry contain 30 gm of sugar per liter. Usually a standard drink contains approximately 14 gm of pure alcohol. Burning 1 gm of alcohol produces more calories than burning carbohydrate. Therefore, drinking one can of beer provides approximately 100 calories (6).

2.4 ALCOHOL METABOLISM

Genetic aspects of alcohol metabolism are discussed in detail in Chapter 5. This section provides only a brief overview. Most of the alcohol is absorbed from the stomach and only a small amount is excreted in breath. Age, gender, race, body weight, and amount of food consumed all affect alcohol metabolism. After consumption approximately 20% of alcohol is absorbed by the stomach, and the rest is absorbed from the small intestine. When alcohol is consumed on an empty stomach, blood alcohol levels peak between 15 and 90 minutes after drinking. Food substantially slows the absorption of alcohol and can even reduce the rate of absorption of alcohol for 4 to 6 hours. Sipping alcohol instead of drinking also slows absorption. In one study, 10 healthy men drank a moderate dosage of alcohol (0.8 gm of alcohol per kg of body weight) in the morning after an overnight fast or immediately after breakfast (two cheese sandwiches, one boiled egg, orange juice, and fruit yogurt). Subjects who drank alcohol on an empty stomach felt more intoxicated than the subjects who drank the same amount of alcohol after breakfast. The average peak blood alcohol in subjects who drank on an empty stomach was 104 mg/dL. In contrast, the average peak blood alcohol in subjects who drank alcohol after eating breakfast was 67 mg/dL. The time required to metabolize all alcohol was on average 2 hours shorter in subjects who drank alcohol after eating breakfast compared to subjects who drank on an empty stomach. (7).

A small amount of alcohol is metabolized by the enzyme present in the gastric mucosa, and also a small amount of alcohol is metabolized by the liver before it can enter the main bloodstream. Then the rest of the alcohol enters the systematic circulation. After drinking the same amount of alcohol, a man would have a lesser peak blood alcohol level compared to a woman with the same body weight. This gender difference in the blood alcohol level is related to the different body water content between a male and a female. Women also metabolize alcohol more slowly than men because the concentration of alcohol dehydrogenase is usually lower in women compared to men. Hormonal changes play a role in the metabolism of alcohol in women, although this finding has been disputed in the medical literature. Some publications indicated that women metabolized alcohol at a higher rate during the luteal phase of the menstrual cycle (19 to 22 days of the cycle), but a few days before getting a period a woman's alcohol metabolism may slow down (8).

Alcohol metabolism by the liver usually follows zero-order kinetics, although at very low (<20 mg/dL) or high (>300 mg/dL) concentrations, alcohol elimination may follow first-order kinetics. Several enzyme systems are involved in the metabolism of ethanol (ethyl alcohol or commonly termed *alcohol*), namely alcohol dehydrogenase (ADH), microsomal ethanol oxidizing system (MEOS), and catalase (9). The first and most important of these, alcohol dehydrogenase, is a family found primarily in hepatocytes.

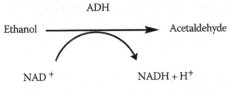

Nicotinamide adenine dinucleotide (NAD)

At least five classes of ADH are found in humans. ADH activity is greatly influenced by the frequency of ethanol consumption. Adults who consume two to three alcoholic beverages per week metabolize ethanol at a rate much lower than alcoholics. For medium-sized adults, the blood ethanol level declines at an average rate of 15 to 20 mg/dL/h or a clearance rate of ~3 oz of ethanol/hour.

The major drug metabolizing family of enzymes found in the liver is the cytochrome P450. This mixed function oxidase plays a minor role in alcohol metabolism, although in alcohol abusers metabolism through CYP2E1 may be significant.

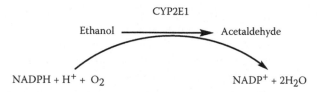

The acetaldehyde produced due to metabolism of alcohol regardless of pathway is subsequently converted to acetate as the result of the action of mitochondrial aldehyde dehydrogenase (ALDH2). Acetaldehyde is fairly toxic compared to ethanol and must be metabolized fast.

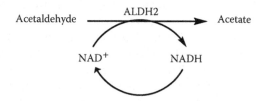

Acetate or acetic acid then enters the citric acid cycle, which is a normal metabolic cycle of living cells, and is converted into carbon dioxide and water.

2.5 USE AND ABUSE OF ALCOHOL: U.S. STATISTICS

Moderate consumption of alcohol has many health benefits, but alcohol abuse is hazardous to the health. According to a survey by the U.S. government of adults age 18 and over, approximately 50% of adults are current drinkers (at least 12 drinks in the past year), 14% are infrequent drinkers (1 to 11 drinks in the past years), 14% were former drinkers, and approximately 25% of adults call themselves lifetime abstainers. Of all drinkers, 70% considered themselves light drinkers, 23% are moderate drinkers, and only 7% identified themselves as heavy drinkers (10). Although the majority of Americans drink sensibly and per capita consumption of alcohol from all alcoholic beverages in 2007 was 2.31 gallons (approximately 24 beers a year per person) according to the National Institute on Alcohol Abuse and Alcoholism, approximately 8% of Americans are alcohol dependent, which is associated with a heavy burden on society. Healthy People 2010 set a national objective of reducing per capita annual alcohol consumption to no more than 1.96 gallons alcohol (11) because the average total societal cost due to alcohol abuse as percent gross domestic product (GDP) in high-income countries including the United States is approximately 1%. This is a high toll for a single factor and an enormous burden on public health (12). According to a 2008 report by Ting-Kai Li, Director of the National Institute on Alcohol Abuse and Alcoholism (NIAAA), alcohol-related problems cost the United States an estimated $185 billion annually with almost half the costs from lost productivities due to alcohol-related disabilities. In the United States over 18 million people age 18 and older suffer from alcohol abuse or dependency, and only 7% of these people receive any form of treatment. In addition, heavy drinkers who are not alcoholics but are at high risk of developing alcohol-related physical or mental damage are seldom identified. The highest prevalence of alcohol dependency in the United States is observed among younger people, ages between 18 and 24. The World Health Organization (WHO) lists alcohol as one of the world's leading causes of disability (13).

According to studies conducted by the Centers for Disease Control and Prevention, alcohol abuse kills approximately 75,000 Americans each year and shortens the life of alcoholics by an average of 30 years. In 2001, 34,833 Americans died from cirrhosis of the liver, a major complication of alcohol abuse, and another 40,933 died from car crashes and other alcohol-related fatalities. Men accounted for 72% of deaths due to alcohol abuse, and 6% of those people were 21 years old or younger (14). In 2005, cirrhosis of the liver was the 12th leading cause of death in the United States claiming 28,175 deaths. Among all cirrhosis-related deaths 45.9% were alcohol related (15). California is the largest alcohol market in the United States. Californians consumed almost 14 billion alcoholic drinks in 2005 which resulted in an estimated 9,439 deaths and 921,029 alcohol-related problems such as crime and injury. The economic burden was estimated to be $38.5 billion, of which $5.4 billion was for medical and mental health spending, $25.3 billion due to loss of work, and another $7.8 billion in criminal justice spending (16). In the United Kingdom, alcohol consumption was responsible for 31,000 deaths in 2005. The UK National Health Service spent an estimated 3 billion pounds in 2005 to 2006 for treating alcohol-related illness and disability. Alcohol consumption was responsible for approximately 10% of disabilities (male, 15%; female, 4%) (17).

2.6 MODERATE, HEAVY, AND BINGE DRINKING

Moderate alcohol consumption is defined by the U.S. government as follows (18):

Men: No more than two standard alcoholic drinks per day
Women: No more than one standard alcoholic drink per day
Adults over 65 (both male and female): No more than one drink per day

Drinking more than recommended can cause serious problems because the health benefits of drinking in moderation disappear fast when consuming more than three to four drinks a day. For all practical purposes, the National Institute on Alcohol Abuse and Alcoholism sets the upper limit of drinking as up to 14 drinks per week for men (no more than four drinks per occasion) and up to 7 seven drinks per week for women (no more than three drinks per occasion). Individuals whose drinking exceeds these guidelines are at increased risk for adverse health effects. Hazardous drinking is defined as the quantity or pattern of alcohol consumption that places individuals at high risk from alcohol-related disorders. Usually hazardous drinking is defined as 21 or more drinks per week by men or more than 7 drinks per occasion at least three times a week. For women, more than 14 drinks per week or drinking more than 5 drinks on one occasion at least three times a week is considered hazardous drinking (19). Alcohol abuse is a leading cause of mortality and morbidity internationally and is ranked by the WHO as one of the top five risk factors for disease burden. Without treatment, approximately 16% of all hazardous or heavy alcohol consumers will progress to becoming alcoholics (20). Heavy consumption of alcohol not only leads to increased domestic violence, decreased productivity, and increased risk of motor vehicle as well as job-related accidents, but also to increased mortality from cirrhosis of the liver, stroke, and cancer. Alcohol overdose may also cause fatality.

Binge drinking is defined as heavy consumption of alcohol within a short period of time with an intention to become intoxicated. Although there is no universally accepted definition for binge drinking, usually consumption of five or more drinks by males and four or more drinks by females is considered "binge drinking," and such a drinking pattern always results in blood alcohol levels above 0.08%, the legal limit for driving. Despite having a legal drinking age of 21 in the United States, binge drinking is very popular among college students. In one study, the authors found that 74.4% of binge drinkers consumed beer exclusively or predominately, and 80.5% of binge drinkers consumed at least some beer. Wine accounted for only 10.9% of binge drinks consumed (21).

2.7 BLOOD ALCOHOL CONCENTRATIONS AND NUMBER OF DRINKS

Blood alcohol level depends on number of alcoholic drinks consumed, gender of the person, body weight, and age, as well as genetic makeup. It is usually assumed that drinking one drink per hour not exceeding two drinks on one occasion and up to one drink for women should be safe because blood alcohol level should be significantly below the accepted legal limit for driving in the United States, which is 0.08% whole blood alcohol (80 mg/dL). DWI stands for "driving with impairment." The charge differs from state to state in the United States and includes driving under the influence (DUI). In some states DWI stands for "driving while intoxicated." Although impairment may also be drug related, alcohol is the major cause of DWI not only in the United States but worldwide. Alcohol-related motor vehicle accidents kill approximately 17,000 Americans annually and are associated with more than $51 billion in total costs annually. There is a strong correlation between binge drinkers and alcohol-impaired drivers in the United States. In one study, the authors found that overall 84% of all alcohol-impaired drivers are binge drinkers. Nonheavy drinkers are also involved in alcohol-related motor vehicle accidents (22). Currently, in all states in the United States, the legal limit for driving is 0.08% alcohol in whole blood. Serum concentration of alcohol is more than whole blood concentration, and in order to calculate whole blood concentration of alcohol, the measured serum concentration must be multiplied by a factor that is generally considered to be 0.85.

Although in the United Kingdom and Canada, the legal limit for driving is also 0.08%, in other countries, lower levels of alcohol are mandated as the acceptable upper limit for driving under the influence of alcohol. In Switzerland, Denmark, Italy, the Netherlands, Austria, Australia, China, Thailand, and Turkey, the upper limit is 0.05% alcohol. In Japan, the upper acceptable limit is only 0.03%, and in

TABLE 2.2

Legal Blood Alcohol Content (BAC) Limits of Driving in Various Countries in the World

Legal Limit for Driving	Countries
0.08% BAC	United States, Mexico, United Kingdom, Ghana, Kenya, New Zealand, Ireland, Malta, Singapore, Seychelles, Uganda, Zambia
0.05% BAC	Austria, Belgium, Bulgaria, Costa Rica, Cambodia, Denmark, Finland, Greece, Hong Kong, Israel, Malaysia, Peru, Portugal, Serbia, Spain, South Africa, Switzerland, Thailand, Tanzania, Turkey
0.03% BAC	India, Nepal, Georgia, Japan, Russia
0.02% BAC	China, Poland, Norway, Sweden, Estonia
Zero tolerance (0% BAC)	Saudi Arabia, United Arab Emirates, Brazil, Bangladesh, Hungary, Jordan, Czech Republic, Kuwait, Pakistan, Iran

certain countries such as various Middle Eastern countries, Hungary, Romania, and Georgia, there is a zero tolerance for blood alcohol in drivers. Legal limits for driving in various countries are listed in Table 2.2.

Although the legal limit of blood alcohol for adult drivers in the United States is 0.08%, some driving impairment may occur even at lower blood alcohol levels. There is general agreement that some impairment of the ability to drive takes place at a blood alcohol level of 0.05%. Even a blood alcohol level of 0.03% affects some cognitive functions that rely on perception and processing of visual information (23). Low blood alcohol level of 0.05% usually produces more relaxation and more social interactions with other individuals. However, intoxication can be encountered at a blood alcohol level of 0.1%, and levels over 0.5% are potentially lethal. The drunkest reported driver in Sweden had a blood alcohol level of 0.55% (24).

2.7.1 WIDMARK FORMULA

In 1932, the Swedish scientist Eric P. Widmark developed a formula that is still used today for calculation of the amount of alcohol ingested and for assessing the concentration of alcohol prior to a blood alcohol analysis (25). The Widmark formula suggests estimating blood alcohol level on a given amount of alcohol administration knowing the subject's body weight and gender.

By using this formula, one can estimate the amount of alcohol consumed by a person:

$$A = C \times W \times r$$

where A represents total amount of alcohol consumed by the person in grams, C is the blood alcohol concentration in grams per liter, W is the body weight of the person expressed in kilograms, and r is a constant assumed to be roughly 0.7 for men and 0.6 for women. A commonly used form of the formula to calculate a blood alcohol concentration from the amount of alcohol consumed by the individual, the body weight, and gender is as follows:

$$C = (A/W \times r) - 0.015\,t$$

where t represents time passed since the beginning of drinking.

In the United States, one standard drink of alcohol has 0.6 ounce of alcohol, and weight of a person is expressed in pounds. However, blood alcohol concentration is expressed as mg per 100 mL of whole blood. Taking into account all these factors, this formula can be modified for calculating blood alcohol concentration as follows:

$$C = (\text{Total amount of alcohol consumed in ounces} \times 5.14/\text{weight in pounds} \times r) - 0.015\,t$$

C is the blood alcohol in percent. Assuming each drink contains 0.6 ounce of alcohol, this equation can be further modified to

$$C = (\text{Number of drinks} \times 3.1/\text{Weight in pounds} \times r) - 0.015\,t$$

The blood alcohol level in women would be higher than that in men of the same weight. These calculations show that regardless of gender, drinking five beers or any five drinks within a 2-hour time frame would result in a blood alcohol level much higher than the legal limit for driving.

2.7.2 Alcoholic Odor in Breath and Endogenous Alcohol Production

Alcohol is almost odorless and the alcoholic smell perceived by people is due to the presence of many complex organic volatile compounds found in alcoholic beverages. Wine aroma is attributed to a large range of molecules from different chemical families including esters, aldehydes, ketones, terpenes, tannins, and sulfur compounds. Some of these compounds originate from grapes, and others are formed during fermentation or aging. In general, more volatile substances are present in white wine compared to red wine (26). Therefore, there is no correlation between blood alcohol level and alcoholic odor. Such odor may also be present in an individual drinking nonalcoholic beer.

In general the human body does not produce enough endogenous alcohol to reach a measurable blood alcohol level. There are reports of measurable endogenous ethanol production in patients with liver cirrhosis. In one report, after a meal in such patients, negligible alcohol levels of 11.3 mg/dL (0.01%) and 8.2 mg/dL (0.008%) were detected in two out of eight patients. Small intestinal bacterial overgrowth generates such small amounts of endogenous alcohol. Patients with cirrhosis of the liver often have small intestinal bacterial overgrowth (27). However, postmortem production of alcohol due to fermentation of sugar by bacteria is well documented. Toxicological analysis of a specimen from a 14-year-old child reveled high amounts of alcohol both in blood and tissue, but ethyl glucuronide, a metabolite of alcohol, was not detected in the liver tissue. The authors concluded that postmortem alcohol in that child was due to the action of the bacterial strain *Lactococcus garvieae* in blood of the deceased which is capable of producing alcohol from glucose (28). In another article the authors observed the presence of ethyl glucuronide in postmortem blood of 93 cases with information of antemortem blood alcohol but detected no ethyl glucuronide in 53 cases where there was no indication of antemortem alcohol or use of alcohol by the deceased. The authors concluded that the presence of ethyl glucuronide in the postmortem blood is a marker of antemortem ingestion of alcohol (29).

2.8   BENEFITS OF MODERATE DRINKING

Benefits of drinking in moderation include reduced risk of coronary heart disease, better survival chance after a heart attack, reduced risk of stroke, reduced risk of developing diabetes, reduced risk of forming gallstones, reduced risk of developing arthritis, reduced risk of developing age-related dementia and Alzheimer's disease, reduced risk of certain types of cancer, increased longevity, and possibly less chance of getting a common cold.

2.8.1   MODERATE ALCOHOL CONSUMPTION AND REDUCED RISK OF HEART DISEASE

The relationship between alcohol consumption and coronary heart disease was examined in the original Framingham Heart Study initiated in 1948 with a 24-year follow-up examination using 2,106 males and 2,639 females. The alcohol consumption showed a U-shaped curve with reduced risk of developing heart disease with moderate drinking but high risk of developing such diseases with heavy drinking. Smoking is a risk factor for developing coronary heart disease, but moderate alcohol consumption may also provide some protection against coronary heart disease among smokers (30). Smokers who smoke one pack of cigarettes per day have twice the risk of developing coronary heart disease than nonsmokers. Alcohol consumption actually lowered the incidence of coronary heart disease in the Framingham study, but when greater than two drinks of alcohol per day were consumed, a rise in mortality from cancer and stroke was observed (31). In the American Cancer Society prospective study among 276,802 American men over a period of 12 years, the authors determined that the relative risk (RR) total mortality was 0.88 for the occasional drinker, 0.84 for those drinking one drink per day, but 1.38 in people drinking six or more drinks per day compared to nondrinkers. However, RR of death from coronary heart disease was lower than one in all groups of drinkers compared to nondrinkers. The RR is defined as the ratio of the chance of a disease developing among members of a population exposed to a factor compared with a similar population not exposed to the factor. Interestingly, the risk of cardiovascular disease was mostly reduced in people who consumed one alcoholic drink per day (RR: 0.79). This group also demonstrated the lowest of all mortality (RR: 0.84) among all groups studied (32).

It is beneficial to drink one drink per day or at least six drinks per week to reduce the risk of coronary heart disease and heart attack. In a study using 18,445 males (aged 40 to 84 years) and a 7-year follow-up, it was revealed that when those individuals consuming one drink per week or less increased their consumption between more than one to six drinks per week, a further 29% reduction in the risk for developing cardiovascular disease was observed compared to individuals who did not increase their alcohol consumption. The authors concluded that among men with initial low alcohol consumption (one or less drink per week), a subsequent moderate increase in alcohol consumption may lower their risk of developing coronary heart disease (33). Diabetic patients are at a higher risk of developing cardiovascular disease. Moderate consumption of alcohol can help these patients to lower their chances to get heart disease. In the Physician's Health Study using a total of 87,938 U.S. physicians (2,970 diagnosed with diabetes mellitus), the authors observed that weekly consumption of alcohol reduced the risk of heart disease by 33% while daily consumption of alcohol reduced the risk by 58% among diabetics. For nondiabetics, weekly consumption of alcohol reduced the risk of heart disease by 18%, while daily consumption of alcohol reduced the risk by 40% (34). Interestingly, women may get the beneficial effects of alcohol from consuming lower amounts and less frequently than men. In one study with 28,448 women and 25,052 men between 50 and 65 years of age who were free from cardiovascular diseases at enrollment in the study, during a 5.7-year follow-up, the authors observed that women who consumed alcohol at least 1 day per week had lower risk of coronary heart disease than those who drank alcohol less than 1 day a week. However, little difference was found between women who consumed at least one drink per week compared to women who consumed two to four drinks per week, five to six drinks per week, or seven drinks per week. For men the lowest risk was found in individuals who consumed one drink per day. The authors conclude that for women alcohol consumption can reduce the risk of heart disease and the frequency of drinking may not be an important factor, but for men the drinking frequency and not the alcohol intake is the determining factor in preventing heart disease (35). Heart failure or congestive heart failure is another potentially lethal heart disease. Moderate alcohol consumption not only can reduce the risk of myocardial infarction but can also provide protective effects against heart failure. In the Cardiovascular Health Study using 5,595 subjects, the authors observed that the risk of heart failure was reduced by 18% in individuals who drank 1 to 6 drinks per week and 34% in individuals who drank 7 to 13 drinks per week. In addition, the authors observed that moderate alcohol consumption lowered the risk of heart failure even in individuals who experienced heart attack (36). Another published report based on a study using 1,154 participants (580 men and 574 women) in Winnipeg, Manitoba, Canada, indicated that the well-established relationship between reduced risk of cardiovascular disease and moderate consumption of alcohol may not be evident until middle age (35 to 49) or older (50 to 64) in men. However, women may benefit from moderate consumption of alcohol at a much younger (18 to 34) age. The beneficial effects of alcohol consumption are negated when alcohol is consumed in a heavy episodic drinking (eight or more drinks per occasion) pattern, especially for middle aged and older men (37).

There are several hypotheses on how moderate drinking can reduce the risk of developing heart disease (Table 2.3). Many studies have demonstrated increased

TABLE 2.3
Hypotheses by which Moderate Alcohol Consumption Reduces Risk of Heart Disease

Mechanism	Comments
Increasing high-density lipoprotein cholesterol	Direct effect of ethanol
Decreasing low-density lipoprotein cholesterol	Direct effect of ethanol
Reduced plaque formation	Ethanol and other polyphenolic compounds present in wine
Reduces risk of blood clotting	Inhibit platelet aggregation
Reduces level of fibrinogen	Complex mechanism
Antioxidant	Various polyphenolic compounds present in wine, especially red wine

high-density lipoprotein cholesterol (HDL-cholesterol) levels in drinkers than in nondrinkers. The Honolulu Heart Study showed that men who drank alcoholic beverages had higher blood levels of HDL cholesterol than nondrinkers. Gordon et al. reviewed data from 10 different studies including the Honolulu Heart Study and observed that there was a positive correlation between amounts of alcohol consumed and serum level of HDL cholesterol. In the male population between ages 50 and 69, the average HDL-cholesterol level was 41.9 mg/dL in people who consumed no alcohol, 47.6 mg/dL in people consuming up to 16.9 gm of alcohol per day (a single drink is 14 gm of alcohol), 50.7 mg/dL in people consuming between 16.9 and 42.2 gm of alcohol per day (one to three drinks a day), and 55.3 mg/dL in people drinking between 42.3 and 84.5 gm of alcohol (three to six drinks) per day. Interestingly, in the Albany, Framingham, and San Francisco studies, the highest levels of HDL cholesterol were 54.6 mg/dL, 50.1 mg/dL, and 57.8 mg/dL (HDL cholesterol levels among nondrinkers were 46.3 mg/dL, 41.4 mg/dL, and 44.4 mg/dL), respectively, in men between ages 50 and 69 who consumed the highest amounts of alcohol per day (42.3 to 85.5 gm/day or approximately three to six drinks per day) (38). In another study, the authors observed that the HDL-cholesterol level in blood was increased by up to 33% in social drinkers versus nondrinkers. A small experiment also revealed an average 15% reduction in HDL-cholesterol levels among social drinkers who abstained from alcohol for a 2-week period (39). In females light drinking (one drink or less a day) was associated with lower blood levels of low-density lipoprotein (LDL) cholesterol and higher levels of HDL cholesterol (40). Another article also demonstrated that serum HDL cholesterol was higher in drinkers than nondrinkers in all age groups (20 to 69 years) of men and women, and the atherogenic index (calculated by using serum total cholesterol and HDL-cholesterol concentrations) was also lower in drinkers than nondrinkers in all age groups of both men and women (41). Alcohol also diminishes thrombus formation on damaged walls of the coronary artery, and the plausible mechanism of inhibition of platelet aggregation is through inhibiting phospholipase A2 (42).

2.8.2 WINE VERSUS OTHER ALCOHOLIC BEVERAGES ON PREVENTING HEART DISEASE

Studies have indicated that the level of increase in HDL cholesterol in blood may explain 50% of the protective effect of alcohol against cardiovascular disease, and the other 50% may be partly related to inhibition of platelet aggregation thus reducing blood clot formation in coronary arteries. It has been suggested that although alcohol can increase the HDL cholesterol level and also can inhibit platelet aggregation, polyphenolic compounds found abundantly in red wine can further reduce platelet activity via mechanisms other than alcohol. In addition, these polyphenolic compounds found in red wine can also increase the level of vitamin E, an important antioxidant, thus providing further protection against various diseases. Therefore, it appears that red wine is more protective against cardiovascular disease than other alcoholic beverages (43). It has been postulated that resveratrol, a polyphenolic compound abundant in red wine (compared to white wine, beer, or spirits), plays an important role as an antioxidant and inhibitor of platelet aggregation which may explain more cardioprotection from consuming red wine (44).

2.8.3 CONSUMPTION OF ALCOHOL AND REDUCED RISK OF STROKE

Another beneficial effect of consuming alcohol in moderation is dramatic reduction in the risk of having a stroke among both men and women regardless of age or ethnicity. The Copenhagen City Heart Study with 13,329 eligible men and women aged between 45 and 84 years with 16 years of follow-up indicated a U-shaped relation between intake of alcohol and risk of stroke. People who consume low to moderate amounts of alcohol experience protective effects of alcohol against stroke, but heavy consumers of alcohol are at higher risk of suffering from a stroke than moderate drinkers or nondrinkers. For moderate drinkers of wine, monthly drinking of alcohol reduced the risk of stroke by 17%, weekly drinking reduced the risk by 41%, and daily drinking reduced the risk by 30%. There was no association between risk of stroke and drinking beer or spirits (45). In the second examination of the Copenhagen City Heart Study with 5,373 men and 6,723 women and 16 years of follow-up, it was observed that at a high stress level weekly total consumption of 1 to 14 drinks compared with no consumption of alcohol was associated with 43% lower risk of stroke in both men and women, but no clear association was observed between risk of stroke and moderate consumption of alcohol in individuals who were at lower stress levels. In addition, this study reported that only drinking beer or wine reduced the risk of stroke in individuals with high stress levels. It was suggested that alcohol may also alter psychological responses to stress in addition to modifying physiological responses (46).

2.8.4 MODERATE CONSUMPTION OF ALCOHOL AND TYPE 2 DIABETES

Moderate consumption of alcohol reduces the risk of developing Type 2 diabetes. Based on 15 studies conducted in the United States, Finland, the Netherlands,

Germany, the United Kingdom, and Japan, with 369,862 men and women and an average follow-up of 12 years, light drinkers (less than half a drink per day or 6 gm of alcohol) had a 13% lower chance of developing Type 2 diabetes while moderate drinkers (half a drink to four drinks per day, 6 to 12 gm of alcohol per day) had a 30% lower risk of developing Type 2 diabetes compared to nondrinkers. It made little difference whether an individual consumes beer, wine, or spirit, and it is better to consume alcohol frequently (such as daily or several times in a week) rather than occasionally. In contrast, heavy consumption of alcohol (more than three and a half drinks per day or 48 gm of alcohol per day) did not have any protective effect against developing Type 2 diabetes. Moreover, these individuals are at slightly higher risk of developing (4% more) Type 2 diabetes than nondrinkers (47). The Finnish twin study where twins with different drinking patterns (22,778 twins) were followed up for 20 years found that moderate alcohol consumption (half a drink to two drinks, 5 to 29.9 gm/day for men; half to one and a half drinks, 5 to 19.9 gm/day for women) was associated with lower risk of developing Type 2 diabetes than light drinking (less than half a drink, less than 5 gm/day). Overweight subjects (body mass index equal to greater than 25 kg/m^2) receive more beneficial effects from moderate alcohol consumption as the risk of developing diabetes was 30% lower in overweight men and 40% lower in overweight women than in nondrinkers. On the other hand, binge drinking and high alcohol consumption may increase the risk of Type 2 diabetes in women, especially lean women, but affected men to a lesser extent (48). Balinus et al., based on reviewing 20 studies, observed a U-shaped relationship between alcohol consumption and risk of developing Type 2 diabetes, where moderate alcohol consumption decreases the risk but heavy alcohol consumption increases the risk. Compared to lifetime abstainers from alcohol, the alcohol protected individuals most from developing diabetes (17% lower risk) who drank an average 22 gm of alcohol per day (one and a half drinks). For women the most protection was observed (40% lower risk) in individuals who consumed 24 gm of alcohol per day. Drinking became deleterious among men who consumed over 60 gm of alcohol per day (four and half drinks) and among women who consumed over 50 gm of alcohol (almost four drinks) a day (49).

2.8.5 MODERATE ALCOHOL CONSUMPTION AND LOWER RISK OF DEMENTIA AND ALZHEIMER'S DISEASE

Moderate alcohol consumption can dramatically reduce the risk of age-related dementia and developing Alzheimer's disease. In a French study using 3,777 community residents aged 65 years or older, it was shown that the subjects who drank three to four glasses of alcoholic beverages (mostly wine) per day (318 subjects) had 82% lower risk of developing senile dementia and 75% lower risk of getting Alzheimer's disease compared to nondrinkers (971 subjects) (50). Mitchell et al. demonstrated that brain cultures preconditioned with moderate alcohol concentrations are resistant to neurotoxic Alzheimer's amyloid-beta peptides. The mechanism of neuroprotection by moderate levels of alcohol is probably related to early increases in NR (N-methyl-D-aspartate receptor) subunits concomitant with enhancement of synaptic localization and activity of NMDA receptors (*N*-methyl-D-aspartate iono-tropic glutamate receptor) (51). However, chronic abusers of alcohol are at higher risk of developing memory loss, dementia, and lack of appropriate motor control due to alcohol-related brain damage. Younger people, especially underage drinkers, are at higher risk of alcohol-related brain damage because alcohol has a detrimental effect on the developing adolescent brain.

2.8.6 MODERATE ALCOHOL CONSUMPTION AND CANCER

Moderate consumption of alcohol may reduce the risk of certain types of cancer. It has been suggested that moderate drinking facilitates elimination of *Helicobacter pylori* (*H. pylori*), a bacteria found in the gut that causes chronic atrophic gastritis (CAG) and gastric cancer. Gao et al., using 9,444 subjects aged 50 to 74, observed that moderate drinkers (less than 60 gm of alcohol per week or four drinks per week) had 29% lower chance of developing CAG than nondrinkers. Both beer and wine drinking provided protection against CAG. In addition to facilitating elimination of *H. pylori*, another mechanism may also contribute to reducing the risk of CAG in moderate drinkers (52). In the California Men's Health Study using 84,170 men ages between 45 and 69, consumption of one or more drinks per day was associated with approximately 60% reduced lung cancer risk in smokers. Even heavy smokers ben-efitted from consuming red wine in moderation. No clear association was observed between moderate drinking and alcohol in individuals who consumed white wine, beer, or other liquors (53). In another study, the author observed that although mod-erate consumption of wine (one drink or less per day) was associated with approxi-mately 23% reduced risk of developing lung cancer, moderate consumption of beer (one or more per day) increases the risk of developing lung cancer by 23% only in men but not in women (54). Jiang et al. reported that intake of beer and wine but not spirits (hard liquors) can reduce the risk of bladder cancer by up to 32% in people drinking beer and wine compared to people who are nondrinkers (55). Consumption of up to one drink per day reduced the risk of head and neck cancer in both men and women, but consuming more than three alcoholic beverages increased the risk of developing this cancer (56). In Italian studies the authors observed that moderate consumption of alcohol reduced the risk of developing renal cell carcinoma (kidney cancer) in both males and females (57).

2.8.7 MODERATE ALCOHOL CONSUMPTION AND PROLONGING LIFE?

Because moderate consumption of alcohol can prevent many diseases including the number one killer, cardiovascular disease, it is expected that moderate drink-ers may live longer than lifetime abstainers of alcohol. Freiberg et al. using 10,576 African American and 105,610 Caucasian postmenopausal women and an 8-year follow-up demonstrated that moderate drinking (one to less than seven drinks per week) was associated with lower mortality among both hypertensive and nonhyper-tensive Caucasian women, but among African American women only hypertensive women received benefits from moderate drinking. Even current drinking of only one drink per month or more increased longevity among Caucasian hypertensive and nonhypertensive women as well as among African American hypertensive women. Low mortality was also observed among African American nondrinking women with normal blood pressure (58). Klatsky et al. studied 10-year mortality in relation to alcohol in 8,060 subjects and observed that persons who consumed two or fewer drinks daily fared best and had 50% reductions in mortality rate compared to non-drinkers. The heaviest drinkers (six or more drinks a day) had a doubled mortality rate compared to moderate drinkers, while people who drink three to five drinks per day had a similar mortality rate to nondrinkers. Therefore, consuming two or less drinks per day is the best practice (59). In the Physician's Health Study using 22,071 male physicians in the United States between ages 40 and 84 with no history of myocardial infarction, stroke, or cancer, and 10-year follow-up, the authors observed that men who consumed two to six drinks per week had the most favorable results (20% to 28% lower mortality rate than people who consumed one drink per week). In contrast, people who consumed more than two drinks per day had an approximately 50% chance of higher mortality than people who consumed just one drink per week (60). A study from the Netherlands reported that in the presence of stress, moderate drinkers are less likely to be absent from work than nondrinkers (61). A 9-year prospective study indicated that moderate consumption of alcohol was associated with the most favorable health scores, indicating that these people in general enjoy better overall health quality than abstainers (62).

2.8.8   MODERATE ALCOHOL CONSUMPTION AND ARTHRITIS

Moderate alcohol consumption reduces the risk of developing rheumatoid arthritis. Results from two Scandinavian studies indicated that among moderate drinkers, the risk of rheumatoid arthritis was significantly reduced (40% to 50%). Smokers had a higher risk of developing rheumatoid arthritis. The authors concluded that smokers should be advised to quit smoking in order to reduce the risk of developing arthritis, but moderate drinkers should not be discouraged from sensible alcohol consumption (63). Moderate alcohol consumption not only reduces the risk of developing rheumatoid arthritis but also may slow the progression of the disease. Nissen et al. reported in a study using 2,908 patients suffering from rheumatoid arthritis that occasional or daily consumption of alcohol reduces the progression of the disease based on radiological studies (x-ray). The best results were observed in male patients (64).

2.8.9   CAN MODERATE DRINKING PREVENT COMMON COLD?

In one study, the authors observed that smokers are at greater risk of developing the common cold than are nonsmokers. Moderate alcohol consumption reduced the incidence of the common cold among nonsmokers but had no protective effects against the common cold in smokers (65). In a large study using 4,272 faculty and staff of five Spanish universities as subjects, the investigators observed that total alcohol intake from drinking beer and spirits had no protective effect against the common cold, whereas moderate wine consumption was associated with reduced risk of common cold. When individuals consumed 14 or more glasses of wine per week, the relative risk of developing the common cold was reduced by 40% compared to that of teetotalers. It was also observed that consumption of red wine provided superior protection against the common cold. The authors concluded that wine drinking, especially drinking red wine, may have a protective effect against the common cold (66).

2.9 HAZARDS OF HEAVY DRINKING AND ALCOHOL ABUSE

As mentioned earlier in the chapter, alcohol abuse costs the U.S. economy approximately $185 billion annually and many deaths due to traffic accidents (13). In 2005, cirrhosis of the liver was the 12th leading cause of death in the United States, claiming 28,175 deaths. Among all cirrhosis-related deaths, 45.9% were alcohol related (67). California is the largest alcohol market in the United States, and Californians consumed almost 14 billion alcoholic drinks in 2005 which resulted in an estimated 9,439 deaths and 921,029 alcohol-related problems such as crime and injury. The economic burden was estimated to be $38.5 billion of which $5.4 billion was for medical and mental health spending, $25.3 billion due to loss of work, and another $7.8 billion in criminal justice spending (68). In the United Kingdom, alcohol consumption was responsible for 31,000 deaths in 2005, and the National Health Services spent an estimated 3 billion pounds in 2005 and 2006 for treating alcohol-related illness and disability. Alcohol consumption was responsible for approximately 10% of disabilities (male, 15%; female, 4%) (69).

Many studies demonstrated the harmful effects of alcohol on a variety of organ systems including liver, heart, brain, immune system, endocrine system, and bones. Alcoholic liver disease and alcoholic cirrhosis of the liver cost many lives every year worldwide. Major adverse effects of chronic alcohol consumption include decreased life span, increased risk of violent behavior, alcohol liver disease including cirrhosis of the liver, mood disorder, brain damage, damage to the heart, increased risk of stroke, damage to the immune system, damage to the endocrine system, bone damage, and increased risk of various cancers. Drinking during pregnancy is associated with poor outcomes in pregnancy including fetal alcohol syndrome.

2.9.1 Chronic Alcohol Abuse and Reduced Life Span

Although moderate drinking is associated with increased longevity, heavier drinking is associated with all causes of decreased longevity compared to abstaining. Heavy consumption of alcohol reduced longevity dependent on two factors: frequency of drinking and number of drinks consumed on one occasion. Even occasional heavy drinking may be detrimental to health. Dawson reported an increased risk of mortality among individuals who usually drink more than five drinks per occasion but who drank even less than once a month (70). Irregular heavy drinking even once a month (five or more drinks per occasion) increases the risk of heart disease rather than protecting the heart as observed in moderate drinkers. The cardioprotective effect of moderate drinking also disappears when average light to moderate drinking is mixed with occasional heavy drinking episodes (71).

In a British study, the authors, based on an investigation of 5,766 men ages 35 to 64 with 21 years of follow-up, observed that consuming between 15 and 21 standard drinks increases the risk of all causes of mortality in these drinkers compared to moderate drinkers (up to 14 standard drinks per week) and nondrinkers by 34% while drinking more than 35 standard drinks per week increases the risk by 49%. The authors further observed that men drinking 35 or more standard drinks per week had double the risk of stroke compared to nondrinkers. The authors concluded that in general the overall association between alcohol consumption and mortality is unfavorable for men drinking 22 standard alcoholic drinks per week or more (72). Using over 43,000 participants and a 14-year follow-up when 2,547 people died, Breslow and Graubard observed that men who consumed five or more drinks on a drinking day had a 30% higher risk of mortality from heart disease, over 50% from cancer, and over 40% from all causes of mortality compared to individuals who drank one drink on a drinking day. The risk of mortality was also increased to some extent with just two drinks per day or more for males. Women drinkers who consumed alcohol (two drinks or more in a session) more than in moderation (one drink or less a day) also showed all-cause higher mortality than moderate drinkers. Among men both quantity and frequency of drinking were significantly associated with mortality from cardiovascular disease, cancer, and other causes, but among women the quantity of alcohol was more important. Women who drank more than in moderation showed higher risk of mortality from cancer than men (73). The London-based Whitehall II cohort study using 10,308 government employees between the ages of 35 and 55 with an 11-year follow-up also concluded that optimal drinking is once or twice a week to daily consumption of one drink or less. People who consume twice a day or more had an increased risk of mortality compared to those drinking once or twice a week (74). Binge drinking is also dangerous. In one study, the authors observed that (based on a population of 1,641 men who drank beer) the risk of death in men who drank six or more bottles of beer on one occasion was almost three times higher than for those who consumed less than three bottles on one occasion (75). Another study based on 13,251 adults also reported that individuals who drank five or more drinks on one occasion were nearly twice as likely to die from injuries than persons who drank fewer than five drinks in a single occasion. Persons drinking nine or more drinks on a single occasion are 3.3 times more likely to die from injuries than people consuming less than five drinks (76).

Other than increasing mortality from various diseases, alcohol abuse is also associated with increased risk of suicide, accidents, and violent crimes. Based on a survey of 31,953 school students, Schilling et al. observed that both drinking while depressed and episodic heavy drinking were associated with self-reported suicide attempts in adolescents (77). Swahn et al. in their study reported that in a high-risk school district in the United States, 35% of seventh graders reported alcohol abuse starting at age 13 or younger. Preteen alcohol users were more involved in violent behavior than nondrinkers. Early alcohol use was also associated with higher risk of suicide attempts among these adolescents (78).

2.9.2 ALCOHOL ABUSE AND VIOLENT BEHAVIOR

Many investigators reported a close link between violent behavior, homicide, and alcohol intoxication. Studies conducted on convicted murderers suggest that about half of them were under heavy influence of alcohol at the time of the murder (79). Alcohol may induce aggression and violent behavior by disrupting normal brain function when consumed in high dosages. By impairing the normal information-processing capability of the brain, alcohol can make a person misjudge a perceived threat and react more aggressively than warranted. Serotonin, a neurotransmitter, is considered a behavioral inhibitor. Alcohol abuse may lead to decreased serotonin activity causing aggressive behavior. High testosterone concentrations in criminals have been associated with violent crimes. Adolescents and young adults with higher levels of testosterone compared to the general population are more often involved in heavy drinking and consequently violent behavior. Young men who exhibit antisocial behavior often "burn out" with older age due to decreased levels of testosterone and increased levels of serotonin. By modulating serotonin and testosterone concentration, alcohol may exert its effect in inducing aggressive and violent behavior when consumed in excess (80).

2.9.3 Alcohol Abuse and Liver Disease

The liver is one of the largest and most complex organs, that synthesizes important proteins vital for life, stores some nutrients, and metabolizes drugs and toxins including alcohol, thus protecting the body from harmful effects of drugs and toxins. Alcohol-induced liver disease can be classified under three categories: fatty liver, alcoholic hepatitis, and cirrhosis of the liver. Heavy drinking for as little as a few days may produce fatty changes in the liver (steatosis) which are reversed after abstinence. However, drinking heavily for a longer period may cause more severe alcohol-related liver injuries such as alcoholic hepatitis and cirrhosis of the liver. The diagnosis of alcoholic hepatitis is a serious medical condition because approximately 70% of such patients may progress to liver cirrhosis, a major cause of death worldwide. Although fatty liver may develop in approximately 90% of alcoholics, only 10% to 35% of them develop alcoholic hepatitis and 10% to 20% of them develop liver cirrhosis. In the United States it is estimated that there are 2 million people who are suffering from alcohol-related liver diseases. Liver cirrhosis is the seventh leading cause of death among young and middle-aged adults and approximately 10,000 to 24,000 deaths from liver cirrhosis annually may be attributable to alcohol abuse (81). The risk of developing alcoholic hepatitis and liver cirrhosis depends on several factors including amount of alcohol consumed per day, length of heavy drinking, gender, ethnicity and genetic predisposition, nutritional status, obesity, type of alcoholic beverage consumed, and the presence of hepatitis C.

The amount of alcohol consumed is a determining factor in developing alcoholic hepatitis and liver cirrhosis. In one report the authors commented that cirrhosis of the liver does not develop below a lifetime ingestion of 100 kg of alcohol (one standard drink is approximately 14 gm of alcohol; therefore lifetime consumption of 7,143 drinks). This amount corresponds to an average of five drinks a day for about 4 years. The authors also commented that consuming alcohol with food lowers the risk of developing cirrhosis of the liver compared to alcohol being consumed on an empty stomach (82). Although only a small percentage of alcoholics develop alcoholic hepatitis and cirrhosis of the liver, other alcohol-related liver damage occurs at a much lower intake of alcohol. In general, the threshold of alcohol-induced liver toxicity is considered to be 40 gm of alcohol per day (approximately three drinks a day) for men and 30 gm (more than just two drinks) or more alcohol a day for women. But in one report, the authors concluded, based on a study of 6,917 subjects, that risk of any alcohol-induced liver damage (noncirrhotic liver damage) may have a threshold of just 30 gm or more alcohol consumption (just over two standard drinks) per day for both males and females, and the risk increases with increasing daily consumption. Drinking outside mealtime and drinking multiple different alcoholic beverages increased the alcohol-induced liver damage (83). However, another study indicated that above a threshold of 7 to 13 drinks per week for women and 14 to 27 drinks per week for men there is a risk of developing some alcohol-related liver problem. The greater sensitivity of women toward alcohol toxicity may be related to genetic predisposition of the metabolism pattern of alcohol in women where more oxidative by-products of alcohol are formed compared to that in men. Consumption of coffee may protect males against alcohol-induced liver damage but no such data are currently available for females (84).

Although fatty liver is common in heavy drinkers, alcohol abuse over a decade is usually needed for developing alcoholic hepatitis and liver cirrhosis. Almost a decade of alcohol abuse is needed for development of severe alcohol-related liver diseases. Daily alcohol consumption of three to six drinks for males and two to three drinks for females over a period of 12 years would most likely cause alcoholic liver diseases. However, this lower amount of alcohol consumption may cause alcoholic liver disease in certain ethnic populations. One Chinese study conducted using 1,300 alcohol drinkers indicated that the risk threshold was only 20 gm of alcohol daily (one and half drinks) for 5 years with a greater risk when alcohol is consumed on an empty stomach, especially hard liquors (spirits). In addition, obese people showed more morbidity from alcohol-related liver diseases (85).

Hepatitis C is a liver disease caused by the Hepatitis C virus. This virus can spread through sharing needles or other equipment for injecting illicit drugs in one's body and also through sexual contact with infected partners. It has been estimated that approximately 4 million Americans are infected with the Hepatitis C virus and between 10,000 and 12,000 die annually. Hepatitis C infection is common among alcohol abusers and this infection may even accelerate alcohol-related liver diseases including cirrhosis of the liver and liver cancer. How much alcohol consumption is safe in a person with hepatitis C infection has not been clearly established. In one study the authors observed that moderate alcohol consumption between 31 and 50 gm per day (two and half drinks to three and half drinks) for males and 21 to 50 gm per day (one and half drinks to three and half drinks) for females could adversely affect the progression of liver damage (86).

The mechanism of alcohol-induced liver disease is complex. While in moderate drinkers alcohol is mostly metabolized by alcohol dehydrogenase in the liver, in alcoholics *CYP2E1*, a member of the cytochrome P-450 drug-metabolizing family of enzymes in the liver, becomes activated. In this process free oxygen radicals are generated causing oxidative damage to liver cells. In addition, acetaldehyde, a toxic product of alcohol metabolism, if not removed quickly by further metabolism, may cause liver toxicity. In alcoholics, due to the tremendous burden of alcohol on the liver for metabolism, both acetaldehyde and NADH (nicotinamide adenine dinucleotide hydrogen) occur leading to oxidative stress on the liver and increased production of fatty acid. Metabolism of fatty acid is also impaired causing fatty acids to build up in the liver which are eventually turned into fat (triglycerides) by the liver. Fatty liver with more alcohol consumption may proceed to liver cirrhosis. Another mechanism of liver damage by alcohol is the excess cytokine production by Kupffer cells of the liver due to release of bacterial endotoxin in the blood by the action of excess alcohol on bacteria present in the gut.

2.9.4 HEAVY ALCOHOL CONSUMPTION AND BRAIN DAMAGE

Although alcohol can cause relaxation and mild euphoria with moderate consumption, these pleasurable effects of alcohol are reversed with increasing blood alcohol level above 100 mg/dL (0.1%). Alcohol has more damaging effects on the adolescent brain than on the adult brain. Onset of drinking at an early age (13 or earlier) has devastating effects on the brain as well as on the life of the person, and such effects follow the person throughout his or her life. Early onset of drinking is also linked to greater risk of alcohol dependence in adult life. Although thiamine deficiency is one of the major factors involved in alcohol-related brain damage, both alcohol and its toxic metabolite acetaldehyde also have direct toxic effects on neurons. Underage drinkers are also susceptible to immediate ill effects of alcohol use such as blackouts, hangover, and alcohol poisoning compared to their adult counterparts. These individuals are also at higher risk of neurodegeneration, impairment of functional brain activity, and neurocognitive deficits. Underage drinking is associated with brain damage causing impaired intellectual development and such damage carries through to adulthood. Because adolescent drinking induces brain structure abnormalities, these changes lead to poor memory, impaired study habits, poor ability to learn, and poor academic performance (87). Using data from 8,661 respondents in a survey of a 10-year study, Harford et al. concluded that education beyond high school has a protective effect against alcohol abuse and dependence. In addition, people who do not attend college may have a higher risk of alcohol abuse than people who attend college (88). Studies also show that children of alcoholics constitute a population at risk for skipping school days, poor performance, and dropping out of school. Children of alcoholics also have a higher incidence of repeating a grade (89).

There is a difference between how alcohol damages the male versus female adolescent brain and the extent of damage. In general a female adolescent brain is more vulnerable to alcohol exposure than a male brain. Adolescents with alcohol abuse disorder have smaller prefrontal cortex volumes compared to healthy adolescents. The prefrontal cortex is located in the cortical region of the frontal lobe and is a crucial area of the brain that is involved in planning complex cognitive behavior such as learning, critical thinking, working with information held mentally, rational judgment, expression of personality, and appropriate social behavior. Unfortunately, the prefrontal cortex volume of alcohol-dependent female adolescents was smaller than that of their male counterparts. Consistent with adult literature, alcohol use during adolescence is associated with prefrontal volume abnormality including differences in white matter, but girls are more affected than boys by adverse effects of alcohol (90).

Although the onset of drinking at an early age carries a much higher risk of alcohol dependence and brain damage with long-lasting effects into adulthood, the onset of drinking at age 21 followed by chronic abuse of alcohol can also cause significant damage to the human brain. The two major alcohol-related brain damages are alcoholic Korsakoff's syndrome and alcoholic dementia. Korsakoff's syndrome is a brain disorder caused by deficiency of thiamine, and major symptoms are severe memory loss, false memory, lack of insight, poor conversation skills, and apathy. Some heavy drinkers may also have a genetic predisposition to developing this syndrome. In Korsakoff's syndrome, loss of neurons is a common feature including micro bleeding in certain regions of gray matter (91). When Wernicke's encephalopathy accompanies Korsakoff's syndrome in an alcoholic, it is called Wernicke–Korsakoff syndrome. Wernicke's encephalopathy and Korsakoff's syndrome are two related diseases, both caused by thiamine deficiency, but clinical symptoms may be different. Alcoholics with Korsakoff's syndrome always have severe amnesic syndrome but may not have classical symptoms of Wernicke's encephalopathy which include ophthalmoplegia, ataxia, and confusion. However, patients with Wernicke–Korsakoff syndrome show most of the symptoms found in both diseases. Damage to the anterior nucleus of the thalamus is commonly found in patients with Korsakoff's syndrome but may also be present in patients suffering from Wernicke's encephalopathy. The anterior nucleus of the thalamus is involved in learning and memory as well as the alertness of an individual. The Royal College of Physicians in London recommends that patients admitted to the hospital who show evidence of chronic misuse of alcohol and poor diet should be treated with B vitamins (92). Paparrigopoulos et al. reported a case where a 52-year-old man with a 10-year history of heavy alcohol abuse was admitted to the hospital and was treated aggressively for Wernicke–Korsakoff syndrome with 600 mg per day oral thiamine supplement in addition to 300 mg of thiamine that was delivered intravenously every day, and he was fully recovered 2 months after therapy (93).

Other than developing Korsakoff's syndrome or Wernicke–Korsakoff syndrome, thiamine deficiency in chronic alcohol abusers is a major cause of alcohol-induced brain damage. Thiamine is a cofactor required by three enzymes involved in carbohydrate metabolism, and brain cells can only use sugar as a fuel. In addition, intermediate products of carbohydrate metabolism pathways are needed for generation of essential molecules for cellular functions, and a reduction in thiamine can interfere with many of these important biochemical processes. Chronic alcohol consumption can result in thiamine deficiency by causing inadequate nutritional thiamine uptake, reduced absorption of thiamine from the gastrointestinal tract, and impaired thiamine utilization by the cells (94). Another study indicated that although thiamine deficiency causes neurodegeneration (loss of neurons) in the brain, alcohol potentiates this effect directly because alcohol can cross the blood-brain barrier and diffuse in the brain (95).

Prefrontal white matter is most severely affected in alcoholics, and there is a correlation between amount of loss and daily consumption of alcohol. Loss of white matter is a major cause of cognitive impairments in alcoholics (96). Significant loss of neurons has also been documented in the cortex, hypothalamus, and cerebellum of alcoholics. The types of neurons that are damaged in chronic alcohol users are larger neurons from the frontal cortex. These neurons are also damaged in patients with Alzheimer's disease. However, there is no direct link between alcoholic brain damage and Alzheimer's disease. Alzheimer patients are more impaired on recalling names, recognition memory, and orientation, while subjects with alcohol-induced dementia were impaired in fine motor control, initial letter fluency, and free recall (97).

Chronic abuse of alcohol results in brain damage to both males and females, but women are more susceptible to alcohol-induced brain damage than men. For the same mean daily alcohol consumption, blood alcohol levels in females may be higher than in males because the female's body burns alcohol slower than the male's. Hommer et al., based on a study of 43 alcoholic men and women and comparing them with 39 healthy controls, demonstrated that alcoholic women had a significantly smaller volume of gray and white matter than healthy subjects. Although alcoholic men also had lower amounts of gray matter and white matter compared to healthy controls, the difference in magnitude was smaller in men than in women. Direct comparison of alcoholic men and women showed that the proportion of the intracranial contents occupied by gray matter was smaller in alcoholic women than in alcoholic men when all other factors were adjusted. In addition, the magnitude of difference of brain volume between alcoholic women and nonalcoholic women was greater than the magnitude of difference of brain volume between alcoholic men and nonalcoholic men. The authors concluded that female brains are more susceptible to alcohol-related damages than male brains (98).

Binge drinkers, both males and females, are at higher risk of developing alcohol-related brain damage. Chronic exposure to the high amounts of alcohol that are ingested during binge drinking leads to stimulation of N-methyl-D-aspartate (NMDA) and calcium receptors, which results in increased release of glucocorticoids (stress molecules such as cortisol that affect carbohydrate metabolism). NMDA-mediated mechanisms and glucocorticoid actions on the hippocampus are associated with brain damage. In addition, ethanol withdrawal becomes more difficult for binge drinkers (99).

Alcohol-related brain damage and loss of cognitive functions may be reversible at least in part, if the brain damage is not permanent and the alcoholic can successfully complete a rehabilitation program and practices complete abstinence. Chronic alcoholism is often associated with brain shrinkage, but this may be reversed at least in part when abstinence is maintained as demonstrated by Trabert et al. based on a study using 28 male patients with severe alcohol dependence. Even with 3 weeks of abstinence, increased brain tissue densities were observed in these subjects (100). Asada reported a case report where a 42-year-old patient was unable to perform his office duties because of slowly progressive amnesia. The initial evaluation of the patient indicated severe verbal memory loss and an early stage of Alzheimer's disease was suspected because the patient did not disclose his habit of heavy alcohol consumption and no thiamine deficiency was found. Later the patient disclosed his habit of heavy alcohol consumption in the past, and with complete abstinence, his memory and cognitive functions improved markedly. Initial studies with FDG-PET (fluorodeoxyglucose-positron emission tomography), an advanced imaging technique, indicated that glucose metabolism was slower in the brain of the patient, and glucose is the only fuel brain cells can use. A 5-year follow-up study using PET imaging indicated that glucose metabolism in the brain was recovered to the normal level and the patient showed dramatically improved cognitive functions (101).

2.9.5 HEAVY ALCOHOL CONSUMPTION AND RISK OF HEART DISEASE AND STROKE

If consumed in moderation alcohol can reduce the risk of heart disease and stroke, but if consumed chronically in excess it increases risk of both heart disease and stroke. Drinking more than three drinks a day (any type of beverage) may be harmful to the heart. Chronic alcohol abuse for several years may result in the following serious medical conditions (102) including alcoholic cardiomyopathy and heart failure, systematic hypertension, heart rhythm disturbances, and hemorrhagic stroke.

Alcoholics who consume 90 gm or more of alcohol a day (seven to eight drinks) for 5 years are at risk of developing alcoholic cardiomyopathy, and if they continue drinking alcohol, cardiomyopathy may proceed to heart failure, a potentially fatal medical condition. This distinct form of heart failure (congestive heart failure) is responsible for 21% to 36% of all cases of nonischemic heart failure. Without complete abstinence, 50% of these patients will die within the next 4 years of developing heart failure (103).

A stroke may be evident when the blood supply in a particular part of the brain is interrupted or decreased, depriving brain cells from supply of glucose, oxygen, and essential nutrients. Within a few minutes brain cells start dying and if not treated soon, stroke may lead to severe brain damage, paralysis, or even death. Controlling high blood pressure, abstinence from tobacco abuse, and lowering cholesterol all can reduce the risk of stroke. Hemorrhagic stroke occurs when a blood vessel in the brain ruptures causing interruption in the blood flow to a part of the brain. A blood vessel may rupture from high blood pressure, or a weak spot in the blood vessel wall (aneurysms). Heavy drinking increases risk of stroke and particularly the risk of hemorrhagic stroke. In one study, the authors observed that risk of hemorrhagic stroke increases in an individual drinking 300 gm or more of alcohol weekly (21 or more drinks) (104).

2.9.6 HEAVY ALCOHOL CONSUMPTION AND IMMUNE SYSTEM

Alcohol abuse is associated with increased risk of bacterial infections, and opportunistic infections (including viral infection). The increased risk of infection in alcohol abusers is due to impairment of the immune system by alcohol. Exposure to alcohol can result in reduced cytokine production. Mast cells are important immune cells that are widely distributed in tissues that are in contact with the external environment such as skin, mucosa of lung, and gastrointestinal tract. Mast cells produce a variety of compounds including cytokines, histamine, eicosanoid, and TNF-a (tumor necrosis factor-alpha) that play important roles in defense against bacteria and parasites. Therefore mast cells are considered as the first line of defense against invading bacteria and parasites. Alcohol reduces the viability of mast cells and may cause cell death. Alcohol-induced reduction of the viability of mast cells could contribute to an impaired immune system associated with alcohol abuse (105). Alcohol also accelerates disease progression in patients with HIV infection because of immunosuppression. In one study using 231 patients with HIV infection who were undergoing antiretroviral therapy, the authors observed that even consumption of two or more drinks daily can cause a serious decline in CD4+ cell count (higher CD4+ counts indicates good response to therapy) (106).

Adult respiratory distress syndrome (ARDS) is a severe form of lung injury. Approximately 200,000 individuals develop ARDS in the United States each year, and nearly 50% of these patients have a history of alcohol abuse. The mortality from ARDS is high (over 40%) and for alcohol abusers approximately 65%. In ARDS survivors, alcohol abuse was also associated with longer stay under ventilation in intensive care units. Alcohol impairs immune function and decreases the level of pulmonary antioxidants; thus it may cause ARDS (107).

2.9.7 ALCOHOL ABUSE AND DAMAGE TO ENDOCRINE SYSTEM INCLUDING REPRODUCTIVE SYSTEM AND BONE

The hypothalamus, located deep within the brain, is the control center for most of the body's hormonal system. The hypothalamus, the pituitary gland, and adrenal glands function together as a well-coordinated unit controlling the hormonal balance of the body. The hypothalamus secretes corticotropin-releasing factor that through a complex mechanism stimulates the adrenal gland to secrete glucocorticoid hormones that influence carbohydrate, lipid, protein, and nucleic acid metabolism, as well as play a vital role in the cardiovascular system, with bone development, and with immune function. The major circulating glucocorticoid hormone in humans is cortisol. Alcohol abuse may lead to a disease known as pseudo-Cushing's syndrome that is indistinguishable from Cushing's syndrome, which is characterized by excess production of cortisol causing high blood pressure, muscle weakness, diabetes, obesity, and a variety of other physical disturbances. Diminished sexual function in alcoholic men has been described for many years. Administration of alcohol in healthy young male volunteers caused a diminished level of testosterone. Even drinking three or more drinks a day may cause significant problems in women, including delayed ovulation or failure to ovulate and menstrual problems, but such problems were not noticed in women who consumed two or fewer drinks a day. This may be related to alcohol-induced estrogen levels in women. Alcoholic women often experience reproductive problems. However, these problems may resolve when a woman practices abstinence from alcohol. To form healthy bone calcium, phosphorus and an active form of vitamin D are essential. Chronic consumption of alcohol may reduce bone mass through a complex process of inhibition of hormonal balance needed for bone growth including testosterone in men, which is diminished in alcoholics. Alcohol abuse may also interfere with pancreatic secretion of insulin, causing diabetes (108).

2.9.8 ALCOHOL ABUSE RELATED TO HIGHER RISK OF CANCER

Although moderate drinking reduces the risk of certain cancers, chronic abuse of alcohol increases cancer risk. Cancer kills an estimated 526,000 Americans annually, next only to heart disease. Cancers of the lung, large bowel, and breast are most common in the United States, and approximately 2% to 4% of all cancer cases may be linked to alcohol abuse. Epidemiological research demonstrated a dose-dependent relation between consumption of alcohol and certain types of cancers; as alcohol consumption increases so does the risk of cancer. The strongest link was found between alcohol abuse and cancer of the mouth, pharynx, larynx, and esophagus. An estimated 75% of all esophageal cancers are attributable to chronic alcohol abuse, while nearly 50% of cancers of the mouth, pharynx, and larynx are associated with chronic heavy consumption of alcohol. Prolonged drinking may result in alcoholic liver disease and cirrhosis of the liver, and such disease can progress to liver carcinoma (liver cancer). There are weak links between alcohol abuse and cancer of the colon, stomach, lung, and pancreatic cancer (109). Disease of the pancreas (pancreatitis) and gallstones are common among alcohol abusers. In alcoholics endotoxin may be released from gut bacteria by action of excess alcohol, and such a process may trigger progression of acute pancreatitis into chronic pancreatitis. Chronic pancreatitis may lead to pancreatic cancer (110). Pancreatic cancer is related to a high mortality rate.

The relation between moderate alcohol consumption and risk of breast cancer is debatable because there are conflicting reports in the medical literature. One Spanish study using 762 women between 18 and 75 years of age showed that even one drink a day may increase the risk of breast cancer, and women who consumed 20 gm or more alcohol a day (one and a half drinks or more) have a 70% higher chance of developing breast cancer than nondrinkers (111). In contrast, another study reported that women who consumed 10 to 12 gm wine per day (one glass of wine) had lower risk of developing breast cancer compared to nondrinkers. However, risk of breast cancer increases in women who drink more than one drink per day (112). Nagata et al., based on a review of 11 reports on the association between alcohol consumption and risk of breast cancer, concluded that epidemiological evidence of the link between alcohol consumption and risk of developing breast cancer remains insufficient (113).

2.9.9 Fetal Alcohol Syndrome

Fetal alcohol syndrome due to prenatal alcohol exposure was first reported by Jones and Smith in 1973 (114). Since then many publications have documented a teratogenic effect of alcohol in both human and animal studies. This syndrome is the most common noninherited (nongenetic) cause of mental retardation in the United States. "Fetal alcohol spectrum disorders" was a term described in 2004 to convey that exposure of the fetus to alcohol produces a continuum of effects and that many babies who do not fulfill all criteria for diagnosis of fetal alcohol syndrome may nevertheless be profoundly impacted negatively throughout their lives due to exposure to alcohol. Therefore, fetal alcohol spectrum disorders include a wide range of permanent birth defects due to maternal consumption of alcohol during pregnancy, which also includes all serious complications found in babies born with fetal alcohol syndrome. The other medical terminology used that was related to birth defects

TABLE 2.4

Abnormalities Present in Babies with Fetal Alcohol Syndrome

Abnormality	Comments
Growth retardation	Low birth weight, lack of weight gain over time, low weight-to-height ratio
Facial abnormalities	Small head circumference, small eye opening, small midface, flat upper lip, low nasal bridge, short nose
Neurodevelopmental problems	Abnormalities of central nervous system, small head size at birth, low or impaired motor skills, hearing loss
Behavioral and cognitive problems	Mental retardation, learning disability, poor memory, language deficiency, poor judgement, problem with reasoning and math
Cardiac malfunctions	Atrial septal defect, ventricular septal defect, and related malfunctions

found in babies where the mother consumed alcohol during pregnancy include partial fetal alcohol syndrome, fetal alcohol effect, alcohol-related neurodevelopmental disorders, and alcohol-related birth defects. Approximately 1 to 4.8 of every 1,000 children born in the United States has fetal alcohol syndrome while as many as 9.1 babies out of 1,000 babies born has fetal alcohol spectrum disorder. This is an alarming statistic because nearly 1 in every 100 babies born in the United States is born with fetal alcohol spectrum disorders. Therefore, fetal alcohol spectrum disorders are a major public health issue affecting up to 1% of the U.S. population (115). Recent school studies indicate that the prevalence of fetal alcohol syndrome in the United States is at least 2 to 7 per 1,000 babies and the current prevalence of fetal alcohol spectrum disorders may be as high as 2% to 5% in the United States among the school population. Such prevalence of alcohol-related complications in newborns is higher among school populations than among the general population (116). Abnormalities present in babies with fetal alcohol syndrome are listed in Table 2.4.

2.10 CONCLUSIONS

Although moderate alcohol consumption has many health benefits, heavy consumption of alcohol is detrimental to health. In moderation alcohol use can increase longevity and reduce risk of heart disease, stroke, and certain type of cancers. In addition, moderate alcohol consumption has neuroprotective effects and reduces the risk of dementia including Alzheimer's type. However, alcohol consumption in excess produces intoxication, withdrawal, brain trauma, central nervous system infection, hypoglycemia, hepatic failure, and Marchiafava-Bignami disease. Nutritional deficiency due to alcohol abuse also causes pellagra and Wernicke–Korsakoff disorder. Additionally, alcohol is a neurotoxin and in sufficient dosage can cause lasting dementia (117). It is generally accepted that no more than one drink per day for a female and no more than two drinks per day for a male below 65 years of age is safe. However, both males and females over 65 must not consume more than one drink per day. There is no safe limit of alcohol during pregnancy, and in order to avoid fetal alcohol syndrome a pregnant woman must practice total abstinence.

REFERENCES

1. Dudley R. 2004. Ethanol, fruit ripening and the historical origins of human alcoholism in primate frugivory. *Integra Comp Biol* 44 (4): 315–323.
2. Hanson D. History of alcohol and drinking around the world. http://www2.postdam.edu/hansondj/Controversies/1114796842 html. Accessed August 23, 2010.
3. Vallee BL. 1998. Alcohol in the Western World. *Sci Am* 278 (6): 80–85.
4. Loyola Marymount University, Los Angeles. Heads up: History of Alcohol Use http://www.lmu.edu/pages25071.aspx. Accessed August 24, 2010.
5. Kerr WC, Greenfield TK, Tujague J, Brown SE. 2005. A drink is a drink? Variation in the amount of alcohol contained in beer, wine and spirits drinks in a US methodological sample. *Alcohol Clin Exp Res* 29 (11): 2015–2021.
6. Liber CS. 2003. Relationship between nutrition, alcohol use and liver disease. *Alcohol Res Health* 27 (3): 220–231.
7. Jones AW, Jonsson KA. 1994. Food-induced lowering of blood ethanol profiles and increased rate of elimination immediately after a meal. *J Forensic Sci* 39 (4): 1084–1093.
8. Gill J. 1997. Women alcohol and the menstrual cycle. *Alcohol Alcohol* 32 (4): 435–441.
9. Zakhari S. 2006. Overview: How is alcohol metabolized by the body? *Alcohol Res Health* 29 (4): 245–254.
10. Summary of Health Statistics for U.S. adults: National interview survey 2008. National Center for Health Statistics, Hyattsville, MD.
11. LaValle R, Williams GD, Ti H. 2009, September. Surveillance Report #87. Apparent per capita alcohol consumption: National, state, regional trends 1977–2007. National Institute on Alcohol Abuse and Alcoholism, Bethesda, MD.
12. Mohapatra S, Patra J, Popova S, Duhig A, et al. 2010. Social cost of heavy drinking and alcohol dependence in high income countries. *Int J Pub Health* 55 (3): 149–157.
13. National Institute on Alcohol Abuse and Alcoholism (NIAAA): FY 2009 Presidents' Budget Request for NIAAA-Directors' Statement before the House Subcommittee on Labor-HHS Appropriations-March 5, 2008. Presented by Ting-Kai Li, M.D. http://www.niaaa.nih.gov/AboutNIAAA/CongressionalInformation/Testimony/statement.html. Accessed August 24, 2010.
14. Msnbc.com News update June 25, 2005. No author listed. Alcohol linked to 75,000 U.S. deaths a year. http://www.msnbc.msn.com/id/6089353. Accessed August 24, 2010.
15. Yoon YH, Yi H. 2008, August. Surveillance Report #83. Liver cirrhosis mortality in the United States, 1970–2005. National Institute on Alcohol Abuse and Alcoholism, Bethesda, MD.
16. Rosen SM, Miller TR, Simon M. 2008. The cost of alcohol in California. *Alcohol Clin Exp Res* 32 (11): 1925–1936.
17. Balakrishnan R, Allender S, Scarborough P, Webster P, et al. 2009. The burden of alcohol related ill health in the United Kingdom. *J Pub Health* (Oxford) 31 (3): 366–373.
18. U.S. Department of Agriculture and U.S. Department of Health and Human Services. 2005. In: *Dietary Guidelines for Americans*. Chapter 9—Alcoholic Beverages. Washington, DC: U.S. Government Printing Office; pp. 43–46. Available at http://www.health.gov/DIETARYGUIDELINES/dga2005/document/html/chapter9.htm. Accessed July 1, 2009.
19. Reid MC, Fiellin DA, O'Connor PG. 2008. Hazardous and harmful alcohol consumption in primary care. *Arch Int Med* 159 (8): 1681–1689.
20. Coulton S. 2008. Alcohol misuse. *Clinical Evidence* (online journal) August 27; pii: 1017.
21. Naimi TS, Brewer RD, Miller JW, Okoro C, et al. 2004. What do binge drinkers drink? Implications for alcohol control policy. *Am J Prev Med* 33 (3): 188–193.

22. Flowers NT, Naimi TS, Brewer RD, Elder RW, et al. 2008. Patterns of alcohol consumption and alcohol-impaired driving in the United States. *Alcohol Clin Exp Res* 32 (4): 639–644.
23. Breitmeier D, Seeland-Schulze I, Hecker H, Schneider U. 2007. The influence of blood alcohol concentrations around 0.03% on neuropsychological functions—a double blind, placebo controlled investigation. *Add Biol* 12 (2): 183–189.
24. Jones AW. 1999. The drunkest driver in Sweden: blood alcohol concentration of 0.545%. *J Stu Alcohol* 60 (3): 400–406.
25. Brouwer IG. 2004. The Widmark formula for alcohol quantification. *SADJ: South African Dental Assoc J* 59 (10): 427–428.
26. Torrens J, Riu-Aumatell M. Lopez-Tamames E, Buxaderas S. 2004. Volatile compounds of red and white wines by head-space-solid phase microextraction using different fibers. *J Chromatogr Sci* 42 (6): 310–316.
27. Madrid AM, Hurtado C, Gatica S, Chacon I, et al. 2002. Endogenous ethanol production in patients with liver cirrhosis, motor alteration and bacterial overgrowth. *Revista Medica de Chile* 130 (12): 1329–1334.
28. Appenzeller BM, Schuman M, Wennig R. 2008. Was a child poisoned by ethanol? Discrimination between ante-mortem and postmortem formation. *Int J Legal Med* 122 (5): 429–434.
29. Hoiseth G, Karinen R, Christophersen AS, Olsen L, et al. 2007. A study of ethyl glucuronide in post mortem blood as a marker of ante-mortem ingestion of alcohol. *Forensic Sci Int* 165 (1): 41–45.
30. Friedman LA, Kimball AW. 1986. Coronary heart disease mortality and alcohol consumption in Framingham. *Am J Epidemiology* 124 (3): 481–489.
31. Castelli WP. 1990. Diet, smoking, and alcohol: influence on coronary heart disease risk. *Am J Kid Dis* 16 (4 Suppl 1): 41–46.
32. Boffetta P, Garfinkel L. 1990. Alcohol drinking and mortality among men enrolled in an American Cancer Society prospective study. *Epidemiology* 1 (5): 342–348.
33. Sesso HD, Stampfer MJ, Rosner B, Hennekens CH, et al. 2000. Seven year changes in alcohol consumption and subsequent risk of cardiovascular disease in men. *Arch Int Med* 160 (17): 2605–2612.
34. Ajani UA, Gaziana JM, Lotufo PA, Hennekens CH, et al. 2000. Alcohol consumption and risk of coronary heart disease by diabetes status. *Circulation* 102 (5): 500–505.
35. Tolstrup J, Jensen MK, Tjonneland A, Overvad K, et al. 2006. Prospective study of alcohol drinking patterns and coronory heart disease in women and men. *Br Med J* 332 (7552): 1244–1248.
36. Bryson CL, Mukamal KJ, Mittleman MA, Fried LP, et al. 2006. The association of alcohol consumption and incident heart failure: the cardiovascular health study. *J Am Coll Cardiol* 48 (2): 305–311.
37. Snow WM, Murray R, Ekuma O, Tyas SL, et al. 2009. Alcohol use and cardiovascular health outcomes: a comparison across age, gender in the Winnipeg Health and Drinking Survey Cohort. *Age Aging* 38 (2): 206–212.
38. Gordon T, Ernst N, Fisher M, Rifkind BM. 1981. Alcohol and high density lipoprotein cholesterol. *Circulation* 64 (3 Part 2, Suppl): III 63–67.
39. Hulley SB, Gordon S. 1981. Alcohol and high density lipoprotein cholesterol: casual inference from diverse study designs. *Circulation* 64 (3 Part 2, Suppl): III 57–63.
40. Wakabayashi I, Araki Y. 2009. Association of alcohol consumption with blood pressure and serum lipid in Japanese female smokers and nonsmokers. *Gender Med* 6 (1): 290–299.
41. Wakabayashi I, Araki Y. 2010. Influence of gender and age on relationship between alcohol drinking and atherosclerosis risk factor. *Alcohol Clin Exp Res* 34 (Suppl 1) S54–S60.

42. Rubin R. 1999. Effect of ethanol on platelet function. *Alcohol Clin and Exp Res* 23 (6): 1114–1118.
43. Ruf JC. 2004. Alcohol, wine and platelet function. *Biol Res* 37: 209–215.
44. Wu JM, Wang ZR, Hsieh TC, Bruder JL, et al. 2001. Mechanism of cardioprotection by resveratrol, a phenolic antioxidant present in red wine. *Int J of Mol Med* 8 (1): 3–17.
45. Truelsen T, Gronbaek M, Schnohr P, Boyen G. 1998. Intake of beer, wine and spirits and risk of stroke: the Copenhagen city heart study. *Stroke* 29 (12): 2467–2472.
46. Nielsen NR, Truelsen T, Barefoot JC, Johnsen SP, et al. 2005. Is the effect of alcohol on risk of stroke confined to highly stressed persons? *Neuroepidemiology* 25 (3): 105–113.
47. Koppes LL, Dekker JM, Hendriks HF, Bouter LM, et al. 2005. Moderate alcohol consumption lowers the risk of Type 2 diabetes: a meta-analysis of prospective observational studies. *Diabetes Care* 28 (3): 719–725.
48. Carrison SW, Hammar N, Grill V, Kaprio J. 2003. Alcohol consumption and the incidence of Type 2 diabetes: 20 years follow up of the Finnish twin cohort study. *Diabetes Care* 26: 2785–2790.
49. Balinus DO, Taylor BJ, Irving H, Roereke M, et al. 2009. Alcohol as a risk factor for Type 2 diabetes: a systematic review and meta-analysis. *Diabetes Care* 32: 2123–2132.
50. Orgogozo JM, Dartigues JF, Lafont S, Letenneur L, et al. 1997. Wine consumption and dementia in the elderly: a prospective study in Bordeaux area. *Revue Neurologique* (Paris) 153: 185–192.
51. Mitchell RM, Neafsey EJ, Collins MA. 2009. Essential involvement of the NMDA receptor in ethanol preconditioning–dependent neuroprotection from amyloid-β *in vitro*. *J Neurochem* 111: 580–588.
52. Gao L, Weck MN, Stegmaier C, Rothenbacher D, et al. 2009. Alcohol consumption and chronic atrophic gastritises: population based study among 9,444 older adults from Germany. *Int J Cancer* 125: 2918–2922.
53. Chao C, Slezak JM, Caan BJ, Quinn VP. 2008. Alcoholic beverage intake and risk of lung cancer: the California Men's Health Study. *Cancer Epidemiol Biomark Prev* 17: 2692–2699.
54. Chao C. 2007. Association between beer, wine, and liquor consumption and lung cancer risk: a meta-analysis. *Cancer Epidemiol Biomark Prev* 16: 2436–2447.
55. Jiang X, Castelao KE, Cortessis VK, Ross RK, et al. Alcohol consumption and risk of bladder cancer in Los Angeles County. *Int J Cancer* 121: 839–845.
56. Freedman ND, Schatzkin A, Leitzmann MF, Hollenbeck MF, et al. 2007. Alcohol and head and neck cancer risk in a prospective study. *B J Cancer* 96: 1469–1474.
57. Pelucchi C, Galeone C, Montella M, Polesel J, et al. 2008. Alcohol consumption and renal cell cancer risk in two Italian case controlled studies. *Ann Oncology* 19: 1003–1008.
58. Freiberg MS, Chang YF, Kraemer KL, Robinon JG, et al. 2009. Alcohol consumption, hypertension, and total mortality among women. *Am J Hyperten* 22: 1212–1218.
59. Klatsky AL, Friedman GD, Siegekaub AB. 1981. Alcohol and mortality: a ten year Kaiser-Permanente experience. *Ann Int Med* 95: 139–145.
60. Camargo CA, Hennekens CH, Gaziano JM, Glynn RJ, et al. 1997. Prospective study of moderate alcohol consumption and mortality in US male physicians. *Arch Int Med* 157: 79–85.
61. Vasse RM, Nijhuis FJ, Kok G. 1998. Association between work stress, alcohol consumption and sickness absence. *Addiction* 93: 231–241.
62. Wiley JA, Camacho TC. 1980. Life-style and future health: evidence from Alameda county study. *Prev Med* 9: 1–21.
63. Kallberg H, Jacobsen S, Bengtsson C, Pedersen M, et al. 2009. Alcohol consumption is associated with decreased risk of rheumatoid arthritis: results from two Scandinavian studies. *Ann Rheumatoid Dis* 68: 222–227.

64. Nissen MJ, Gabay C, Scherer A, Finchk A. 2010. The effect of alcohol on radiographic progression in rheumatoid arthritis. *Arthritis Rheum* 62: 1265–1272.
65. Cohen S, Tyrell DA, Russell MA, Jarvis NJ, et al. 1993. Smoking, alcohol consumption and susceptibility to the common cold. *Am J Public Health* 83 (9): 1277–1283.
66. Takkouch B, Regueira-Mendez C, Garcia-Closas R, Figueiras A, et al. 2002. Intake of wine, beer, and spirits and the risk of clinical common cold. *Am J Epidemiol* 155 (9): 853–858.
67. Yoon YH, Yi H. 2008, August. Surveillance Report #83. Liver cirrhosis mortality in the United States, 1970–2005. National Institute on Alcohol Abuse and Alcoholism, Bethesda, MD.
68. Rosen SM, Miller TR, Simon M. 2008. The cost of alcohol in California. *Alcohol: Clin Exp Res* 32 (11): 1925–1936.
69. Balakrishnan R, Allender S, Scarborough P, Webster P, et al. 2009. The burden of alcohol related ill health in the United Kingdom. *J Pub Health* (Oxford) 31 (3): 366–373.
70. Dawson DA. 2001. Alcohol and mortality from external causes. *J Stu Alcohol Drug* 62 (6): 790–797.
71. Roerecke M, Rehm J. 2010. Irregular heavy drinking occasions and risk of ischemic heart disease: a systematic review and meta-analysis. *Am J Epidemiol* 171 (6): 633–644.
72. Hart CL, Smith GD, Hole DJ, Hawthorne VM. 1999. Alcohol consumption and mortality from all causes, coronary heart disease, and stroke: results from a prospective cohort study of Scottish men with 21 years of follow up. *BMJ* 318 (7200): 1725–1729.
73. Breslow RA, Graubard BI. 2008. Prospective study of alcohol consumption in the United States: quantity, frequency and cause of cause specific mortality. *Alcohol Clin Exp Res* 32 (3): 513–521.
74. Britton A, Marmot M. 2004. Different measures of alcohol consumption and risk of coronary heart disease and all cause mortality: a 11 year follow up of the Whitehall II cohort study. *Addiction* (99): 109–116.
75. Kauhanen J, Kaplan GA, Goldberg DE, Salonen JT. 1997. Beer binging and mortality: results from Kuopio ischemic heart disease risk factor study, a prospective population based study. *BMJ* 315: 846–851.
76. Anda RF, Williamson DF, Remington PL. 1998. Alcohol and fatal injuries among US adults. Finding from the NHANES I Epidemiologic follow up study. *JAMA* 260 (17): 2529–2532.
77. Schilling EA, Aseltine RH, Glanovsky JL, James A, et al. 2009. Adolescent alcohol use, suicidal indention and suicide attempts. *J Adolesc Health* 44: 335–341.
78. Swahn MH, Bossarte RM, Sullivent EE III. 2008. Age of alcohol use initiation, suicidal behavior, and peer and dating violence victimization and perception among high risk seventh grade adolescents. *Pediatrics* 121: 297–305.
79. Palijan TZ, Kovacevic D, Radeljak S, Kovac M, et al. 2009. Forensic aspects of alcohol abuse and homicide. *Collegium Anthropologium* 33: 893–897.
80. Alcohol violence and aggression. Alcohol Alter, National Institute on Alcohol Abuse and Alcoholism. No. 38, October 1997. http://pubs.niaaa.nih.gov/publications/aa38thm. Accessed August 26, 2010.
81. DeBarkey SF, Stinson FS, Grant BF, Dufour MC. 1996. Surveillance report #41. Liver cirrhosis mortality in the United States 1970–1993. National Institute on Alcohol Abuse and Alcoholism, Bethesda, MD.
82. Belentani S, Tribelli C. 2001. Spectrum of liver disease in general population. Lessons from Dionysos study. *J Hepatol* 35: 531–537.
83. Bellentani S, Saccoccio G, Costa G, Tribelli C, et al. 1997. Drinking habits as cofactors of risk of alcohol induced liver damage. The Dionysos study group. *Gut* 42: 845–850.
84. Walsh K, Alexander G. 2000. Alcoholic liver disease. *Postgrad Med* 281: 280–286.

85. Lu X, Luo JY, Tao M, et al. 2004. Risk factors for alcoholic liver disease in China. *World J Gastroenterol* 10: 2423–2426.
86. Hezode C, Lonjon I, Roudot-Thorval F, Pawlotsky JM, et al. 2003. Impact of moderate alcohol consumption on histological activity and fibrosis in patients with chronic hepatitis C, and specific influence of steatosis, a prospective study. *Aliment Pharmacol and Ther* 17: 1031–1037.
87. Zeigler DW, Wang CC, Yoast RA, Dickinson BD, et al. 2005. The neurocognitive effects of alcohol on adolescents and college students. *Prev Med* 40: 23–32.
88. Harford TC, Yi HY, Hilton ME. 2006. Alcohol abuse and dependence in college and noncollege samples: a ten year prospective follow up in a national survey. *J Stud Alcohol* 67: 803–809.
89. Casas-Gil MJ, Navarro-Guzman JL. 2002. School characteristics among children of alcohol parents. *Psychol Rep* 90: 341–348.
90. Medina KL, McQueeny T, Nagel BJ, Hanson KL, et al. 2008. Prefrontal cortex volumes in adolescents with alcohol use disorders: unique gender effects. *Alcohol Clin Exp Res* 32: 386–394.
91. Kopelman MD, Thomson AD, Guerrini I, Marshall EJ. 2009. The Korsakoff syndrome: clinical aspect, psychology and treatment. *Alcohol Alcohol* 44: 148–154.
92. Harper C. 2007. The neurotoxicity of alcohol. *Hum Exp Toxicol* 26 (3): 251–257.
93. Paparrigopoulos T, Tzavellas E, Karaiskos D, Kouzoupis A, et al. 2010. Complete recovery from undertreated Wernicke-Korsakoff syndrome following aggressive thiamine treatment. *In Vivo* 24: 231–233.
94. Martin PR, Singleton CK, Hiller-Sturmhofel S. 2003. The role of thiamine deficiency in alcoholic brain damage. *Alcohol Res Health* 27: 134–142.
95. Ke ZJ, Wang X, Fan Z, Luo J. 2009. Ethanol promotes deficiency-induced neural death: involvement of double stranded RNA activated protein kinase activity. *Alcohol Clin Exp Res* 33: 1097–1103.
96. Mochizuki H, Masaki T, Matsushita S, Ugawa Y, et al. 2005. Cognitive impairment and diffuse white matter atrophy in alcoholics. *Clin Neurophysiol* 116: 223–228.
97. Saxton J, Munro CA, Butters MA, Schramke C, et al. 2000. Alcohol, dementia, and Alzheimer's disease: comparison of neuropsychological profiles. *J Geriat Psychia Neurol* 13: 141–149.
98. Hommer D, Momenan R, Kaiser E, Rawlings R. 2001. Evidence for a gender related effect of alcoholism on brain volumes. *Am J Psychiatry* 158: 198–204.
99. Hunt WA. 1993. Are binge drinkers more at risk of developing brain damage? *Alcohol* 10 (6): 559–561.
100. Trabert W, Betz T, Niewald M, Huber G. 1995. Significant reversibility of alcohol brain shrinkage within 3 weeks of abstinence. *Acta Psychia Scand* 92: 87–90.
101. Asada T, Takaya S, Takayama Y, Yamauchi H, et al. 2010. Reversible alcohol related dementia: a five year follow up using FDG-PET and neuropsychological tests. *Int Med* 49: 283–287.
102. Klatsky AL. 2010. Alcohol and cardiovascular health. *Physiol Behavior* 100: 76–81.
103. Laonigro I, Correale M, Di Biase M, Altomare E. 2009. Alcohol abuse and heart failure. *Eu J Heart Failure* 11: 453–462.
104. Ikehara S, Iso H, Yamagishi K, Yamamoto S. 2009. Alcohol consumption, social support and risk of stroke and coronary heart disease among Japanese men: the JPHC study. *Alcohol Clin Exp Res* 33: 1025–1032.
105. Numi K, Methuen T, Maki T, Lindstedt KA, Kovanen PT, et al. 2009. Ethanol induces apoptosis in human mast cells. *Life Sciences* 85 (19–20): 678–684.
106. Baum MK, Rafie C, Lai C, Sales S, et al. 2010. Alcohol use accelerates HIV disease progression. *AIDS Res Hum Retrovir* 26: 511–518.

107. Boe DM, Vandivier RW, Burnham EL, Moss M. 2009. Alcohol abuse and pulmonary disease. *J Leukocytes Biol* 76: 1097–1104.
108. Emanuele N, Emanuele MA. 1997. Alcohol alters critical hormonal balance. *Alcohol Health Res World* 21: 53–64.
109. Alcohol Alert No. 21, PH 345. 1993, July. National Institute on Alcohol Abuse and Alcoholism, Bethesda, MD.
110. Apte M, Pirola R, Wilson J. 2009. New insights into alcoholic pancreatitis and pancreatic cancer. *J Gastroenterol Hepatol* 24 (3, Suppl 3): S351–S356.
111. Martin-Moreno JM, Boyle P, Gorgojo L, Willett WC, et al. 1993. Alcoholic beverage consumption and risk of breast cancer in Spain. *Cancer Causes Control* 4: 345–353.
112. Bessaoud F, Daures JP. 2008. Pattern of alcohol (especially wine) consumption and breast cancer risk: a case controlled study among population in Southern France. *Ann Epidemiol* 18: 467–475.
113. Nagata C, Mizoue T, Tanaka K, Tsuji I, et al. 2007. Alcohol drinking and breast cancer risk: an evaluation based on a systematic review of epidemiological evidence among Japanese population. *Jpn J Clin Oncol* 37; 568–574.
114. Jones KL, Smith DW. 1973. Recognition of the fetal alcohol syndrome in early infancy. *Lancet* 302: 999–1001.
115. Sampson PD, Streissguth AP, Bookstein FL. 1997. Incidence of fetal alcohol syndrome and prevalence of alcohol related neurodevelopmental disorder. *Teratology* 56: 317–326.
116. May PA, Gossage JP, Kalberg WO, Robinson LK, et al. 2009. Prevalence and epidemiological characteristics of FASD from various research methods with an emphasis on recent in school studies. *Develop Disability Res Rev* 15; 176–192.
117. Brust JC. 2010. Ethanol and cognition: indirect effects, neurotoxicity and neuroprotection: a review. *Int J Environ Res Public Health* 7: 1540–1557.

3 Slate and Trait Markers of Alcohol Abuse

Joshua Bornhorst, Annjanette Stone,
John Nelson, and Kim Light

CONTENTS

3.1 Introduction 48
3.2 Biochemical (Slate) Markers of Chronic Alcohol Use 52
    3.2.1 Short-Term Direct Markers for Alcohol Consumption......................54
        3.2.1.1 Direct Ethanol Measurements 54
        3.2.1.2 Fatty Acid Ethyl Esters (FAEE).................................55
        3.2.1.3 5-Hydroxytryptophol (5-HTOL) 55
        3.2.1.4 Ethylglucuronide (EtG)..55
    3.2.2 Long-Term Alcohol Use Markers 56
        3.2.2.1 General Markers of Nutritional Status and Liver Damage ...56
        3.2.2.2 N-Acetyl-β-Hexosaminidase (β-Hex).............................56
        3.2.2.3 Mean Corpuscular Volume (MCV) 57
        3.2.2.4 Lipid Markers...57
        3.2.2.5 γ-Glutamyltransferase (GGT) 57
        3.2.2.6 Carbohydrate Deficient Transferrin (CDT) 58
        3.2.2.7 Combination of CDT and GGT or γ-CDT............................59
        3.2.2.8 Acetaldehyde..59
        3.2.2.9 Sialic Acid..60
3.3 Genetic Markers of Alcohol Dependence 60
    3.3.1 Specific Markers of Alcohol Susceptibility......................................60
    3.3.2 Alcohol and Aldehyde Dehydrogenases...60
    3.3.3 Neurotransmitter Systems 63
    3.3.4 Dopamine 64
    3.3.5 Dopamine Receptors 64
    3.3.6 Dopamine Transporters 65
    3.3.7 Dopamine Metabolizing Enzymes DβH 65
    3.3.8 Catechol-O-Methyltransferase (COMT)..66
    3.3.9 Monoamine Oxidase (MAO)-A...66
    3.3.10 GABA (γ-Aminobutyric Acid) 66
    3.3.11 Acetylcholine 67
    3.3.12 Glutamate...68
    3.3.13 Serotonin..69
    3.3.14 Neuropeptide Y (NPY)...69

    3.3.15 *ACN9*...70
    3.3.16 Opioids...70
3.4   Pharmacogenetics of Alcohol Abuse Treatment 70
3.5   Advances in Molecular Profiling Technologies of Markers of Alcoholism ... 71
    3.5.1   Transcriptomics 71
    3.5.2   Genomics 72
        3.5.2.1   Metabolomics..75
    3.5.3   Proteomics 76
    3.5.4   Epigenetics..76
3.6   Conclusions...78
References..78

3.1   INTRODUCTION

Slate markers are based on measurable biochemical changes following alcohol abuse. Such markers can be either direct ethanol intake markers such as ethanol and ethanol metabolites, or the elevation of indirect biochemical markers that are representative of past ethanol intake. Trait markers are genetic markers indicating a degree of susceptibility of an individual for alcohol abuse. Although trait markers are generally in research and developmental stages, many slate markers of alcohol abuse are well established and widely available. There is increasing interest in the further development and understanding of clinically useful genetic markers. This chapter reviews both biochemical and genetic markers that could be employed to identify susceptibility to alcohol dependence, identify abuse, assist in the selection of effective therapeutic options, and monitor therapeutic compliance. These markers are summarized in Table 3.1. Please note that unless specified otherwise the term *alcohol* denotes ethanol throughout this chapter.

Alcohol use is widespread in modern society. Alcohol product sales data indicate that for the year 2007 every American 14 years and older accounted for 1.21 gallons of beer, 0.38 gallons of wine, and 0.73 gallons of distilled spirits summing to a total of 2.31 gallons [1]. Consumption (per capita) declined from 1990 through 1997 and 1998 reaching a trough of 2.14 gallons. Consumption on a per capita basis has steadily increased since [1].

It is clear that not all alcohol consumption is associated with negative health and welfare impacts, and it should be noted that the majority of individual consumption may not result in notable adverse consequences. Total consumption volume is usually an indicator of disease linkage, especially for chronic disease states over time. Acute effects of alcohol are more closely associated with the patterns of consumption, although the volume may also be important. For example, consumption of two standard drinks per day may be associated with positive impacts on some disease conditions; however, the same total volume consumed within a few hours is more likely to be associated with acute accident or injury [2,3].

Nevertheless, alcohol consumption has important and significant adverse impacts throughout our society, including at individual, family, community, business, state, and national levels. For example, 39.5% of all individuals dying in a traffic accident during the year 2004 are considered to be deaths related to alcohol; a total of 16,919

TABLE 3.1
Indicators of Alcohol Abuse and Susceptibility

Observational Markers

DSM-IV (*Diagnostic and Statistical Manual of Mental Disorders,*
   fourth edition, American Psychiatric Association) diagnostic criteria
Concurrent use of other substances

"Slate" Markers

Short-Term Markers of Alcohol Use
Serum/plasma/blood/saliva
Ethanol
Fatty acid esters (FAEE)
5-Hydroxytryptophol (5-HTOL)
Ethylglucoronide (EtG)

Long-Term Markers of Alcohol Use
General markers of nutritional status and liver damage
N-Acetyl-β-hexosaminase (β-hex)
Mean corpuscular volume (MCV)
Lipid markers
Gamma-glutamyltransferase (GGT)
Carbohydrate deficient transferrin (CDT)
Combination of CDT and GGT
Acetaldehyde
Sialic acid

"Trait" Genetic Markers of Alcohol Abuse Susceptibility
Alcohol and aldehyde dehydrogenases
Neurotransmitter systems
Dopamine-associated markers
GABA (gamma-aminobutyric acid)
Acetylcholine
Glutamate
Serotonin
Neuropeptide Y (NPY)
ACN9
Opioids

Emerging Techniques for the Identification of Markers
of Alcohol Abuse Susceptibility
Transcriptomics
Genomics
Metabolomics
Proteomics
Epigenetics

people [4]. In addition, 430,000 hospital discharges in the United States during the year 2006 carried a primary diagnosis related to alcohol consumption [5].

Extensive studies exist regarding similarities and differences in alcohol consumption based on gender, age, and ethnic background. Within this context it is clear that on a gram per day (g/d) basis, men consume more alcohol than women [6,7]. Consumption volume tends to increase with age, and heavy drinking is inversely related to educational level [6]. In regards to age, the population of young adults aged 18 to 29 years, while representing 27% of the total U.S. population, was shown to account for roughly 45% of alcohol consumption [7]. Differences in alcohol consumption across the globe are less than might be expected, and perceived differences may have more to do with the type of beverage consumed than the absolute amounts of alcohol [6].

Alcohol use exposes every organ in the body to the drug, and more than 60 disease conditions have been linked to its consumption [2,3]. In addition to the obvious direct alcohol psychopathologies such as abuse, dependence (or addiction), diseases of the liver, and organic brain damage, alcohol use is identified as contributing to a wide variety of cancers including stomach, pancreas, colon, rectum, prostate, salivary gland, ovarium, endometrium, and bladder [2,3]. This contribution of alcohol to these cancers is identified as having a linear dose-response impact whereby the more volume of alcohol consumed, the greater is the relative risk of a negative outcome [2,3].

Using the diagnostic criteria as specified in the *Diagnostic and Statistical Manual of Mental Disorders* (*DSM-IV*), an estimated 19.3 million Americans over the age of 12 years required treatment for alcohol abuse or dependence in 2007 [8]. Briefly, the *DSM-IV* criteria differentiate abuse from dependence on the basis of the frequency of events causing clinically significant impairment or distress within specific domains of everyday life. These domains can be failures to fulfill major obligations at work, school, and so forth; hazardous situations; legal problems resulting from abuse; and continued use despite significant problems. The diagnosis of dependence requires three or more events within a 12-month period along with the addition of evidence of tolerance, withdrawal, or relapse [8,9].

Of those diagnosed with alcohol abuse or dependence by these criteria, only 8.1% actually received treatment at a specialty treatment facility, and 87.4% did not receive any treatment because they did not perceive the need for treatment [8]. Among those who did feel the need for treatment between 2004 and 2007, 42% admitted they were not ready to actually stop alcohol use and 34.5% indicated cost or insurance barriers prevented engagement in treatment [8]. Science continues to identify and develop a more detailed understanding of the biological changes in specific brain regions that underscore the dependence or addiction process [10]. It should be noted that while deemed the most reliable observational criteria for diagnosis of alcohol abuse or dependence, the *DSM-IV* criteria are highly subjective in their determination.

Within the context of alcohol consumption and the correlative aspects that might lead to a marker of alcohol dependence or addiction, the findings of alcohol-associated consequences and concordant use of other substances is promising. In this regard, the use of tobacco is highly correlated with heavy alcohol consumption, as is the appearance of characteristic features of antisocial personality disorder [6]. In addition, the appearance of insomnia when a person stops drinking is highly associated with alcohol dependence diagnosis [6]. Individuals who reported anxiety when they stopped drinking were 12 times as likely to be alcohol dependent as those not reporting anxiety [6]. Thus, the correlation of excess anxiety and insomnia upon alcohol withdrawal may be associated with the factors of vulnerability to addiction. Further research is ongoing in these areas.

Critical to the identification and treatment of problems resulting from alcohol overuse are the identification of objective slate or trait markers or factors highly associated in a specific and selective manner with alcohol-related pathologies, especially alcohol dependence/addiction (i.e., alcoholism). Given the individual and societal costs associated with abuse, there is a clear need for effective testing to conclusively identify alcoholism and alcohol abuse, as well as susceptibility to alcoholism although this important goal continues to remain elusive. A number of policy groups have recommended the implementation of alcohol screening and intervention strategies in routine health-care settings [11]. Overall studies of such programs have shown that these efforts can result in a lowering of disease burden and economic health-care costs [12].

A number of widely available biochemical markers of alcohol use have been characterized. These biochemical markers range from short-term direct markers of ethanol and its metabolites to long-term biochemical markers that can be used to monitor chronic alcohol abuse. In many cases biochemical markers have the advantage that they can be performed relatively rapidly and inexpensively. However, sensitivity and specificity are always critical concerns with these markers. These markers are discussed in Section 3.2.

In addition to biochemical changes or markers associated with alcohol use, genetic testing could potentially be used to identify individuals who are predisposed to developing alcoholism or who have clinically relevant abnormalities of ethanol metabolism. A number of the candidate genes and polymorphisms have been identified based on their involvement in alcohol metabolism or the neuronal response to alcohol. Evaluation of variants of these genes and the development of genetic testing may also be useful from a pharmacogenetic perspective in guiding drug and dose selection for detoxification and abstinence therapy and are further explored in Section 3.3.

Alcohol dependence (AD) is a complex psychiatric disorder modulated by genetic and environmental factors as well as important gene–environment interactions. Although alcoholism does not demonstrate a clear pattern of Mendelian inheritance, several large family, adoption, and twin studies have shown that the genetic contribution to AD etiology is considerable, with heritability estimated at 50% to 64% [13–19]. Recent twin studies have demonstrated that there is no difference in heritability of alcohol dependence between men or women [20]. While some studies indicate that the prevalence of alcohol-associated disorders varies among ethnic minorities in the United States, other studies indicate that overall total alcohol consumption may be similar for different ethnic backgrounds [6,11,21].

Social, cultural, and biological factors all contribute to the differences in alcohol use and abuse among these ethnic groups. For instance, a genetic variant in the aldehyde dehydrogenase 2 gene may provide Asian and Pacific Islander populations with some protection against alcohol consumption and abuse [22]; yet, effective drug and psychosocial treatment options for Asians/Pacific Islanders (as well as other ethnic minorities) are understudied [23,24]. Therefore, it is important to identify genes that predispose individuals from diverse populations to alcoholism, genes that alter treatment response, and genes that interact with other environmental factors.

Elucidation of the significance of potential genetic polymorphisms associated with alcohol abuse is ongoing. Molecular genetics profiling technology is increasingly moving from expensive, single-locus methodologies to cost-effective, genomewide analysis formats that may spur further advances in understanding the impact of genetic variation on ethanol susceptibility. Advances in molecular profiling technologies within the disciplines of genomics, transcriptomics, proteomics, and metabolomics are essential for the development of prevention strategies and personalized treatments of alcoholism and are discussed in Section 3.4.

Perhaps the most exciting area of current research into potential slate and trait markers of alcohol consumption and the onset of alcohol dependence and addiction is in the area of epigenetic alterations. Studies of changes in genetic expression or epigenetics of the individual's expression will likely lead to more specific and selective objective markers. Epigenetics refers to alterations to the chromatin structure or related noncoding sequence modifications that alter the expression of the DNA coding sequences and involve three general mechanisms: histone-based modifications, methylation or demethylation of DNA, and the expression of non-protein-coding RNA transcripts (termed small RNAs) that modify genetic expression. Such alterations may be time and tissue specific and in some cases are shown to be transgenerational [25]. Histone methylation and DNA methylation appear to act in concert, although DNA methylation may be less important as an epigenetic marker in specific circumstances [25]. It has been shown that alcohol-induced hyperacetylation may be identified as an indicator of hepatotoxicity [26]. In addition, the role of genetic background and alcohol-induced alterations of gene expression has led to initial identification of a potential neuroadaptive response pattern triggered by abstinence in mice that could result in specific targets for identification of alcohol dependence and addiction [27].

3.2 BIOCHEMICAL (SLATE) MARKERS OF CHRONIC ALCOHOL USE

There is considerable interest in capabilities to detect individuals who have previously abused alcohol by means of established and widely available biochemical markers. Biomarkers for the detection of chronic alcohol use could be used to screen for heavy alcohol consumption, identify changes in drinking behavior, and monitor therapy of alcoholism [28]. It is proposed that emergency departments would save $3.81 for every dollar invested in biochemical marker screening and intervention programs [29]. The availability of methods for these markers is varied, with some readily available but others requiring "home-brew" testing techniques using mass spectroscopy. Estimates of the relative sensitivities and specificities of many of these markers can be found in Hannksela et al. [28]. Table 3.2 provides a brief summary of the markers discussed in this work.

TABLE 3.2

Selected Markers of Alcohol Consumption

Marker	Abbreviation	Approximate Period of Detection of Prior Alcohol Use	Comments
Ethanol, blood	EtOH	Less than 1 day	Direct marker of intake; elevates anion and osmolar gaps; usually exhibits zero-order elimination
Fatty acid ethyl esters	FAEEs	Less than 1 day in plasma, several months in hair and meconium	Exhibits high sensitivity for exposure; represents a metabolite of ethanol; presence in hair and meconium may be useful in forensic and neonatal exposure applications
5-Hydroxytryptophol, urine	5-HTOL	1–3 days	Often utilized as part of a 5-HTOL/5-HIAA ratio
Ethyl glucuronide, urine	EtG	1 week	A minor metabolite of ethanol, can be detected in hair and tissues; used in forensic investigations
Aminotransferases, plasma	AST/ALT	3 weeks	A De Ritis ratio (AST/ALT) >2 suggests alcoholic liver damage
N-acetyl β- hexosaminidase	β-Hex	1–2 weeks, plasma; 2–4 weeks, urine	Has a half-life of 6 days in plasma; elevated in kidney insufficiency, pregnancy, and diabetes
High-density lipoprotein, plasma	HDL	Long term	Widely available; elevated values associated with alcohol dependence; low specificity
Phosphatidylethanol, blood	PEth	2–3 weeks	A metabolite of ethanol with a longer period of detection; poor stability in nonfrozen biological specimens
Gamma-glutamyltransferase, plasma	GGT	2–3 weeks	Nonspecific, but sensitive, commonly utilized marker
Carbohydrate-deficient transferrin, plasma	CDT	2–3 weeks	A relatively specific marker; genetic variants of transferrin can cause false positives
GGT and CDT in combination, plasma	Gamma-CDT	2–3 weeks	Increased sensitivity as compared to GGT or CDT alone without loss of specificity
Acetaldehyde adducts (hemoglobin-associated acetaldehyde)	HAA	Variable, 1–4 months for HAA	Titers of IgA autoantibodies can also be utilized as alcohol exposure markers

continued

TABLE 3.2 (continued)
Selected Markers of Alcohol Consumption

Marker	Abbreviation	Approximate Period of Detection of Prior Alcohol Use	Comments
Mean corpuscular volume	MCV	2–4 months	MCV >100 fl suggestive of alcohol abuse in absence of anemia; B12 and/or folate deficiency, smoking, hypothyroidism, or other liver and hematological disorders can also increase MCV
Sialic acid aminosaccharides, apolipoprotein J, plasma	SIJ, ApoJ	Variable, ~8 weeks for ApoJ	Ethanol exposure decreases sialylation of ApoJ; false positives observed in patients with tumors, diabetes, and cardiovascular disease

3.2.1 Short-Term Direct Markers for Alcohol Consumption

The National Academy of Clinical Biochemistry laboratory medicine practice guidelines recommend that a direct serum ethanol test be available to emergency departments with a turnaround time of less than 1 hour [30]. Indeed, alcohol is the most common of all drugs involved in emergency department visits, with some emergency departments reporting that one third of patients admitted during the evening exhibited substantially elevated blood alcohol levels [31].

3.2.1.1 Direct Ethanol Measurements

Direct measurements of circulating ethanol in serum have the advantages of being widely available and are relatively specific [32]. Commonly used methods include serum determinations by enzyme assay and gas chromatography that can be utilized with turnaround times of less than 1 hour. Blood or serum analysis using the enzyme alcohol dehydrogenase is relatively specific for ethanol with only slight positive interferences observed in the presence of other alcohols such as isopropanol, methanol, and ethylene glycol [33]. Breath analyzers that measure the vapor pressure of ethanol are sometimes used at the bedside but suffer a host of potential interferences [34]. Although osmolar gap calculation is an imperfect method for screening for the presence of ethanol, it still is occasionally utilized in emergency situations [35].

Rapid blood or serum/plasma ethanol testing can be used to investigate alcohol concentrations in individuals with elevated anion and osmolar gaps and may have concurrent poisoning with other alcohols or who exhibit clinical signs of ethanol intoxication consistent with a sedation toxidrome. These signs can include general central nervous system (CNS) depression, coma, respiratory depression, hypotension, and hypothermia. Other potential uses for serum alcohol testing in the emergency room include proving an individual is deemed to be incapacitated in regards to making medical care decisions, confirmation of sobriety prior to transfer, as a potential indication of alcohol dependence, and as an indicator for the provision of counseling [31].

While the direct measurement of alcohol and immediate metabolites in blood, saliva, and breath are clinically useful, these types of measurements have a short window of detection (lasting less than approximately 12 hours) following alcohol ingestion. Ethanol is generally cleared in a linear manner (zero-order kinetics) at a rate of approximately 20 mg/dL/hour [36]. Thus, direct ethanol measurement is not always well suited for the detection of alcohol dependence, although increased alcohol tolerance (reduced functional impairment) and high levels of blood alcohol are often associated with alcohol dependence [31]. Given this short window of detection, other longer-term markers of ethanol are often preferentially utilized for determination of alcohol abuse or dependence.

3.2.1.2   Fatty Acid Ethyl Esters (FAEE)

Fatty acid ethyl esters (FAEE) represent an additional direct alcohol consumption marker with a slightly longer window of detection than ethanol concentration determinations. These products are formed as part of the metabolism of ethanol and are only present in the serum for approximately 24 hours following alcohol ingestion. Fatty acid ethyl esters (FAEE) represent an interesting potential long-term alcohol consumption marker. A number of FAEE species are incorporated into hair where they can be measured after extraction by gas chromatography-mass spectrometry [37]. This is a potentially powerful method with 90% specificity and sensitivity [28]. Hair samples are useful in forensic and clinical toxicology [38,39]. Its concentration in hair appears to be accumulative over time and can be a measure of chronic alcohol consumption. Furthermore, this marker can be measured both in neonatal hair and meconium samples to investigate prenatal alcohol exposure [40].

3.2.1.3   5-Hydroxytryptophol (5-HTOL)

Alcohol and acetaldehyde affect the metabolism of serotonin resulting in an increased production the serotonin metabolite 5-hydroxytryptophol (5-HTOL). This is excreted in the urine and can be detected for approximately 5 to 15 hours longer than ethanol. This marker is deemed a sensitive and specific marker for ethanol consumption in the preceding 24 hours and may have application in forensic toxicology [41]. The ratio of 5-HTOL to the serotonin metabolite 5-hydroxyindole-3-acetic acid (5-HIAA) is also useful in the detection of recent alcohol intake [42].

3.2.1.4   Ethylglucuronide (EtG)

Ethylglucuronide/ethylsulfate is a marker garnering increasing interest which can be detected for a period of a few days following alcohol ingestion [43,44]. Ethylglucuronide is a minor metabolite of ethanol resulting from conjugation with glucuronic acid. This can be detected in the blood for ~36 hours and can be detected for several days in urine and tissues for several days following cessation of alcohol intake [44]. EtG is also present in hair and is a promising marker for postmortem investigations of alcohol use [45].

3.2.2 LONG-TERM ALCOHOL USE MARKERS

Long-term markers for alcohol use/abuse which remain elevated for long periods of time can be used to investigate the role of alcohol as an etiological factor for other disorders, to initiate detoxification therapy, to treat dependency and motivate patients to modify drinking habits, and to monitor abstinence [28,45]. Thus, markers of chronic alcohol intake that can be detected over long periods of time following drinking cessation are often preferentially utilized. Several biochemical markers are available to identify excessive or chronic alcohol use. Some long-term markers of alcohol use include mean corpuscular volume (MCV) or markers of liver function such as γ-glutamyltransferase (EGT), and carbohydrate deficient transferrin (CDT). In general, long-term markers are responsive to alcohol intake in the preceding weeks to months.

3.2.2.1 General Markers of Nutritional Status and Liver Damage

Almost all alcoholics with liver disease or alcoholic hepatitis suffer from dietary imbalance and protein calorie malnutrition (PCM). Reduced albumin and prealbumin concentrations have been used as markers for kwashiorkor PCM. A number of significant vitamin and mineral deficiencies are also commonly observed in alcohol dependent individuals [46]. Serum uric acid is often elevated in cases of alcohol dependence [45].

Serum IgG, IgM, and IgA gamma globulins are often elevated in response to antigenic stimulation. Twofold elevations of IgA are observed in 90% of alcoholism cases, IgG and IgM can also be elevated albeit at a lower frequency. Elevations of gamma globulins are nonspecific, which limits their diagnostic value [47,48].

Liver damage is associated with chronic alcohol use. Total bilirubin is elevated in 60% of patients with alcoholic hepatitis, and prolonged alkaline phosphatase and prolonged prothrombin times are often concurrently observed [46]. Hyaluronic acid is a marker of liver fibrosis and cirrhosis and is often elevated in cases of severe alcohol abuse [45]. The serum transaminases aspartate (AST) and alanine (ALT) aminotransferase can also be elevated in liver damage but appear to have limited sensitivity for alcohol abuse [49]. Obesity tends to increase overall ALT ratios, limiting the specificity of using elevated ALT values alone as a marker of alcohol abuse [45]. A De Ritis ratio (AST/ALT) of >2 may be suggestive of an alcohol origin of liver disease. Most patients with nonalcoholic liver disease have ratios of less than one, except for individuals with nonalcoholic steatohepatitis [45]. Overall, the sensitivities and specificities of AST and ALT are generally considered to be somewhat lower than those of some of the other markers described later [28].

3.2.2.2 N-Acetyl-β-Hexosaminidase (β-Hex)

N-Acetyl-β-hexosaminidase is an enzyme elevated in heavy drinkers. Alcohol intake reduces biliary excretion of β-hexosaminidase, which can be measured in the serum or urine and has a serum half-life of 1 week [50]. Levels return to baseline in plasma after about 10 days of cessation making it a potential intermediate term marker of alcohol consumption. While some studies have shown it to be a sensitive marker of heavy drinking, other disorders such as diabetes and hypertension also cause elevations of β-Hex [51].

3.2.2.3   Mean Corpuscular Volume (MCV)

Although the exact mechanism remains unknown, red blood cell size (MCV) increases in an apparently dose-dependent relationship with intensity of alcohol intake. An increase in MCV can be noted in patients with <40 g/day consumption of alcohol [52]. Following cessation of alcohol consumption, a significant number of enlarged red blood cells remain in circulation for 2 to 4 months. The sensitivity of MCV for alcohol dependence has been estimated at ~40%. Increased MCV can be associated with thyroid disease, folate deficiency, blood loss, pharmaceutical intake, nonalcoholic liver diseases, and various hematological disorders. Thus while about 4% of all adults exhibit elevated MCV only about 65% of those cases can be attributed to alcohol intake [53,54]. In the absence of anemia an MCV of >100 fL is strongly suggestive of high alcohol intake in males [45].

3.2.2.4   Lipid Markers

Concentrations of high-density lipoprotein (HDL) increase in cases of prolonged alcohol consumption. Although the diagnostic use of HDL is limited due to a number of confounding factors, HDL can be elevated in cases of low daily amounts of chronic intake [55]. HDL may be useful in the identification of patients in the early phase of alcohol dependence who do not yet have significant liver disease [47]. Chronic alcohol users also can exhibit increased serum triglyceride concentrations.

The phospholipid phosphatidylethanol (PEth) is formed after alcohol consumption and can be detected in whole blood up to 3 weeks after cessation of drinking [56]. While PEth appears to be sensitive to heavy alcohol use, specimens have poor stability and must be kept frozen at −80°C to prevent degradation prior to analysis [45].

3.2.2.5   γ-Glutamyltransferase (GGT)

Enzymatic determination of serum enzyme activities of γ-glutamyltransferase (GGT) is perhaps the most commonly utilized traditional marker for chronic alcohol consumption. In general, a positive correlation exists between GGT serum enzyme activity and alcohol consumption [47]. This biliary canalicular enzyme is induced by alcohol consumption and is increased in cases of acute hepatocellular damage. Increased activity elevations are seen in cases of severe alcoholic liver disease. Binge drinking seems to elevate GGT activity more than mild chronic drinking. However, lower levels of daily alcohol consumption (<40 g/day) and moderate routine alcohol consumption can both result in elevated GGT activities [57,58].

GGT is deemed to be one of the more sensitive markers for alcoholism. Overall estimates of sensitivity vary widely ranging from 40% to 80% in alcohol-dependent individuals [47]. In cases of heavy alcohol consumption GGT is typically 10-fold higher than normal and returns to normal approximately 2 weeks after alcohol intake cessation. The sensitivity of GGT for alcohol consumption has been shown to be higher in men than in women [57]. Other individuals who tend to experience mild to moderate increases in average GGT activities include obese individuals, smokers, older individuals, postmenopausal women as compared to premenopausal women, and women taking oral contraceptives [59,60].

The existence of a number of relatively common causes of significantly increased GGT activity other than alcohol consumption limits its specificity and utility for screening for chronic alcohol consumption. Elevated GGT activity is used as a marker of liver disease as almost all liver disorders, especially those involving intrahepatic or posthepatic liver consumption. The induction and/or release of GGT can occur upon drug ingestion (including barbiturates, tricyclic antidepressants, warfarin, and monoamine oxidase inhibitors). Furthermore, nonalcoholic fatty liver disease and viral liver infection can cause elevations of serum GGT activity [28].

3.2.2.6 Carbohydrate Deficient Transferrin (CDT)

Quantification of the relative amount of carbohydrate deficient transferrin (CDT) in the serum provides a biomarker for the detection of heavy alcohol use in preceding weeks [61,62]. Transferrin is an iron transport glycoprotein involved in iron transport and is produced in hepatocytes. Transferrin exists as a heterogenous population of isoforms that differ in the number of attached charged carbohydrate sialic acid chains. Chronic alcohol (ethanol) use markedly increases the concentrations of the asialo and disialo CDT isoforms. The half-life of these marker CDT isoforms is approximately 2 weeks. Alcohol intake has little or no effect on the concentrations of the trisialo, tetrasialo, and pentasialo transferrin isoforms. Thus, the detection of specific CDT isoforms relative to total transferrin concentrations can be utilized to monitor sustained alcohol intake.

There are several analytical methods cleared by the U.S. Food and Drug Administration (FDA) for detecting CDT. These methods include immunoassay, high-performance liquid chromatography (HPLC) methods, and capillary gel electrophoresis methods [63]. Consumption of greater than 50 g of ethanol per day (roughly four to five drinks/day) for a period of 1 to 2 weeks is required to cause a significant measurable increase in the serum CDT fraction by these methods.

A number of clinical uses have been proposed for CDT. While false-positives can result from this test and it may not be suitable for general population-based screening, this test can be used in the initial evaluation of patients for whom there is suspicion of alcohol abuse or in patients with disorders that are often associated with alcohol abuse such as liver disease, pancreatitis, or depression. The specificity of this test makes it suitable for follow-up testing to investigate abnormalities of other biomarkers for alcohol abuse. Finally this test has been used to monitor patients who are considered to be at high risk for excessive alcohol use or alcohol abuse relapse.

As a biomarker for heavy alcohol consumption, the use of %CDT compares favorably with γ-glutamyltransferase and mean corpuscular volume, demonstrating equal or superior sensitivity and specificity in a number of studies. A study of 119 normal and 46 pathological (heavy alcohol use) samples correctly classified all samples. However, meta-analysis has shown that commonly used methods demonstrate clinical sensitivity of approximately 40% at 95% clinical specificity for substantial alcoholic intake [64,65].

Although elevated levels of CDT are deemed to be fairly specific for sustained alcohol intake, a number of factors can elevate CDT. Patients suspected of having autosomal congenital disorders of glycosylation may exhibit elevation of CDT isoforms in the absence of alcohol intake. In addition, the presence of rare genetic variants of transferrin (including D, B1, B2) may interfere with CDT analysis. Advanced liver damage (including biliary cirrhosis, hepatocellular carcinoma, and severe chronic viral hepatitis) can increase observed relative CDT levels, leading to an estimated reduction in specificity to 70% to 80% for the exclusion of heavy alcohol use. Other potential confounding factors include the presence of monoclonal antibodies, pancreatic or kidney transplantation, and immunosuppressive or anti-epileptic drug therapy.

CDT has been demonstrated to be most sensitive in males older than 40 years. Sensitivity of the CDT assay may be decreased in females, especially in those administered hormone replacement or hormone contraceptives, and pregnant females. Furthermore, increased body mass index has been demonstrated to decrease overall sensitivity for excessive alcohol intake while smoking increases diagnostic sensitivity [61,66,67].

3.2.2.7 Combination of CDT and GGT or γ-CDT

Although no single marker gives ideal sensitivity and specificity, increased performance can be achieved by the use of multiple markers for alcohol use. One marker combination that has been shown to exhibit superior prediction of heavy chronic alcohol use is that of GGT and CDT, which is often referred to as γ-CDT. A γ-CDT equation has been generated that includes contributions by both GGT and CDT markers and has been shown to correlate to alcohol consumption and is consistently elevated in patients who consume >40 g/day of ethanol. In a comparison of 257 male alcoholics versus 362 moderate to occasional social drinkers, γ-CDT was able to differentiate the populations with 75% sensitivity and 93% specificity [62].

3.2.2.8 Acetaldehyde

Acetaldehyde adducts are produced when acetaldehyde formed during alcohol metabolism conjugates to various circulating proteins [47]. The half-life of these adducts is dependent on the circulating protein. One potential useful acetaldehyde adduct is hemoglobin-associated acetaldehyde that can be detected for 28 days following alcohol ingestion [68]. Diagnostically useful autoantibodies are often produced against these adducts and can also be used to detect heavy alcohol intake [32]. The concentrations of free and bound acetaldehyde in whole blood or whole-blood-associated acetaldehyde have been used as markers of heavy alcohol consumption [69]. Hemoglobin-bound acetaldehyde will continue to increase in the presence of alcohol intake over the course of the 4-month life span of red blood cells in a process analogous to the well-known accumulation of hemoglobin A1C in the presence of glucose. This process makes whole-blood-associated acetaldehyde a potentially attractive monitor of cumulative longer-term alcohol consumption in individuals in alcohol treatment programs.

3.2.2.9   Sialic Acid

Sialic acid is a potential marker for alcohol intake. Total serum sialic acid levels have been shown to be increased in social drinkers. While total serum sialic acid may have sensitivities and specificities approaching CDT, it appears to take longer to return to baseline following cessation potentially limiting its utility as a marker. The presence of increased sialic acid aminosaccharides was demonstrated to be elevated in the serum of heavy alcohol users and has potential as a marker of alcohol abuse in the preceding 2 to 5 weeks [70]. The level of sialylation of the glycoprotein apolipoprotein J has also been proposed as a marker of heavy alcohol intake [71].

3.3   GENETIC MARKERS OF ALCOHOL DEPENDENCE

3.3.1   SPECIFIC MARKERS OF ALCOHOL SUSCEPTIBILITY

In this section, a summary of the best characterized genetic markers of alcohol susceptibility is presented. Each section focuses on genetic variations in a particular enzymatic pathway and reviews what is known about the resulting impact on alcohol-related diseases. Comprehensive reviews exist for each of these markers and should be consulted for more specific details [72–77].

Genotyping, which includes the identification of polymorphic gene variants such as single nucleotide polymorphisms (SNPs), has experienced rapid improvement in regard to throughput rates, reduction of costs, increased accuracy, and simplicity of operation [78]. All SNP genotyping methods share two common steps: sample preparation for interrogating an SNP, followed by measurement of the allele-specific product using a variety of physical methods. Commonly utilized SNP genotyping methods include quantitative real-time PCR, rapid DNA sequencing, and oligonucleotide microarrays [79]. Table 3.3 provides a summary of some of the polymorphisms discussed in this work.

3.3.2   ALCOHOL AND ALDEHYDE DEHYDROGENASES

Metabolism of ethanol is a two-stage process of elimination: The first stage involves oxidation of ethanol to acetaldehyde by the liver enzymes from the alcohol dehydrogenase (ADH) family. Acetaldehyde is a toxic by-product that may contribute to the addictive process [80]. In the second stage, aldehyde dehydrogenases (ALDHs) oxidize the acetaldehyde intermediate to NADH and acetate which are subsequently oxidized to carbon dioxide for elimination. Linkage and association studies have identified the genes involved in this two-stage process as major contributors to the pathobiology of alcoholism or as important for possible pharmacotherapeutic approaches to alcohol dependence treatment [81–84]. Several of the genes that encode ADH and ALDH enzymes exhibit functional polymorphisms that contribute to the interindividual variation in alcohol metabolizing capacity. Seven known *ADH* genes cluster on a 365 kb region of chromosome 4q23: Class I *ADH* (1A, 1B, 1C), *ADH4*, *ADH5*, *ADH6*, and *ADH7* [85,86]. The Class I *ADH* genes, mainly expressed in the liver, are responsible for about 70% of the total metabolism of ethanol [87]. Polymorphisms in *ADH1A*, *ADH1B*, and *ADH1C* genes produce enzymes with

TABLE 3.3

Selected Polymorphisms from Genes Associated with Alcohol Dependence

CHR	Gene Mutation	Protein Residue	Enzyme Affected	Effect of Polymorphism on Enzyme	Reference
4q23	ADH1B*2	His47Arg	ADH1B	*2 allele has faster conversion of alcohol to acetaldehyde; protective effect against alcoholism in several populations	Crabb et al., 2004 [89]; Whitfield et al., 2002 [90]
4q23	ADH1B*3	Arg370Cys	ADH1B	*3 allele has a protective effect against alcoholism in African Americans and Native Americans	Osier et al., 1999 [95]
9q21.13	ALDH1A1*2 ALDH1A1*3	17bp Del 3bp Del (promoter)	ALDH1A	Modulates expression of ALDH1A gene; protective effect against alcohol dependency in African Americans and Native Americans	Spence et al., 2003 [99]; Ehlers et al., 2004 [100]
11q22-q23	ANKK1 "TaqIA" rs1800497 C > T	Glu713Lys	DRD$_2$ (dopamine receptor)	T-allele affects substrate binding; specificity of the gene product is affected	Le Foll et al., 2009 [103]
11p15.5	DRD$_4$ VNTR (Exon 3)	2–10 repeats of 48bp	DRD$_4$ (dopamine receptor)	7-repeat VNTR decreases expression of gene; reduced sensitivity to dopamine as compared to the 2- and 4-repeat VNTR	Du et al., 2010 [122]
5p15.33	DAT1 VNTR	3'UTR VNTR	SLC6A3 (dopamine transporter)	May lower the expression rate of DAT; allele 9 carriers associated with alcohol dependence	Köhnke et al., 2005 [129]; Samochowiec et al., 2006 [126]
9q34.2	DβH*444 rs1108580 G > A (Exon 2)	Silent mutation (Glu > Glu)	Dopamine	A-allele associated with low DβH and alcoholism (note: effect may result from linkage disequilibrium with a functional polymorphism in close proximity)	Köhnke et al., 2006a [134]; Zabetian et al., 2003 [135]
22q11.2	COMT rs4680 A > G	Val$_{158/108}$Met	Dopamine	Met$_{158/108}$ enzyme has lower activity; linked to decreased dopamine in frontal cortex, higher alcohol consumption in men	Kauhanen et al., 2000 [140]; Köhnke, 2008 [76]

continued

TABLE 3.3 (continued)

Selected Polymorphisms from Genes Associated with Alcohol Dependence

CHR	Gene Mutation	Protein Residue	Enzyme Affected	Effect of Polymorphism on Enzyme	Reference
Xp11	$MAO\text{-}A\text{-}\mu VNTR$	30bp repeat (promoter)	Dopamine	3,5 VNTR copies are associated with lower transcriptional efficiency; VNTR polymorphism associated with dependence and antisocial behavior	Contini et al., 2006 [146]
5q34	$GABA_A$ 1236 C > T	α6 Pro385Ser	$GABA_A$	T-allele (Ser_{385}) linked with susceptibility to alcohol dependence	Radel et al., 2005 [167]
7q33	$CHRM2$ rs1824024 G > T	(intron 4)	nAChRs	T-allele linked with susceptibility to alcohol dependence	Luo et al., 2005 [183]
9q34.3 12p12	$NMDAR1$ $NMDA2B$	(intron)	Glutamate	Increased expression of NMDA receptor influences EtOH tolerance and dependence	Schumann et al., 2003 [192] Wernicke et al., 2003 [191]
6q13-15	$5\text{-}HTR1B$	G861C	Serotonin autoreceptor	Increased risk of EtOH antisocial alcoholism	Lappalainen et al., 1998 [200]
17q11	$5\text{-}HTT$	5-HTTLPR	Serotonin transporter 5-HTT linked polymorphic region	Association between severe alcohol dependence and the S allele of the 5-HTTLPR	Pinto et al., 2008 [198]
7p15 4q32 4q32 4q32	NPY $NYP1R$ $NYP2R$ $NYP5R$	Leu7Pro(NYP) and multiple others	Neuropeptide Y Receptor Receptor Receptor	Associated with increased EtOH dependence and withdrawal; may also influence cocaine addiction	Wetherill et al., 2008 [202]; Vengeliene et al., 2008 [201]
7q21	$ACN9$	Multiple	ACN9	Associated with increased EtOH dependence	Dick et al., 2008 [252]
8q11.2	$OPRK1$	Multiple	Opioid receptor (Kappa)	Associated with increased EtOH dependence	Haile et al., 2008 [206]; Edenberg et al., 2008 [253]
1p36.3	$PDYN1$	Multiple	Opioid receptor (ligand)		
6q24-25	OPMR1	A118G variant Asn40Asp	Opioid receptor (mu)	Associated with customizing naltrexone EtOH treatment	Van der Zwaluw et al., 2007 [210]; Ooteman et al., 2009 [208]

Notes: GWAS, genome-wide association study; SNP, single-nucleotide polymorphism; CHR, chromosome; MAF, minor allele frequency; Freq, frequency.

kinetic properties that differ from the reference proteins [85]. The *ADH1B*2* allele (His47Arg), which is common in Asian populations and responsible for "facial flushing," has a K_m almost 18 times greater than the *ADH1B*1* allele, therefore, it rapidly converts ethanol to aldehyde: K_m (ethanol) = concentration of ethanol equivalent to a 50% enzyme capacity) [85]. The *ADH1B*2* allele is protective against alcoholism in East Asian, Chinese, European, Jewish, and African populations [88–92]. The *ADH1B*3* allele (Arg370Cys), which is common in African and Native American Indian populations, has a K_m that is 400 times greater than the reference allele. The *ADH1B*3* allele has also shown a protective effect against alcoholism in select African American and Native American populations [93,94]. The *ADH1C* gene has three polymorphic alleles: *ADH1C*1* (reference allele), *ADH1C*2* (Arg272Gln, Ile350Val), and *ADH1C*3* (Thr352). In one study, *ADH1C*1* has been associated with protection from alcohol dependence in Native Americans [95].

The human aldehyde dehydrogenase (ALDH) family of enzymes is encoded by 19 putatively functional genes and 3 pseudogenes located on several different chromosomes. Two genes, *ALDH1* (located on 9q21.13; cytosolic isoenzyme) and *ALDH2* (located on 12q24.2; mitochondrial isoenzyme), encode enzymes significantly associated with acetaldehyde oxidation in the liver [96]. The *ALDH2*2* (Glu487Lys) polymorphism located in the mutant ADLH2 enzyme is prevalent in many Asian ethnic groups [97] and produces acetaldehyde accumulation with an alcohol-flushing reaction [98] similar to the *ADH1B*2* allele (also see www.aldh.org). Another gene from the ALDH superfamily has two promoter variants, *ALDH1A1*2* (a 17bp deletion) and *ALDH1A1*3* (a 3bp deletion), with protective effects against alcohol dependence in African American and Native American populations [99,100].

Thus far, association studies have not attempted a comprehensive genotyping approach on functional polymorphisms in both *ADH* and *ALDH* genes for alcohol dependence in multiple ethnic populations. Recently, Kuo et al. [101] tested polymorphisms in seven *ADH* genes, along with the *ALDH1A1* and *ALDH2* genes, for association with alcohol dependence in the Irish Affected Sibling Pair Study of Alcohol Dependence. In this study, numerous SNPs in the *ADH* gene were associated with AD; however, the authors concluded that they did not have a full coverage set of SNPs in the alcohol metabolism genes to adequately test for associations in AD [101]. Microarray or whole-genome sequencing strategies can overcome this limitation. Another large case-control study found associations between AD and *ADH5* genotypes; also, diplotypes of *ADH1A, ADH1B, ADH7*, and *ALDH2* were linked to AD in European Americans or African Americans [102]. Likewise, these authors suggested that a dense set of genetic markers is essential for comprehending the complex interactions among the alcohol-metabolizing enzymes and AD.

3.3.3 NEUROTRANSMITTER SYSTEMS

Neurotransmitters located in the brain are endogenous chemicals responsible for transmission of signals from neurons to target cells across synapses. Major neurotransmitter systems include the dopamine, noradrenaline, cholinergic, and serotonin systems; they are linked to the behavioral aspects of ethanol dependence. A synopsis of select neurotransmitters and the genes encoding them is described in this section: dopamine receptors and transporters, dopamine metabolizing enzymes, γ-aminobutyric acid, acetylcholine, glutamate, serotonin, neuropeptide Y, ACN9, and opioids. A brief discussion of individual genetic variants in genes from each major neurotransmitter system is included. In this review chapter, focus is placed on the mesolimbic dopamine system because it has been extensively studied in relationship to alcoholism.

3.3.4 DOPAMINE

Dopamine is the neurotransmitter used by the mesolimbic, nigrostriatal, and tuberoinfundibular pathways of the brain. The mesolimbic dopaminergic pathway or "reward" pathway and the mesocortical pathway are associated with memory, motivation, emotional response, desire/reward, addiction, and hallucinations (if not working properly). Dysfunctions in the dopaminergic systems are involved in several pathological conditions including Parkinson's disease, Tourette's syndrome, drug addiction, and hyperactivity disorders [103]. Therefore, it is highly likely that the dopaminergic system may contribute to alcoholism predisposition and vulnerability.

3.3.5 DOPAMINE RECEPTORS

Dopamine receptors, metabotropic G-protein-coupled receptors, include the D_1 family (DRD$_1$, DRD$_5$) and the D_2 family (DRD$_2$, DRD$_3$, DRD$_4$). DRD$_1$ is involved in the rewarding/reinforcing effects of drugs of abuse, but few genetic studies have been completed because functional polymorphisms in the *DRD₁* gene have not been elucidated. *DRD₁* does contain a 5′ UTR polymorphism with an A > G transition on chromosome 5q35.1 [103]. Batel et al. [104] found a specific haplotype (rs686-T-rs4532*G) within *DRD₁* that is significantly linked to alcohol dependence [104].

By contrast, DRD$_2$ is the most studied dopamine receptor in relation to alcohol-associated disorders. 10kb downstream from the *DRD₂* gene is TaqIA, a C > T substitution actually located in an exon of the *ANKK1* gene [105]. This polymorphism causes a nonsynonymous coding change (Glu713Arg) that has been investigated in 40 case-control studies and several meta-analyses involving alcoholism (see Table 1 in Le Foll et al. [103]. A meta-analysis by Smith et al. [106] of genotyping data from 44 studies of the *Taq1* polymorphism found a significant association with alcohol dependency in persons with one or two copies of the A1 allele [106]. A similar meta-analysis by Le Foll et al. [103] also detected a modest association of the A1 allele and alcohol dependency. Because these results conflict with other studies [107–109], one possible explanation for the discrepancies is that the A1 allele is in linkage disequilibrium with other influential genes near the chromosomal region of *DRD₂* such as the *ANKK1* gene. The amino acid substitution may change the substrate-binding specificity of *ANKK1* which is involved in signal transduction [105]. This possibility is also supported by the COGA study on *ANKK1* and alcoholism that found a strong association with the 5′ linkage disequilibrium block of this gene; this region does not contain TaqIA [73].

The dopamine receptor DRD_3 contains a functional polymorphism (Ser9Gly), producing an allele (Gly) that is more sensitive to dopamine [110]. A small case-control study by Limosin et al. [111] found an increased risk of impulsiveness, an addiction trait, in French alcoholics who are heterozygous for the Ser9Gly mutation (also known as the *Bal*I polymorphism due to a restriction site produced by the variant) [111]. Numerous studies have failed to find any association with alcohol dependence and DRD_3 variants [112–114].

DRD_4 receptors, G-coupled receptors encoded by the DRD_4 gene, have been linked to several psychiatric disorders (schizophrenia, bipolar disorder), neurological disorders (Parkinson's disease), and addictive behaviors (including novelty or thrill seeking). DRD_4 has a variable number of tandem repeat (VNTR) polymorphisms on exon 3 that may affect gene expression by binding nuclear factors [115]. The DRD_4 VNTR can vary from 2 to 10 48-bp repeats. Alleles with less than seven repeats are "short alleles" and alleles with more than seven repeats are "long alleles." Novelty seeking, which is found in some alcoholic personalities, has been associated with the DRD_4 VNTR long allele [116–118]. Several older studies tried to find an association of the DRD_4 VNTR with alcoholism but failed [119–121]. A study by Du et al. reported an association between *DRD4* VNTR and alcoholism in Mexican Americans [122].

DRD_5 receptors, G-coupled receptors that stimulate adenylyl cyclase, are similar in structure and function to the D_1 receptors [123]. There are several functional polymorphisms located in DRD_5 [124], but genetic studies in relation to alcohol have not been attempted due to the complexity of the gene location [125]. Nevertheless, advanced molecular technologies make the analysis of this candidate locus feasible, and polymorphisms in *DRD* may contribute to disorders influenced by the dopaminergic system.

3.3.6 DOPAMINE TRANSPORTERS

One member of the SLC6 gene family of neurotransmitter transporters is the sodium-dependent dopamine transporter (DAT, alias SLCA3). Expression of this transporter is influenced by a VNTR polymorphism located in the 3′ untranslated region of the dopamine transporter (*DAT1*) gene, found on chromosome 5q15.3 [76]. Several case-control studies and family-based association studies have reported conflicting associations with the *DAT1* VNTR polymorphism and alcohol-related phenotypes; however, most of these studies had very small sample sizes [120,126–131]. A case-control study using a German population by Köhnke et al. found a significant association between the diagnosis of alcoholism and prevalence of the nine-repeat allele in the *DAT1* gene [129].

3.3.7 DOPAMINE METABOLIZING ENZYMES DBH

Dopamine-β-hydroxylase (DβH) catalyzes the conversion of dopamine to norepinephrine; several polymorphisms in the *DβH* gene have been the focus of addiction research for many years. Two studies did not find an association between the *DβH*-1021 C > T polymorphism and alcoholism or severe withdrawal symptoms

[132,133]. The $D\beta H$ *444 G > A polymorphism, found in exon 2, was associated with alcoholism [134] but is probably linked with another functional polymorphism close to $D\beta H$ on chromosome 9q34 [134–136].

3.3.8 CATECHOL-O-METHYLTRANSFERASE (COMT)

Catechol-O-methyltransferase (COMT) is a major enzyme located in the frontal cortex of the brain that metabolizes epinephrine, norepinephrine, and dopamine. COMT metabolizes dopamine to the metabolite homovanillic acid (HVA). The *COMT* gene is found on chromosome 22q11, a locus previously linked to schizophrenia [137]. It contains the functional polymorphism $Val_{158/108}Met$ in soluble COMT (*S*-COMT) or $Val_{108}Met$ in membrane-bound COMT (*MB*-COMT), an enzyme that is expressed in brain neurons [138,139]. The $Val_{158/108}$ enzyme has more activity than the $Met_{158/108}$ form and is linked with decreased amounts of dopamine in the frontal cortex [14]. The $Met_{158/108}$ allele is associated with higher consumption of alcohol in men [140] and higher anxiety levels in women [141]. In a genetic substudy from the COMBINE Alcoholism Treatment Study, the relationship between candidate loci and drug metabolism or alcohol addiction was explored [142,143]. In the *COMT* gene, the $Val_{158}Met$ polymorphism was associated with brain endogenous opioid function, response to stress and anxiety, and differences in cognitive and emotional processes [144].

3.3.9 MONOAMINE OXIDASE (MAO)-A

Monoamine oxidase (MAO) enzyme, which also catalyzes dopamine to HVA, has a repeat polymorphism in the MAO-A gene (located on chromosome Xp11) that affects the transcriptional activity of the gene. Alleles with three or five copies of the repeat have diminished transcriptional efficiency as compared to alleles with three and a half or four copies [145]. Numerous association studies have found a positive relationship between MAO-A-μVNTR genotype and alcohol dependence or antisocial behavior linked to alcohol dependence [146–148]. These results have been contradicted by other case-control studies that did not find a positive association between MAO-A-μVNTR genotype and alcohol dependence phenotype [149–152].

3.3.10 GABA (γ-AMINOBUTYRIC ACID)

The chief inhibitory neurotransmitter in the central nervous system, γ-aminobutyric acid (GABA), is an amino acid responsible for regulating neuronal excitability. The neurotransmitter GABA binds to specific receptors located in the plasma membrane of neuronal cells. Two main classes of GABA receptors have been well characterized: ionotropic $GABA_A$ and $GABA_C$ receptors, and metabotropic $GABA_B$ receptors [153–156]. $GABA_A$ receptors are a family of ligand-activated chloride ion channels composed of multiple receptor subunits: (α 1–6), β 1–3), γ 1–3), δ, ε, θ, π, and ρ 1–3), and their splice variants [153,157,158]. Most of the receptor subunits of $GABA_A$ have been assigned to chromosomes 4, 5, 15, and X [74,159]. When the neurotransmitter GABA binds to $GABA_A$, the conformational structure of the receptor is changed, and the membrane pore opens for the flow of chloride ions. $GABA_B$ receptors, widely distributed throughout the human brain, are G-protein coupled receptors that stimulate the opening of potassium channels. $GABA_C$ receptors, which differ in complexity of structure, abundance, distribution, and function from $GABA_A$ and $GABA_B$ receptors, can be found in retinal, hippocampus, spinal cord, and pituitary tissues [160,161].

The GABA neurotransmitter system interacts with the corticomesolimbic dopamine system (CMDS) in the brain to reinforce the effects of alcohol. New pharmacological treatments that target the inhibitory and excitatory modulators of the CMDS may prevent relapse or reduce heavy drinking [160]. Two clinical trials on the anticonvulsant drug Topiramate, which modulates the CMDS by facilitating GABA function through interaction with the $GABA_A$ receptor, have significantly increased abstinence from alcohol and reduced heavy drinking in alcohol-dependent subjects [160,162].

Tabakoff et al. [163] used a genomics approach in the rat model to identify a list of candidate genes that influence human consumption of alcohol. They compiled a PubMed literature search on the functionality of the candidate gene products and determined that "defined pathways are linked to presynaptic GABA release, activation of dopamine neurons, and postsynaptic GABA receptor trafficking, in brain regions including the hypothalamus, ventral tegmentum and amygdale" [163]. Then, a custom genotyping microarray of 1,350 single nucleotide polymorphisms (SNPs) was designed from the list of candidate genes (i.e., an "addiction array") and used to conduct genetic association studies with two different human populations. SNPs in two genes (*GAD1, MPDZ*) were significantly associated with alcohol consumption in the Montreal, Canada, population, whereas SNPs in four genes (*CHRM5, GABRB2, MAPK1, PPP1R1B*) were associated with alcohol consumption in the Sydney, Australia, population. An interesting outcome from this study is the observation that genetic factors contributing to alcohol consumption may differ from genetic factors associated with alcohol dependence in humans.

Chromosomal region 5q34 in humans contains a $GABA_A$ receptor gene cluster implicated in the genetics of alcoholism [74,164]. Previous association studies in humans have implicated *GABRA₆*, *GABRB₂*, and *GABRG₂* (major $GABA_A$ receptor subunit genes in the human genome located on 5q34) with alcohol dependence [164–166]. The α6 Pro385Ser amino acid polymorphism, located within the $GABA_A$ subunit cluster on 5q34, was implicated by haplotype-based localization with susceptibility to alcohol dependence [74,167]. Radel et al. discovered two other polymorphisms localized on 5q34, the *GABRA6* 1519T and the *GABRB2* 1412T alleles, which were associated with alcohol dependence in both Finns and Southwestern Native American populations [167].

3.3.11 ACETYLCHOLINE

Acetylcholine (ACh) is a neurotransmitter found in the brain and autonomic nervous system; it is the natural agonist of muscarinic and nicotinic receptors. Muscarinic receptors (mAChRs) are G protein-coupled acetylcholine receptors found in the plasma membrane of neurons throughout the central nervous system. Nicotinic acetylcholine receptors (nAChRs) are receptor ion channels found in the autonomic nervous system. Several genetic studies have shown that common genetic factors in genes may contribute to alcohol and nicotine co-addiction [168–171]. Because neuronal nicotinic receptors (nAChRs) are a common site of action of nicotine and alcohol, candidate gene and whole genome association studies have targeted genes coding the nAChRs [172].

Human clinical research studies have implicated two genes that encode the α4 and β2 nAChR subunits (*CHRNA4* and *CHRNB2*) with alcohol and tobacco dependence [173–175]. However, most genetic association studies examining single nucleotide polymorphisms within *CHRNA4* and *CHRNB2* genes have not found evidence for an association between nicotine or alcohol addictions and these genetic variants [176–178]. Because several neuronal pathways (opioidergic, serotonergic, glutamatergic, dopaminergic, GABAergic, and cholinergic systems) with connections to nAChRs are modified by nicotine and alcohol use, research continues with candidate gene and whole genome association studies to understand the comorbidity of these two drugs [172] and to search for effective drug treatments. In a recent review of neuronal nicotinic acetylcholine receptors, Chatterjee et al. [179] suggest that two FDA-approved nAChR ligands, varenicline and mecamylamine, may be effective drugs to treat not only smoking cessation but also alcohol use disorders [179].

The cholinergic muscarinic receptor 2 (*CHRM2*) gene is located on the long arm of chromosome 7 where a locus for genetic susceptibility was found in two separate genome-wide linkage studies [180,181]. The Collaborative Study of the Genetics of Alcoholism (COGA) found three SNPs in *CHRM2* that increase risk of alcohol dependence [182]. One of these SNPs, rs1824024, was implicated in another association study of *CHRM2* and alcohol dependence in European American and African American populations [183]. In a Korean case-control study, SNP rs1824024 was significantly associated with the Alcohol Use Disorders Identification Test (AUDIT) and the Alcohol Dependence Scale (ADS), instruments used to measure the severity of symptoms in alcohol dependence subjects [184]. Recent studies associate the *CHRM2* gene with brain oscillations and cognition [185], alcohol and drug comorbidity [186], and major depression in women [187].

3.3.12 GLUTAMATE

Glutamate is the main excitatory neurotransmitter. Ethanol potently inhibits glutamate receptor activity. There are many locations of various glutamate receptor subunits throughout the genome for NMDA (*N*-methyl-D-aspartic acid) receptor genes that can be found on chromosomes 9q, 16p, 12p, 17q, 19p, and 19q [188]. A calcium-conducting glutamate receptor known as the *N*-methyl-D-aspartate receptor (NMDAR) is believed to be involved in the evolution of alcohol addiction. Ethanol has an inhibitory effect on this receptor which is attenuated upon prolonged ethanol exposure. Increased receptor expression is believed to underlie the development of ethanol tolerance and dependence as well as acute and delayed signs of withdrawal, particularly agitation and delirium tremens [189]. Protein kinases such as Fyn and Src are important in phosphorylation of NR2 subunits (A and B) that upregulate channel function, regulating NMDA receptors, and controlling signal pathways

[190]. Several allelic variants of the *NMDAR1* and the *NMDAR2B* receptor genes have been linked with alcoholism and related traits [191], but other studies have not consistently linked an association [192]. Withdrawal from chronic alcohol usage leads to increased activity in the NMDA receptor and increased influx of calcium which are attributed to neurotoxicity and neuronal cell death. Topiramate, which can decrease release of dopamine and also antagonize the glutamate activity, has been shown to be efficacious to placebo in the management of alcohol dependence [193].

3.3.13 SEROTONIN

Serotonin, also known as 5-hydroxytryptamine (5-HT), is a monoamine neurotransmitter that is derived from tryptophan. Serotonin is found in the central nervous system, the gastrointestinal tract, and the dense granules of platelets. The serotonin transporter (5-HTT) has received attention as a marker for alcohol dependency risk because of evidence supporting a relationship between serotonergic function and alcohol consumption [194]. Halliday has suggested that chronic alcohol intake may exert neurotoxic effects upon neurotransmitter systems such as the serotonergic system [195]. A serotonin deficiency state has been postulated to account for an alcohol-dependent individual's undercontrol of intrusive alcohol-related thoughts and inability to restrain impulses to engage in drinking behaviors [196]. Several studies showed inconsistent data on the potential association of the serotonin transporter genotype with the risk of developing alcoholism, with differences in data possibly stemming from numerous environmental and biochemical variables [197]. A variant in the gene that encodes for the 5-HT transporter, the *5-HTTLPR* polymorphism, has been associated with alcohol consumption. Expression of the gene stems from either short (S) alleles of 14 copies or long (L) 16 copies of a 20–23 base pair repeated sequence, with the S variant associated with reduced transporter protein expression and increased alcohol dependence [198]. In a study of 176 treatment-seeking men and women with alcohol dependence, Thompson et al. used genotyping methods to examine the association between *5-HTTLPR* genotype and obsessive compulsive alcohol craving but did not find a significant relationship to the particular variant allele (La), which is associated with greater 5-HTT transcription and alcohol craving [199]. An older study by Lappaleinan et al. examined both Finnish and Southwestern American Indian subjects and noted significance of the *HTR1B-G861C* (a serotonin autoreceptor) as well as a dinucleotide repeat locus, *D6S284,* closely linked to antisocial alcoholism [200]. The end result of many studies is suggestive that there are most likely several, if not many, genetic factors such as single nucleotide repeats or polymorphisms of the serotonin neurotransmitter system that play a role in alcohol dependence.

3.3.14 NEUROPEPTIDE Y (NPY)

Neuropeptide Y (NPY) is a 36-amino-acid peptide widely expressed in the mammalian nervous system, with high levels in brain regions such as the hypothalamus, in particular the arcuate and the paraventricular nuclei, the hippocampal formation, the amygdala, and the septum. Neuropeptide Y has been demonstrated in animal and human studies, either with NPY variation or NPY receptors, to be associated with alcohol dependence and alcohol withdrawal symptoms. Hypoactivity of the NPY system has been shown to promote alcohol consumption [201]. Genetic variations in NPY receptors appear to be associated with alcohol dependence, alcohol withdrawal, cocaine dependence, and concurrent alcohol and cocaine dependence [202].

3.3.15   ACN9

The *ACN9* gene is located on chromosome 7q in humans. *ACN9* was originally identified in respiratory-competent yeast mutants that were not able to use acetate for carbon assimilation, and there are homologs of the *ACN9* gene proteins that are also found not only in humans, but in the mouse, and in nematodes as well [203]. The ACN9 protein is a mitochondrial intermembrane space protein that is involved in gluconeogenesis and the assimilation of ethanol or acetate into carbohydrate [204]. The Collaborative Study on the Genetics of Alcoholism (COGA) is a multisite collaboration aimed at identifying genes contributing to alcohol dependence in which in a sampling of 105 pedigrees, chromosome 7 provided the strongest evidence of linkage to alcohol dependence [76]. Furthermore, four of the eight significant single nucleotide polymorphisms identified in this project as being associated with alcohol dependence were located in or near the *ACN9* gene.

3.3.16   OPIOIDS

The opioid neurotransmitter system has been associated with alcoholism, due to the fact that certain opioid antagonists, such as naltrexone, have been used to clinically manage alcohol dependence and withdrawal [205]. While no conclusive association has been made between the mu and delta opioid receptor genes and alcohol dependence, the genetic variants of the kappa receptor gene (*OPRK1*) and a gene encoding for an opioid ligand, prodynorphin (*PDYN*), have been strongly associated with a risk of alcohol dependence [76]. Earlier studies in rodents demonstrated that alcohol consumption increases levels of beta-endorphin that binds to the opioid receptor [206]. It should be noted that some variants of the mu opioid receptor *OPRM1* gene alter the efficacy of naltrexone [205]. A factor that may be of importance in the opioid system is that there are polymorphisms in genes that can be utilized not only in the evaluation of alcohol dependence, such as *OPRK1* and *PDYN*, but also in the treatment of alcoholism, such as *OPRM1*. This is discussed in more detail below.

3.4   PHARMACOGENETICS OF ALCOHOL ABUSE TREATMENT

Several classes of drugs have been utilized for alleviation of alcohol withdrawal symptoms and dependency treatment [205]. Pharmacogenetic intra-individual genetic variation may impact metabolism and response to these drugs and represents an emerging area of study. Benzodiazepines, for example, are utilized to treat alcohol withdrawal symptoms. As the metabolism is mediated by cytochrome P450 isozymes, pharmacogenetic variation in these isozymes may substantially affect drug kinetics resulting in altered efficacy or potential toxicity [207].

Some of the genes described previously in this chapter may also be relevant to potential pharmacologic treatment of dependence. These include genetic variations in the *ADH* and *ALDH* genes which can affect response to the alcohol dependence treatment drug disulfiram. This drug inhibits ALDH and leads to the accumulation of the ethanol metabolite acetaldehyde, which contributes to a "sick" feeling on the part of the individual (M81).

The drug naltrexone is sometimes utilized in order to attempt to prevent the development of alcohol dependence. The *OPMR1* gene encodes the mu opioid receptor, which is the primary site of action for the most commonly used opioids. At the *OPRM1* gene a functional receptor polymorphism (A118G) has been identified that can alter β-endorphin binding to the receptor and has been linked to increased efficacy of naltrexone treatment among alcohol-dependent patients [208]. β-Endorphin is approximately three times more potent against the A118G variant receptor than at the most common allelic form in agonist-induced activation of G protein-coupled potassium channels [209]. While a literature review of the *OPRM1* gene concluded that clinical studies do not unequivocally support an association between polymorphisms in *OPRM1* and alcohol dependence, the discovery of such polymorphisms in the opioid system is promising in tailoring treatment for alcohol dependency [210].

3.5 ADVANCES IN MOLECULAR PROFILING TECHNOLOGIES OF MARKERS OF ALCOHOLISM

Advances in molecular profiling technologies within the disciplines of genomics, transcriptomics, proteomics, and metabolomics will likely play an essential role in the development of prevention strategies and personalized treatments of alcoholism, as well as the identification of new therapeutic targets. While these techniques are not in current clinical use, they may ultimately substantially change our approach to testing for trait markers of alcoholism. This section will provide an overview of the sophisticated array-based "omics" technologies used in the context of large epidemiological studies to expand the genetic knowledge of alcoholism.

3.5.1 TRANSCRIPTOMICS

Two relatively new technologies, next-generation sequencing and microarray systems, have changed the way complex genetic disorders are studied. Entire "transcriptomes" from large numbers of tissues can be analyzed simultaneously by either sequencing or microarray technology [217]. Gene expression profiling, using high-quality RNA extracted from human postmortem brain tissue, has been applied to several neurological and psychiatric disorders including Alzheimer's disease [211–213], schizophrenia [212,214], and major depression [215,216]. Numerous expression studies in postmortem brains of alcohol abusers versus matched controls have identified differentially expressed genes involved in "myelination, ubiquitination, apoptosis, cell adhesion, neurogenesis, and neural disease" [217]. Data from these transcriptional studies indicate that multiple genes in multiple systems are affected by alcohol abuse. Subsequent informatics-based analyses of the function of candidate gene products found in whole-genome gene expression studies may lead to the discovery of important genetic/biochemical pathways that contribute to alcohol consumption and dependence [163]. Recently, the field of transcriptomics has been transformed by RNA sequencing of alternate splice variants [218], novel transcripts from gene fusion events [219], and new classes of noncoding RNAs (ncRNAs) using next-generation sequencing [220,221].

3.5.2 GENOMICS

The complex etiology of alcoholism and hundreds of genes are involved in multiple molecular networks within neuronal cells. Identification of genes associated with alcohol dependence can be accomplished by candidate gene studies, whole-genome association (WGA) and whole-genome linkage studies. In whole-genome linkage studies, broad chromosomal regions are linked to a disease using a panel of polymorphisms tested for meiotic linkage in a large family-based set of DNAs [75]. Next, a candidate gene study using microarray or another molecular approach is conducted on the region of interest found in the whole-genome linkage study. One example of a large-scale family-based data set is the Collaborative Study on the Genetics of Alcoholism (COGA) conducted by the National Institute of Alcohol Abuse and Alcoholism (NIAAA), whose primary goal is the identification of genes contributing to alcoholism susceptibility (see the following National Institutes of Health link: http://www.niaaa.nih.gov/ResearchInformation/ExtramuralResearch/SharedResources/projcoga.htm). COGA has more than 228 publications listed at www.niaaagenetics.org; results from their whole-genome linkage studies have been summarized in previous publications [222,223], and a brief synopsis follows. Linkage analysis on the COGA study using a large panel of 1,717 SNPs found additional regions on chromosome 1 for alcohol dependence, chromosome 10 and 13 for illicit drug dependence, and chromosomes 2, 10, and 13 for comorbidity of alcohol and drug dependence [224].

Genome-wide association studies (GWAS) use dense microarray panels with more than 1 million polymorphisms and copy number variant markers run on unrelated individuals from case-control studies of complex diseases. A summary table of the models used and the major findings obtained by applying microarray technologies to GWAS studies of alcoholism is presented (see Table 3.4). The first study listed in Table 3.4 is a GWAS using case-control samples from the COGA families [223]. After genotyping almost 1.2×10^6 markers on 1,192 cases and 692 controls, a focused genotyping panel with 199 of the most promising SNP candidates was analyzed in 262 pedigrees and 59 trios. None of the SNPs met genome-wide significance, but a cluster of genes on chromosome 11 was associated with alcohol dependence.

Another study summarized in Table 3.4 is a pooling-based genome-wide association study of alcohol addiction (AD) and nicotine addiction (ND) in Australian and Netherland populations [225]. The Australian GWAS found three SNPs for comorbid AD/ND that achieved genome-wide significance: rs7530302 near *MARK1*, rs1784300 near *DD6*, and rs12882384 in *KIAA1409*. After performing a meta-analysis of the

TABLE 3.4

Genome-Wide Association Studies in Alcohol Dependence

Biological Model	Genotyping Platform	Number of Cases/Controls in Discovery Panel	Replication Panel(s)	Novel Genes Reported	Authors	Reference
COGA: Family case-control study on alcohol dependence	Illumina Human Hap1M v3 (1,199,187 markers)	1,192/692	199 SNPs in 262 pedigrees and 59 trios	Chrm11: *SLC22A18, PHLDA2, NAP1L4, SNORA54, CARS, OSBPL5* **Replicated from previous GWAS publications:** *CPE, DNASE2B, SLC10A2, ARL6IP5, ID4, GATA4, SYNE1, ADCY3*	Edenberg et al. [223]	**Alcohol Clin Exp Res** 34(5): 840–852, 2010
Australian Twin Registry: large sibship study (**BIGSHIP**), Alcohol Extreme discordant and concordant study (**EDAC**), and Nicotine Addiction Genetics (**NAG**) study	Illumina Human Hap1M v3 (1,199,187 markers) or **Human Hap1M v1** (1.05 × 10^6 markers)	1,224/1,162 (AD*) 1,273/1,113 (ND*) 599/488 (AD and ND)*	1738 cases/1802 controls Replication panel Netherlands Study of Depression and Anxiety (**NESDA**) and Netherlands Twin Registry (**NTR**)	ND: *ARHGAP10* AD and ND: *MARK1, near DDX6, KIAA1409*	Lind et al. [225]	**Twin Res Hum Genet** 13(1): 10–29, 2010

continued

TABLE 3.4 (continued)
Genome-Wide Association Studies in Alcohol Dependence

Biological Model	Genotyping Platform	Number of Cases/ Controls in Discovery Panel	Replication Panel(s)	Novel Genes Reported	Authors	Reference
German Addiction Research Network (GARN) for cases; PopGen, KORA-gen, Heinz RECALL study for controls	Illumina Human Hap 550 (550 × 10³ markers)	487/1,358	1,024 cases/996 controls	PECR, CAST, ERAP1, PPP2R2B, ESR1, CCDC41, ADH1C, GATA4, CDH13	Treutlein et al. [226]	Arch Gen Psychiatry 66(7): 773–784, 2009
Study of Addiction, Genetics, and Environment (SAGE): COGA, FSCD, COGEND	Human Hap1M v1 (1.05 × 10⁶ markers)	1,897/1,932	258 COGA Pedigree cases; 487cases/1,358 controls	None	Bierut et al. [222]	Proc Natl Acad Sci 107(11): 5082–5087, 2010

Notes: *AD, alcohol dependence; ND, nicotine dependence; AD and ND, comorbidity for alcohol and nicotine dependence.

10,000 most significant AD and ND polymorphisms from both populations, they found a gene network (ion-channels, cell adhesion molecules) that may be associated with AD, ND, or comorbid AD/ND.

A third GWAS study summarized in Table 3.4 is from a population of 487 German male inpatients with alcohol dependence and an age onset of younger than 28 years, along with 1,358 population-based controls [226]. All samples were genotyped using the Illumina Human Hap 550 Beadchip containing more than 550,000 markers. A replication panel of 139 SNPs was genotyped in a follow-up study of 1,024 cases and 996 controls. Fifteen SNPs in the replication panel showed genome-wide significance (just like the GWAS); nine SNPs were located in genes previously associated with alcohol dependence (*PECR, CAST, ERAP1, PPP2R2B, ESR1, CCDC41, ADH1C, GATA4,* and *CDH13*). An additional two SNPs had genome-wide significance and are located in chromosomal region 2q35, implicated by linkage and animal studies with alcohol phenotypes. These two SNPs are near the peroxisomal trans-2-enoyl-CoA reductase (*PECR*) gene, a possible candidate for further genomic tests.

Finally, the last study mentioned in Table 3.4 is a large GWAS of 1,897 alcohol-dependent cases and 1,932 alcohol-exposed controls from the Study of Addiction: Genetics and Environment (SAGE) as part of the Gene, Environment Association Studies (GENEVA) consortium [222]. Although 15 SNPs had $p < 10^{-5}$, all of these SNPs failed to meet genome-wide significant thresholds of 5×10^{-8} in replication studies.

Copy number variants (CNVs) are large DNA sequences (>1,000 nucleotides in length) that are duplicated or deleted a variable number of times as copies of a genome relative to a reference genome. These have been associated with susceptibility or resistance to many diseases including CNS disorders like autism, schizophrenia, and idiopathic learning disability [199,227–229]. A European study on schizophrenia (4,345 schizophrenia patients; 35,079 controls) focused on genomic copy number variants in chromosome 16p13.1 using microarray tools [230]. They discovered a copy number duplication located within the 16p13.1 locus that is associated with schizophrenia, alcoholism, dyslexia, and attention deficit hyperactivity disorder in a single Icelandic family. Association statistics were not calculated for this finding due to the small size of the affected family; obviously, a large case-control study of alcoholism using whole-genome copy number variation arrays is needed for replication of this association and to search for novel CNV events.

3.5.2.1 Metabolomics

The National Institute on Alcohol Abuse (NIAAA); National Institute of Diabetes, Digestive, and Kidney Diseases (NIDDK), and Office of Dietary Supplements (ODS) have prioritized the need for research to identify new bio-markers of alcohol use disorders using genomic, proteomic, and metabolomic technologies as evidenced by recent R21 grant opportunities (www.niaaa.nih.gov/Search/Results.aspx?k=proteomics). Current advances in metabolomics, the study of small-molecule metabolite profiles that are the end products of cellular processes [231], offer unique opportunities to discover novel biomarkers for alcohol severity [232]. Traditionally, endogenous metabolites from biological specimens are analyzed by nuclear magnetic resonance (NMR), whereas exogenous metabolites from acute or chronic toxicity are measured with mass spectrometry.

New bioinformatic approaches and computational tools have been developed to integrate the data obtained from metabolomics and other "omics." For example, FDA's National Center for Toxicological Research (Jefferson, Arkansas), Center for Excellence on Metabolomics, uses a software tool known as Scaled-to-Maximum, Aligned, and Reduced Trajectories (SMART) analysis to interpret metabolomic data (see www.fda.gov/nctr).

3.5.3 PROTEOMICS

Proteomics (the large-scale study of proteins, their structures, and their functions) is another approach to search for biomarkers of dependence, consumption, and relapse, in addition to pathologies induced by alcohol. Conventional techniques to study proteins include Western blot, enzyme-linked immunosorbent assay (ELISA) and mass spectrometry. High-throughput proteomic approaches, including protein microarray platforms, matrix-assisted laser desorption/ionization time-of-flight mass spectrometry, liquid chromatography-tandem mass spectrometry, isotope-coded affinity tags, and next-generation sequencing, will facilitate rapid biomarker and drug discoveries. Lai [232a] used laser capture-mass spectrometry/mass spectrometry to profile 322 serum carrier protein-based proteins as potential biomarkers of alcohol dependence. They identified eight novel proteins associated with alcohol abuse: gelsolin, selenoprotein P, serotransferrin, tetranectin, hemopexin, histidine-rich glycoprotein, plasma kallikrein, and vitronectin. Next-generation sequencing identifies and quantifies proteins on a genome-wide scale by parallel processing of clonally amplified molecules; massive numbers of analytes as well as samples are processed simultaneously [233,234]. Since one practical application of proteomics is the identification of potential new drug targets for medication development, the use of next-generation sequencing will accelerate this field of alcohol research [235].

3.5.4 EPIGENETICS

The study of epigenetic changes (DNA methylation, posttranslational histone modifications, and regulation) in both normal and pathological biospecimens on a genome-scale is known as epigenetics. Epigenetic changes can regulate gene expression by modulating the structure of the chromatin without changing the DNA code; aberrant epigenetic mechanisms are implicated in the development of many diseases, including cancer, cardiovascular, immunological, and neuropsychiatric disorders [236–238].

A recent study shows that individuals may have personalized epigenomic signatures that are subject to modification by environmental exposures or aging [236,239]. The authors analyzed global CpG methylation using high-throughput array-based methylation in a small sample set of DNAs from the AGES (Age, Gene/Environment Susceptibility) study in Reykjavik, Iceland. Each participant provided two DNA specimens: at the start of the study and 11 years later. Data from the second collection of specimens had 227 regions with highly variable patterns of methylation that correlated with body mass index, a phenotype associated with disease risk. These experiments lay the groundwork for future studies to identify associations between global methylation signatures and alcohol dependence.

Histone modifications affect the gene-regulatory functions of proteins. Modifications of histones can occur by any of at least six different chemical moieties including methylation, acetylation, phosphorylation, ubiqiutinylation, ADP-ribosylation, and sumoylation [25]. There is clear evidence that alterations in epigenetic regulation through histone modifications contribute to alcohol-related brain damage [240]. The authors demonstrated that a histone deacetylase inhibitor (trichostatin A) reduced suppression seen in neural stem cell differentiation in the presence of ethanol. They postulated that ethanol alters neural differentiation through modifications in histones and other epigenetic dysfunctions. Discovery of posttranslational histone modifications on a global scale is accomplished with a next-generation sequencing (NGS) technique known as ChIP-seq (i.e., chromatin immunoprecipitation followed by sequencing) [220,241]. In their review article on technological advances in genomics, de Magalhaes et al. suggest that next-generation sequencing platforms and their applications like ChIP-seq will form the basis of molecular biology research for many years; therefore, ChIP-seq may become a valuable tool for alcohol research [242].

MicroRNAs, a class of noncoding RNAs transcribed by RNA polymerase II, regulate epigenetic modification by directly targeting histone deacetylases and DNA methyltransferases [243]. More than 500 microRNAs (miRNAs) have been discovered in humans, and control cellular proliferation, differentiation, and apoptosis [244]. Tsang et al. used a computational approach to predict functionality of miRNAs and determined that they may regulate cellular pathways by targeting several members of a specific network [245]. A small subset of miRNAs is sensitive to alcohol (see Table 1 in Pietrzykowski [244]), and a specific microRNA species, miR-9, can cause both gene activation and gene suppression. There are three miR-9 genes that encode for different miR-9 precursors; two of the genes are located on different alcoholism susceptibility loci within chromosomes 1 and 15 [246–250]. Experiments are ongoing to determine if miR-9 is linked to susceptibility to alcoholism. An exciting development in the field of molecular genomics is the development of microarray and sequencing technologies for whole-genome miRNA studies. Screening the entire set of human miRNAs for susceptibility to alcoholism is a feasible option for future research.

An epigenetic model has been applied to fetal alcohol spectrum disorders (FASDs) in an attempt to identify testable hypotheses and to include the alterations that occur subsequent to alcohol use prior to conception or consideration of reproduction [251]. This important aspect of alcohol teratology has been a rare point of discussion that has suffered by a lack of a hypothetical model for rigorous investigation. This recent report now outlines a usable and elegant model that warrants attention.

The availability of a framework for the analysis and mechanistic understanding of the alterations to genetic expression that can be induced by alcohol consumption beyond direct gene mutation or single nucleotide polymorphism analysis adds a further dimension to the realm of slate and trait markers. There is an expansive network for a variety of epigenetic modifications that will need to be identified and associated with their ability to indicate increasing states of alcohol consumption, abuse, and ultimately dependence/addiction. The ultimate future identification of an epigenetic "alcoholism signature" is a distinct possibility.

3.6 CONCLUSIONS

In addition to available biochemical "slate" markers, a number of promising gene variant "trait" markers have been identified and may see more widespread use. Emerging microarray and sequencing technologies, in concert with large data repositories, are making it easier to understand the organization and function of the human genome sequence. Finally, it is becoming clear that in order to more fully understand a complex polygenic disease like alcoholism, which has many environmental effectors, researchers will need to account for both genomic and epigenetic data. These advances may substantially change our use of trait markers associated with alcohol abuse.

REFERENCES

1. LaVallee, R.A., G.D. Williams, and H. Yi, *Surveillance Report #87: Apparent Per Capita Alcohol Consumption: National, State, and Regional Trends: 1970–2007*. 2009, National Institute on Alcohol Abuse and Alcoholism, Division of Epidemiology and Prevention Research, Bethesda, MD.
2. *Global Status Report on Alcohol 2004*. 2004, World Health Organization, Geneva, Switzerland.
3. *WHO Expert Committee on Problems Related to Alcohol Consumption*. 2006, World Health Organization, Geneva.
4. Yi, H., C.M. Chen, and G.D. Williams, *Surveillance Report #76: Trends in Alcohol-Related Fatal Traffic Crashes United States 1982–2004*, in *Alcohol Epidemiologic Data System*. 2006, National Institute on Alcohol Abuse and Alcoholism, Division of Epidemiology and Prevention Research, Bethesda, MD.
5. Chen, C.M., and H. Yi, *Trends in Alcohol Related Morbidity among Short-Stay Community Hospital Discharges, United States, 1979–2006*, in *Alcohol Epidemiological Data System*. 2008, National Institute on Alcohol Abuse and Alcoholism, Division of Epidemiology and Prevention Research, Division of Epidemiology and Prevention Research, Bethesda, MD.
6. Glanz, J., B. Grant, M. Monteiro, et al., WHO/ISBRA Study on State and Trait Markers of Alcohol Use and Dependence: analysis of demographic, behavioral, physiologic, and drinking variables that contribute to dependence and seeking treatment. International Society on Biomedical Research on Alcoholism. *Alcohol Clin Exp Res*, 2002, 26(7): 1047–1061.
7. Greenfield, T.K., and J.D. Rogers, Who drinks most of the alcohol in the U.S.? The policy implications. *J Stud Alcohol*, 1999, 60(1): 78–89.
8. *The NSDUH Report: Alcohol Treatment: Need, Utilization and Barriers*. 2009, Substance Abuse and Mental Health Services Administration, Office of Applied Studies, Rockville, MD.
9. *American Psychiatric Association: Diagnostic and Statistical Manual of Mental Disorders*, Fourth Edition, Text Revision. 2000, American Psychiatric Association, Washington, DC.

10. Koob, G.F., Neurobiological substrates for the dark side of compulsivity in addiction. *Neuropharmacology*, 2009, 56 Suppl 1: 18–31.
11. Schermer, C.R., Feasibility of alcohol screening and brief intervention. *J Trauma*, 2005, 59(3 Suppl): S119–S123; discussion S124–S133.
12. Kraemer, K.L., The cost-effectiveness and cost-benefit of screening and brief intervention for unhealthy alcohol use in medical settings. *Subst Abus*, 2007, 28(3): 67–77.
13. Nurnberger, J.I., Jr., R. Wiegand, K. Bucholz, et al., A family study of alcohol dependence: co aggregation of multiple disorders in relatives of alcohol-dependent probands. *Arch Gen Psychiatry*, 2004, 61(12): 1246–1256.
14. Goldman, D., G. Oroszi, and F. Ducci, The genetics of addictions: uncovering the genes. *Nat Rev Genet*, 2005a, 6(7): 521–532.
15. Prescott, C.A., and K.S. Kendler, Genetic and environmental contributions to alcohol abuse and dependence in a population-based sample of male twins. *Am J Psychiatry*, 1999, 156(1): 34–40.
16. Bienvenu, O.J., D.S. Davydow, and K.S. Kendler, Psychiatric "diseases" versus behavioral disorders and degree of genetic influence. *Psychol Med*, 2011, 41(1): 33–40.
17. Strat, Y.L., N. Ramoz, G. Schumann, et al., Molecular genetics of alcohol dependence and related endophenotypes. *Curr Genomics*, 2008, 9(7): 444–451.
18. Yang, B.Z., H.R. Kranzler, H. Zhao, et al., Association of haplotypic variants in DRD2, ANKK1, TTC12 and NCAM1 to alcohol dependence in independent case control and family samples. *Hum Mol Genet*, 2007, 16(23): 2844–2853.
19. Stacey, D., T.K. Clarke, and G. Schumann, The genetics of alcoholism. *Curr Psychiatry Rep*, 2009, 11(5): 364–369.
20. Heath, A.C., K.K. Bucholz, P.A. Madden, et al., Genetic and environmental contributions to alcohol dependence risk in a national twin sample: consistency of findings in women and men. *Psychol Med*, 1997, 27(6): 1381–1396.
21. Galvan, F.H., and R. Caetano, Alcohol use and related problems among ethnic minorities in the United States. *Alcohol Res Health*, 2003, 27(1): 87–94.
22. Makimoto, K., Drinking patterns and drinking problems among Asian-Americans and Pacific Islanders. *Alcohol Health Res World*, 1998, 22(4): 270–275.
23. Kaskutas, L.A., C. Weisner, M. Lee, et al., Alcoholics anonymous affiliation at treatment intake among white and black Americans. *J Stud Alcohol*, 1999, 60(6): 810–816.
24. Gomberg, E.S., Treatment for alcohol-related problems: special populations: research opportunities. *Recent Dev Alcohol*, 2003, 16: 313–333.
25. Shukla, S.D., J. Velazquez, S.W. French, et al., Emerging role of epigenetics in the actions of alcohol. *Alcohol Clin Exp Res*, 2008, 32(9): 1525–1534.
26. Shepard, B.D., and P.L. Tuma, Alcohol-induced protein hyperacetylation: mechanisms and consequences. *World J Gastroenterol*, 2009, 15(10): 1219–1230.
27. Hashimoto, J.G., M.R. Forquer, M.A. Tanchuck, et al., Importance of genetic background for risk of relapse shown in altered prefrontal cortex gene expression during abstinence following chronic alcohol intoxication. *Neuroscience*, 173: 57–75.
28. Hannuksela, M.L., M.K. Liisanantti, A.E. Nissinen, et al., Biochemical markers of alcoholism. *Clin Chem Lab Med*, 2007, 45(8): 953–961.
29. Gentilello, L.M., B.E. Ebel, T.M. Wickizer, et al., Alcohol interventions for trauma patients treated in emergency departments and hospitals: a cost benefit analysis. *Ann Surg*, 2005, 241(4): 541–550.
30. Wu, A.H., C. McKay, L.A. Broussard, et al., National Academy of Clinical Biochemistry laboratory medicine practice guidelines: recommendations for the use of laboratory tests to support poisoned patients who present to the emergency department. *Clin Chem*, 2003, 49(3): 357–379.
31. Church, A.S., and M.D. Witting, Laboratory testing in ethanol, methanol, ethylene glycol, and isopropanol toxicities. *J Emerg Med*, 1997, 15(5): 687–692.

32. Hietala, J., H. Koivisto, J. Latvala, et al., IgAs against acetaldehyde-modified red cell protein as a marker of ethanol consumption in male alcoholic subjects, moderate drinkers, and abstainers. *Alcohol Clin Exp Res*, 2006, 30(10): 1693–1698.
33. Gadsden, R.H., E.H. Taylor, and S.J. Steindel, Ethanol in biological fluids by enzymatic analysis, in *Selected Methods of Emergency Toxicology*, C. Frings and W.R. Faulkner, Editors. 1986, AACC Press, Washington, DC.
34. Gibb, K.A., A.S. Yee, C.C. Johnston, et al., Accuracy and usefulness of a breath alcohol analyzer. *Ann Emerg Med*, 1984, 13(7): 516–520.
35. Whittington, J.E., S.L. La'ulu, J.J. Hunsaker, et al., The osmolal gap: what has changed? *Clin Chem*, 56(8): 1353–1355.
36. Baselt, R.C., ed. *Disposition of Toxic Drugs and Chemicals in Man*. 5th ed. 2000, CTI, Foster City, CA.
37. Pragst, F., and M. Yegles, Determination of fatty acid ethyl esters (FAEE) and ethyl glucuronide (EtG) in hair: a promising way for retrospective detection of alcohol abuse during pregnancy? *Ther Drug Monit*, 2008, 30(2): 255–263.
38. Pragst, F., and M.A. Balikova, State of the art in hair analysis for detection of drug and alcohol abuse. *Clin Chim Acta*, 2006, 370(1–2): 17–49.
39. Wurst, F.M., M. Yegles, C. Alling, et al., Measurement of direct ethanol metabolites in a case of a former driving under the influence (DUI) of alcohol offender, now claiming abstinence. *Int J Legal Med*, 2008, 122(3): 235–239.
40. Chan, D., D. Caprara, P. Blanchette, et al., Recent developments in meconium and hair testing methods for the confirmation of gestational exposures to alcohol and tobacco smoke. *Clin Biochem*, 2004, 37(6): 429–438.
41. Beck, O., and A. Helander, 5-hydroxytryptophol as a marker for recent alcohol intake. *Addiction*, 2003, 98 Suppl 2: 63–72.
42. Johnson, R.D., R.J. Lewis, D.V. Canfield, et al., Accurate assignment of ethanol origin in postmortem urine: liquid chromatographic-mass spectrometric determination of serotonin metabolites. *J Chromatogr B Analyt Technol Biomed Life Sci*, 2004, 805(2): 223–234.
43. Schmitt, G., P. Droenner, G. Skopp, et al., Ethyl glucuronide concentration in serum of human volunteers, teetotalers, and suspected drinking drivers. *J Forensic Sci*, 1997, 42(6): 1099–1102.
44. Wurst, F.M., G.E. Skipper, and W. Weinmann, Ethyl glucuronide—the direct ethanol metabolite on the threshold from science to routine use. *Addiction*, 2003, 98 Suppl 2: 51–61.
45. Niemela, O., and P. Alatalo, Biomarkers of alcohol consumption and related liver disease. *Scand J Clin Lab Invest*, 2010, 70(5): 305–312.
46. Ropero-Miller, J., and R. Winecker, Alcoholism, in *Clinical Chemistry Theory, Analysis, Correlation*, L. Kaplan, A. Pesce, and S. Kazmierczak, Editors. 2003, Mosby, St Louis, MO.
47. Niemela, O., Biomarkers in alcoholism. *Clin Chim Acta*, 2007, 377(1–2): 39–49.
48. Mendenhall, C.L., Alcoholic hepatitis. *Clin Gastroenterol*, 1981, 10(2): 417–441.
49. Whitfield, J.B., W.J. Hensley, D. Bryden, et al., Some laboratory correlates of drinking habits. *Ann Clin Biochem*, 1978, 15(6): 297–303.
50. Karkkainen, P., and M. Salaspuro, beta-Hexosaminidase in the detection of alcoholism and heavy drinking. *Alcohol Alcohol Suppl*, 1991, 1: 459–464.
51. Javors, M.A., and B.A. Johnson, Current status of carbohydrate deficient transferrin, total serum sialic acid, sialic acid index of apolipoprotein J and serum beta-hexosaminidase as markers for alcohol consumption. *Addiction*, 2003, 98 Suppl 2: 45–50.
52. Koivisto, H., J. Hietala, P. Anttila, et al., Long-term ethanol consumption and macrocytosis: diagnostic and pathogenic implications. *J Lab Clin Med*, 2006, 147(4): 191–196.

53. Savage, D.G., A. Ogundipe, R.H. Allen, et al., Etiology and diagnostic evaluation of macrocytosis. *Am J Med Sci*, 2000, 319(6): 343–352.

54. Wymer, A., and D.M. Becker, Recognition and evaluation of red blood cell macrocytosis in the primary care setting. *J Gen Intern Med*, 1990, 5(3): 192–197.

55. Skinner, H.A., S. Holt, W.J. Sheu, et al., Clinical versus laboratory detection of alcohol abuse: the alcohol clinical index. *Br Med J (Clin Res Ed)*, 1986, 292(6537): 1703–1708.

56. Aradottir, S., G. Asanovska, S. Gjerss, et al., Phosphatidylethanol (PEth) concentrations in blood are correlated to reported alcohol intake in alcohol-dependent patients. *Alcohol Alcohol*, 2006, 41(4): 431–437.

57. Anton, R.F., and D.H. Moak, Carbohydrate-deficient transferrin and gamma-glutamyl-transferase as markers of heavy alcohol consumption: gender differences. *Alcohol Clin Exp Res*, 1994, 18(3): 747–754.

58. Hietala, J., K. Puukka, H. Koivisto, et al., Serum gamma-glutamyl transferase in alcoholics, moderate drinkers and abstainers: effect on gt reference intervals at population level. *Alcohol Alcohol*, 2005, 40(6): 511–514.

59. Puukka, K., J. Hietala, H. Koivisto, et al., Age-related changes on serum GGT activity and the assessment of ethanol intake. *Alcohol Alcohol*, 2006, 41(5): 522–527.

60. Puukka, K., J. Hietala, H. Koivisto, et al., Additive effects of moderate drinking and obesity on serum gamma-glutamyl transferase activity. *Am J Clin Nutr*, 2006, 83(6): 1351–1354; quiz 1448–1449.

61. Arndt, T., Carbohydrate-deficient transferrin as a marker of chronic alcohol abuse: a critical review of preanalysis, analysis, and interpretation. *Clin Chem*, 2001, 47(1): 13–27.

62. Sillanaukee, P., and U. Olsson, Improved diagnostic classification of alcohol abusers by combining carbohydrate-deficient transferrin and gamma-glutamyltransferase. *Clin Chem*, 2001, 47(4): 681–685.

63. Bortolotti, F., F. Tagliaro, F. Cittadini, et al., Determination of CDT, a marker of chronic alcohol abuse, for driving license issuing: immunoassay versus capillary electrophoresis. *Forensic Sci Int*, 2002, 128(1–2): 53–58.

64. Whitfield, J.B., V. Dy, P.A. Madden, et al., Measuring carbohydrate-deficient transferrin by direct immunoassay: factors affecting diagnostic sensitivity for excessive alcohol intake. *Clin Chem*, 2008, 54(7): 1158–1165.

65. Scouller, K., K.M. Conigrave, P. Macaskill, et al., Should we use carbohydrate-deficient transferrin instead of gamma-glutamyltransferase for detecting problem drinkers? A systematic review and meta-analysis. *Clin Chem*, 2000, 46(12): 1894–1902.

66. Legros, F.J., V. Nuyens, M. Baudoux, et al., Use of capillary zone electrophoresis for differentiating excessive from moderate alcohol consumption. *Clin Chem*, 2003, 49(3): 440–449.

67. Hock, B., M. Schwarz, I. Domke, et al., Validity of carbohydrate-deficient transferrin (%CDT), gamma-glutamyltransferase (gamma-GT) and mean corpuscular erythrocyte volume (MCV) as biomarkers for chronic alcohol abuse: a study in patients with alcohol dependence and liver disorders of non-alcoholic and alcoholic origin. *Addiction*, 2005, 100(10): 1477–1486.

68. Peterson, C.M., L. Jovanovic-Peterson, and F. Schmid-Formby, Rapid association of acetaldehyde with hemoglobin in human volunteers after low dose ethanol. *Alcohol*, 1988, 5(5): 371–374.

69. Halvorson, M.R., J.L. Campbell, G. Sprague, et al., Comparative evaluation of the clinical utility of three markers of ethanol intake: the effect of gender. *Alcohol Clin Exp Res*, 1993, 17(2): 225–229.

70. Sillanaukee, P., M. Ponnio, and K. Seppa, Sialic acid: new potential marker of alcohol abuse. *Alcohol Clin Exp Res*, 1999, 23(6): 1039–1043.

71. Ghosh, P., E.A. Hale, and R. Lakshman, Long-term ethanol exposure alters the sialylation index of plasma apolipoprotein J (Apo J) in rats. *Alcohol Clin Exp Res*, 1999, 23(4): 720–725.

72. Dick, D.M., and T. Foroud, Candidate genes for alcohol dependence: a review of genetic evidence from human studies. *Alcohol Clin Exp Res*, 2003, 27(5): 868–879.

73. Dick, D.M., and L.J. Bierut, The genetics of alcohol dependence. *Curr Psychiatry Rep*, 2006b, 8(2): 151–157.

74. Radel, M., and D. Goldman, Pharmacogenetics of alcohol response and alcoholism: the interplay of genes and environmental factors in thresholds for alcoholism. *Drug Metab Dispos*, 2001, 29(4 Pt 2): 489–494.

75. Ducci, F., and D. Goldman, Genetic approaches to addiction: genes and alcohol. *Addiction*, 2008, 103(9): 1414–1428.

76. Kohnke, M.D., Approach to the genetics of alcoholism: a review based on pathophysiology. *Biochem Pharmacol*, 2008, 75(1): 160–177.

77. Crabbe, J.C., Alcohol and genetics: new models. *Am J Med Genet*, 2002, 114(8): 969–974.

78. Kwok, P.Y., Methods for genotyping single nucleotide polymorphisms. *Annu Rev Genomics Hum Genet*, 2001, 2: 235–258.

79. Gut, I.G., Automation in genotyping of single nucleotide polymorphisms. *Hum Mutat*, 2001, 17(6): 475–492.

80. Zakhari, S., Overview: how is alcohol metabolized by the body? *Alcohol Res Health*, 2006, 29(4): 245–254.

81. Williams, J.T., H. Begleiter, B. Porjesz, et al., Joint multipoint linkage analysis of multivariate qualitative and quantitative traits. II. Alcoholism and event-related potentials. *Am J Hum Genet*, 1999, 65(4): 1148–1160.

82. Sherva, R., J.P. Rice, R.J. Neuman, et al., Associations and interactions between SNPs in the alcohol metabolizing genes and alcoholism phenotypes in European Americans. *Alcohol Clin Exp Res*, 2009, 33(5): 848–857.

83. Corbett, J., N.L. Saccone, T. Foroud, et al., A sex-adjusted and age-adjusted genome screen for nested alcohol dependence diagnoses. *Psychiatr Genet*, 2005, 15(1): 25–30.

84. Prescott, C.A., P.F. Sullivan, P.H. Kuo, et al., Genomewide linkage study in the Irish affected sib pair study of alcohol dependence: evidence for a susceptibility region for symptoms of alcohol dependence on chromosome 4. *Mol Psychiatry*, 2006, 11(6): 603–611.

85. Edenberg, H.J., The genetics of alcohol metabolism: role of alcohol dehydrogenase and aldehyde dehydrogenase variants. *Alcohol Res Health*, 2007, 30(1): 5–13.

86. Osier, M.V., A.J. Pakstis, H. Soodyall, et al., A global perspective on genetic variation at the ADH genes reveals unusual patterns of linkage disequilibrium and diversity. *Am J Hum Genet*, 2002, 71(1): 84–99.

87. Lee, S.L., G.Y. Chau, C.T. Yao, et al., Functional assessment of human alcohol dehydrogenase family in ethanol metabolism: significance of first-pass metabolism. *Alcohol Clin Exp Res*, 2006, 30(7): 1132–1142.

88. Thomasson, H.R., H.J. Edenberg, D.W. Crabb, et al., Alcohol and aldehyde dehydrogenase genotypes and alcoholism in Chinese men. *Am J Hum Genet*, 1991, 48(4): 677–681.

89. Crabb, D.W., M. Matsumoto, D. Chang, et al., Overview of the role of alcohol dehydrogenase and aldehyde dehydrogenase and their variants in the genesis of alcohol-related pathology. *Proc Nutr Soc*, 2004, 63(1): 49–63.

90. Whitfield, J.B., Alcohol dehydrogenase and alcohol dependence: variation in genotype-associated risk between populations. *Am J Hum Genet*, 2002, 71(5): 1247–1250; author reply 1250–1251.

91. Hasin, D.S., F.S. Stinson, E. Ogburn, et al., Prevalence, correlates, disability, and comorbidity of DSM-IV alcohol abuse and dependence in the United States: results from the National Epidemiologic Survey on Alcohol and Related Conditions. *Arch Gen Psychiatry*, 2007, 64(7): 830–842.
92. Luczak, S.E., S.H. Shea, L.G. Carr, et al., Binge drinking in Jewish and non-Jewish white college students. *Alcohol Clin Exp Res*, 2002, 26(12): 1773–1778.
93. Wall, T.L., L.G. Carr, and C.L. Ehlers, Protective association of genetic variation in alcohol dehydrogenase with alcohol dependence in Native American Mission Indians. *Am J Psychiatry*, 2003, 160(1): 41–46.
94. Edenberg, H.J., X. Xuei, H.J. Chen, et al., Association of alcohol dehydrogenase genes with alcohol dependence: a comprehensive analysis. *Hum Mol Genet*, 2006, 15(9): 1539–1549.
95. Osier, M., A.J. Pakstis, J.R. Kidd, et al., Linkage disequilibrium at the ADH2 and ADH3 loci and risk of alcoholism. *Am J Hum Genet*, 1999, 64(4): 1147–1157.
96. Ramchandani, V.A., P.Y. Kwo, and T.K. Li, Effect of food and food composition on alcohol elimination rates in healthy men and women. *J Clin Pharmacol*, 2001, 41(12): 1345–1350.
97. Eng, M.Y., S.E. Luczak, and T.L. Wall, ALDH2, ADH1B, and ADH1C genotypes in Asians: a literature review. *Alcohol Res Health*, 2007, 30(1): 22–27.
98. Wall, T.L., M.L. Johnson, S.M. Horn, et al., Evaluation of the self-rating of the effects of alcohol form in Asian Americans with aldehyde dehydrogenase polymorphisms. *J Stud Alcohol*, 1999, 60(6): 784–789.
99. Spence, J.P., T. Liang, C.J. Eriksson, et al., Evaluation of aldehyde dehydrogenase 1 promoter polymorphisms identified in human populations. *Alcohol Clin Exp Res*, 2003, 27(9): 1389–1394.
100. Ehlers, C.L., J.P. Spence, T.L. Wall, et al., Association of ALDH1 promoter polymorphisms with alcohol-related phenotypes in southwest California Indians. *Alcohol Clin Exp Res*, 2004, 28(10): 1481–1486.
101. Kuo, P.H., G. Kalsi, C.A. Prescott, et al., Association of ADH and ALDH genes with alcohol dependence in the Irish Affected Sib Pair Study of alcohol dependence (IASPSAD) sample. *Alcohol Clin Exp Res*, 2008, 32(5): 785–795.
102. Luo, X., H.R. Kranzler, L. Zuo, et al., Diplotype trend regression analysis of the ADH gene cluster and the ALDH2 gene: multiple significant associations with alcohol dependence. *Am J Hum Genet*, 2006, 78(6): 973–987.
103. Le Foll, B., A. Gallo, Y. Le Strat, et al., Genetics of dopamine receptors and drug addiction: a comprehensive review. *Behav Pharmacol*, 2009, 20(1): 1–17.
104. Batel, P., H. Houchi, M. Daoust, et al., A haplotype of the DRD1 gene is associated with alcohol dependence. *Alcohol Clin Exp Res*, 2008, 32(4): 567–572.
105. Neville, M.J., E.C. Johnstone, and R.T. Walton, Identification and characterization of ANKK1: a novel kinase gene closely linked to DRD2 on chromosome band 11q23.1. *Hum Mutat*, 2004, 23(6): 540–545.
106. Smith, L., M. Watson, S. Gates, et al., Meta-analysis of the association of the Taq1A polymorphism with the risk of alcohol dependency: a HuGE gene-disease association review. *Am J Epidemiol*, 2008, 167(2): 125–138.
107. Bolos, A.M., M. Dean, S. Lucas-Derse, et al., Population and pedigree studies reveal a lack of association between the dopamine D2 receptor gene and alcoholism. *JAMA*, 1990, 264(24): 3156–3160.
108. Cook, B.L., Z.W. Wang, R.R. Crowe, et al., Alcoholism and the D2 receptor gene. *Alcohol Clin Exp Res*, 1992, 16(4): 806–809.
109. Heinz, A., T. Sander, H. Harms, et al., Lack of allelic association of dopamine D1 and D2 (TaqIA) receptor gene polymorphisms with reduced dopaminergic sensitivity to alcoholism. *Alcohol Clin Exp Res*, 1996, 20(6): 1109–1113.

110. Jeanneteau, F., B. Funalot, J. Jankovic, et al., A functional variant of the dopamine D3 receptor is associated with risk and age-at-onset of essential tremor. *Proc Natl Acad Sci USA*, 2006, 103(28): 10753–10758.

111. Limosin, F., L. Romo, P. Batel, et al., Association between dopamine receptor D3 gene BalI polymorphism and cognitive impulsiveness in alcohol-dependent men. *Eur Psychiatry*, 2005, 20(3): 304–306.

112. Gorwood, P., F. Limosin, P. Batel, et al., The genetics of addiction: alcohol-dependence and D3 dopamine receptor gene. *Pathol Biol (Paris)*, 2001, 49(9): 710–717.

113. Lee, M.S., and S.H. Ryu, No association between the dopamine D3 receptor gene and Korean alcohol dependence. *Psychiatr Genet*, 2002, 12(3): 173–176.

114. Wiesbeck, G.A., K.M. Dursteler-MacFarland, F.M. Wurst, et al., No association of dopamine receptor sensitivity in vivo with genetic predisposition for alcoholism and DRD2/DRD3 gene polymorphisms in alcohol dependence. *Addict Biol*, 2006, 11(1): 72–75.

115. Schoots, O., and H.H. Van Tol, The human dopamine D4 receptor repeat sequences modulate expression. *Pharmacogenomics J*, 2003, 3(6): 343–348.

116. Golimbet, V.E., M.V. Alfimova, I.K. Gritsenko, et al., Relationship between dopamine system genes and extraversion and novelty seeking. *Neurosci Behav Physiol*, 2007, 37(6): 601–606.

117. Ebstein, R.P., A.H. Zohar, J. Benjamin, et al., An update on molecular genetic studies of human personality traits. *Appl Bioinformatics*, 2002, 1(2): 57–68.

118. Schinka, J.A., E.A. Letsch, and F.C. Crawford, DRD4 and novelty seeking: results of meta-analyses. *Am J Med Genet*, 2002, 114(6): 643–648.

119. Parsian, A., S. Chakraverty, L. Fisher, et al., No association between polymorphisms in the human dopamine D3 and D4 receptors genes and alcoholism. *Am J Med Genet*, 1997, 74(3): 281–285.

120. Sander, T., H. Harms, J. Podschus, et al., Allelic association of a dopamine transporter gene polymorphism in alcohol dependence with withdrawal seizures or delirium. *Biol Psychiatry*, 1997b, 41(3): 299–304.

121. Hutchison, K.E., H. LaChance, R. Niaura, et al., The DRD4 VNTR polymorphism influences reactivity to smoking cues. *J Abnorm Psychol*, 2002, 111(1): 134–143.

122. Du, Y., M. Yang, H.W. Yeh, et al., The association of exon 3 VNTR polymorphism of the dopamine receptor D4 (DRD4) gene with alcoholism in Mexican Americans. *Psychiatry Res*, 2010, 177(3): 358–360.

123. Sidhu, A., Coupling of D1 and D5 dopamine receptors to multiple G proteins: Implications for understanding the diversity in receptor-G protein coupling. *Mol Neurobiol*, 1998, 16(2): 125–134.

124. Cravchik, A., and P.V. Gejman, Functional analysis of the human D5 dopamine receptor missense and nonsense variants: differences in dopamine binding affinities. *Pharmacogenetics*, 1999, 9(2): 199–206.

125. Housley, D.J., M. Nikolas, P.J. Venta, et al., SNP discovery and haplotype analysis in the segmentally duplicated DRD5 coding region. *Ann Hum Genet*, 2009, 73(Pt 3): 274–282.

126. Samochowiec, J., J. Kucharska-Mazur, A. Grzywacz, et al., Family-based and case-control study of DRD2, DAT, 5HTT, COMT genes polymorphisms in alcohol dependence. *Neurosci Lett*, 2006, 410(1): 1–5.

127. Dobashi, I., T. Inada, and K. Hadano, Alcoholism and gene polymorphisms related to central dopaminergic transmission in the Japanese population. *Psychiatr Genet*, 1997, 7(2): 87–91.

128. Muramatsu, T., and S. Higuchi, Dopamine transporter gene polymorphism and alcoholism. *Biochem Biophys Res Commun*, 1995, 211(1): 28–32.

129. Kohnke, M.D., A. Batra, W. Kolb, et al., Association of the dopamine transporter gene with alcoholism. *Alcohol Alcohol*, 2005, 40(5): 339–342.

130. Franke, P., S.G. Schwab, M. Knapp, et al., DAT1 gene polymorphism in alcoholism: a family-based association study. *Biol Psychiatry*, 1999, 45(5): 652–654.

131. Foley, P.F., E.W. Loh, D.J. Innes, et al., Association studies of neurotransmitter gene polymorphisms in alcoholic Caucasians. *Ann N Y Acad Sci*, 2004, 1025: 39–46.

132. Kohnke, M.D., C.P. Zabetian, G.M. Anderson, et al., A genotype-controlled analysis of plasma dopamine beta-hydroxylase in healthy and alcoholic subjects: evidence for alcohol-related differences in noradrenergic function. *Biol Psychiatry*, 2002, 52(12): 1151–1158.

133. Freire, M.T., M.H. Hutz, and C.H. Bau, The DBH-1021 C/T polymorphism is not associated with alcoholism but possibly with patients' exposure to life events. *J Neural Transm*, 2005, 112(9): 1269–1274.

134. Kohnke, M.D., W. Kolb, A.M. Kohnke, et al., DBH*444G/A polymorphism of the dopamine-beta-hydroxylase gene is associated with alcoholism but not with severe alcohol withdrawal symptoms. *J Neural Transm*, 2006a, 113(7): 869–876.

135. Zabetian, C.P., S.G. Buxbaum, R.C. Elston, et al., The structure of linkage disequilibrium at the DBH locus strongly influences the magnitude of association between diallelic markers and plasma dopamine beta-hydroxylase activity. *Am J Hum Genet*, 2003, 72(6): 1389–1400.

136. Cubells, J.F., and C.P. Zabetian, Human genetics of plasma dopamine beta-hydroxylase activity: applications to research in psychiatry and neurology. *Psychopharmacology (Berl)*, 2004, 174(4): 463–476.

137. Badner, J.A., and E.S. Gershon, Meta-analysis of whole-genome linkage scans of bipolar disorder and schizophrenia. *Mol Psychiatry*, 2002, 7(4): 405–411.

138. Lotta, T., J. Vidgren, C. Tilgmann, et al., Kinetics of human soluble and membrane-bound catechol O-methyltransferase: a revised mechanism and description of the thermolabile variant of the enzyme. *Biochemistry*, 1995, 34(13): 4202–4210.

139. Matsumoto, M., C.S. Weickert, M. Akil, et al., Catechol O-methyltransferase mRNA expression in human and rat brain: evidence for a role in cortical neuronal function. *Neuroscience*, 2003, 116(1): 127–137.

140. Kauhanen, J., T. Hallikainen, T.P. Tuomainen, et al., Association between the functional polymorphism of catechol-O-methyltransferase gene and alcohol consumption among social drinkers. *Alcohol Clin Exp Res*, 2000, 24(2): 135–139.

141. Enoch, M.A., K. Xu, E. Ferro, et al., Genetic origins of anxiety in women: a role for a functional catechol-O-methyltransferase polymorphism. *Psychiatr Genet*, 2003, 13(1): 33–41.

142. COMBINE, S.R.G., Testing combined pharmacotherapies and behavioral interventions in alcohol dependence: rationale and methods. *Alcohol Clin Exp Res*, 2003a, 27(7): 1107–1122.

143. COMBINE, S.R.G., Testing combined pharmacotherapies and behavioral interventions for alcohol dependence (the COMBINE study): a pilot feasibility study. *Alcohol Clin Exp Res*, 2003b, 27(7): 1123–1131.

144. Goldman, D., G. Oroszi, S. O'Malley, et al., COMBINE genetics study: the pharmacogenetics of alcoholism treatment response: genes and mechanisms. *J Stud Alcohol Suppl*, 2005b (15): 56–64; discussion 33.

145. Sabol, S.Z., S. Hu, and D. Hamer, A functional polymorphism in the monoamine oxidase A gene promoter. *Hum Genet*, 1998, 103(3): 273–279.

146. Contini, V., F.Z. Marques, C.E. Garcia, et al., MAOA-uVNTR polymorphism in a Brazilian sample: further support for the association with impulsive behaviors and alcohol dependence. *Am J Med Genet B Neuropsychiatr Genet*, 2006, 141B(3): 305–308.

147. Samochowiec, J., K.P. Lesch, M. Rottmann, et al., Association of a regulatory polymorphism in the promoter region of the monoamine oxidase A gene with antisocial alcoholism. *Psychiatry Res*, 1999, 86(1): 67–72.

148. Schmidt, L.G., T. Sander, S. Kuhn, et al., Different allele distribution of a regulatory MAOA gene promoter polymorphism in antisocial and anxious-depressive alcoholics. *J Neural Transm*, 2000, 107(6): 681–689.

149. Saito, T., H.M. Lachman, L. Diaz, et al., Analysis of monoamine oxidase A (MAOA) promoter polymorphism in Finnish male alcoholics. *Psychiatry Res*, 2002, 109(2): 113–119.

150. Zalsman, G., Y.Y. Huang, J.M. Harkavy-Friedman, et al., Relationship of MAO-A promoter (u-VNTR) and COMT (V158M) gene polymorphisms to CSF monoamine metabolites levels in a psychiatric sample of caucasians: A preliminary report. *Am J Med Genet B Neuropsychiatr Genet*, 2005, 132B(1): 100–103.

151. Kohnke, M.D., U. Lutz, W. Kolb, et al., Allele distribution of a monoamine oxidase A gene promoter polymorphism in German alcoholics and in subgroups with severe forms of alcohol withdrawal and its influence on plasma homovanillic acid. *Psychiatr Genet*, 2006b, 16(6): 237–238.

152. Parsian, A., B.K. Suarez, B. Tabakoff, et al., Monoamine oxidases and alcoholism. I. Studies in unrelated alcoholics and normal controls. *Am J Med Genet*, 1995, 60(5): 409–416.

153. Kumar, A., R. Goyal, and A. Prakash, Possible GABAergic mechanism in the protective effect of allopregnenolone against immobilization stress. *Eur J Pharmacol*, 2009, 602(2–3): 343–347.

154. Helm, K.A., R.P. Haberman, S.L. Dean, et al., GABAB receptor antagonist SGS742 improves spatial memory and reduces protein binding to the cAMP response element (CRE) in the hippocampus. *Neuropharmacology*, 2005, 48(7): 956–964.

155. Padgett, C.L., and P.A. Slesinger, GABAB receptor coupling to G-proteins and ion channels. *Adv Pharmacol*, 2010, 58: 123–147.

156. Addolorato, G., L. Leggio, S. Cardone, et al., Role of the GABA(B) receptor system in alcoholism and stress: focus on clinical studies and treatment perspectives. *Alcohol*, 2009, 43(7): 559–563.

157. Olsen, R.W., and W. Sieghart, International Union of Pharmacology. LXX. Subtypes of gamma-aminobutyric acid(A) receptors: classification on the basis of subunit composition, pharmacology, and function. Update. *Pharmacol Rev*, 2008, 60(3): 243–260.

158. Sieghart, W., and G. Sperk, Subunit composition, distribution and function of GABA(A) receptor subtypes. *Curr Top Med Chem*, 2002, 2(8): 795–816.

159. Whiting, P.J., The GABA-A receptor gene family: new targets for therapeutic intervention. *Neurochem Int*, 1999, 34(5): 387–390.

160. Johnston, G.A., M. Chebib, J.R. Hanrahan, et al., Neurochemicals for the Investigation of GABA(C) receptors. *Neurochem Res*, 2010, 35(12): 1970–1977.

161. Rae, C., F.A. Nasrallah, J.L. Griffin, et al., Now I know my ABC. A systems neurochemistry and functional metabolomic approach to understanding the GABAergic system. *J Neurochem*, 2009, 109 Suppl 1: 109–116.

162. Ait-Daoud, N., W.J. Lynch, J.K. Penberthy, et al., Treating smoking dependence in depressed alcoholics. *Alcohol Res Health*, 2006, 29(3): 213–220.

163. Tabakoff, B., L. Saba, M. Printz, et al., Genetical genomic determinants of alcohol consumption in rats and humans. *BMC Biol*, 2009, 7: 70.

164. Loh, E.W., S. Higuchi, S. Matsushita, et al., Association analysis of the GABA(A) receptor subunit genes cluster on 5q33-34 and alcohol dependence in a Japanese population. *Mol Psychiatry*, 2000, 5(3): 301–307.

165. Schuckit, M.A., C. Mazzanti, T.L. Smith, et al., Selective genotyping for the role of 5-HT2A, 5-HT2C, and GABA alpha 6 receptors and the serotonin transporter in the level of response to alcohol: a pilot study. *Biol Psychiatry*, 1999, 45(5): 647–651.

166. Sander, T., D. Ball, R. Murray, et al., Association analysis of sequence variants of GABA(A) α6, β2, and γ2 gene cluster and alcohol dependence. *Alcohol Clin Exp Res*, 1999, 23(3): 427–431.

167. Radel, M., R.L. Vallejo, N. Iwata, et al., Haplotype-based localization of an alcohol dependence gene to the 5q34 {gamma}-aminobutyric acid type A gene cluster. *Arch Gen Psychiatry*, 2005, 62(1): 47–55.

168. Swan, G.E., D. Carmelli, and L.R. Cardon, The consumption of tobacco, alcohol, and coffee in Caucasian male twins: a multivariate genetic analysis. *J Subst Abuse*, 1996, 8(1): 19–31.

169. Swan, G.E., D. Carmelli, and L.R. Cardon, Heavy consumption of cigarettes, alcohol and coffee in male twins. *J Stud Alcohol*, 1997, 58(2): 182–190.

170. True, W.R., H. Xian, J.F. Scherrer, et al., Common genetic vulnerability for nicotine and alcohol dependence in men. *Arch Gen Psychiatry*, 1999, 56(7): 655–661.

171. Hettema, J.M., L.A. Corey, and K.S. Kendler, A multivariate genetic analysis of the use of tobacco, alcohol, and caffeine in a population based sample of male and female twins. *Drug Alcohol Depend*, 1999, 57(1): 69–78.

172. Schlaepfer, I.R., N.R. Hoft, A.C. Collins, et al., The CHRNA5/A3/B4 gene cluster variability as an important determinant of early alcohol and tobacco initiation in young adults. *Biol Psychiatry*, 2008, 63(11): 1039–1046.

173. Steinlein, O., T. Sander, J. Stoodt, et al., Possible association of a silent polymorphism in the neuronal nicotinic acetylcholine receptor subunit α4 with common idiopathic generalized epilepsies. *Am J Med Genet*, 1997, 74(4): 445–449.

174. Phillips, H.A., I. Favre, M. Kirkpatrick, et al., CHRNB2 is the second acetylcholine receptor subunit associated with autosomal dominant nocturnal frontal lobe epilepsy. *Am J Hum Genet*, 2001, 68(1): 225–231.

175. Kent, L., F. Middle, Z. Hawi, et al., Nicotinic acetylcholine receptor α4 subunit gene polymorphism and attention deficit hyperactivity disorder. *Psychiatr Genet*, 2001, 11(1): 37–40.

176. Li, M.D., J. Beuten, J.Z. Ma, et al., Ethnic- and gender-specific association of the nicotinic acetylcholine receptor α4 subunit gene (CHRNA4) with nicotine dependence. *Hum Mol Genet*, 2005, 14(9): 1211–1219.

177. Lueders, K.K., S. Hu, L. McHugh, et al., Genetic and functional analysis of single nucleotide polymorphisms in the β2-neuronal nicotinic acetylcholine receptor gene (CHRNB2). *Nicotine Tob Res*, 2002, 4(1): 115–125.

178. Silverman, M.A., M.C. Neale, P.F. Sullivan, et al., Haplotypes of four novel single nucleotide polymorphisms in the nicotinic acetylcholine receptor β2-subunit (CHRNB2) gene show no association with smoking initiation or nicotine dependence. *Am J Med Genet*, 2000, 96(5): 646–653.

179. Chatterjee, S., and S.E. Bartlett, Neuronal nicotinic acetylcholine receptors as pharmacotherapeutic targets for the treatment of alcohol use disorders. *CNS Neurol Disord Drug Targets*, 2010, 9(1): 60–76.

180. Foroud, T., H.J. Edenberg, A. Goate, et al., Alcoholism susceptibility loci: confirmation studies in a replicate sample and further mapping. *Alcohol Clin Exp Res*, 2000, 24(7): 933–945.

181. Reich, T., H.J. Edenberg, A. Goate, et al., Genome-wide search for genes affecting the risk for alcohol dependence. *Am J Med Genet*, 1998, 81(3): 207–215.

182. Wang, J.C., A.L. Hinrichs, H. Stock, et al., Evidence of common and specific genetic effects: association of the muscarinic acetylcholine receptor M2 (CHRM2) gene with alcohol dependence and major depressive syndrome. *Hum Mol Genet*, 2004, 13(17): 1903–1911.
183. Luo, X., H.R. Kranzler, L. Zuo, et al., CHRM2 gene predisposes to alcohol dependence, drug dependence and affective disorders: results from an extended case-control structured association study. *Hum Mol Genet*, 2005, 14(16): 2421–2434.
184. Hun Jung, M., B. Lae Park, B.C. Lee, et al., Association of CHRM2 polymorphisms with severity of alcohol dependence. *Genes Brain Behav*, 2010, 10(2): 253–256.
185. Rangaswamy, M., and B. Porjesz, Uncovering genes for cognitive (dys)function and predisposition for alcoholism spectrum disorders: a review of human brain oscillations as effective endophenotypes. *Brain Res*, 2008, 1235: 153–171.
186. Dick, D.M., J.C. Wang, J. Plunkett, et al., Family-based association analyses of alcohol dependence phenotypes across DRD2 and neighboring gene ANKK1. *Alcohol Clin Exp Res*, 2007, 31(10): 1645–1653.
187. Comings, D.E., S. Wu, M. Rostamkhani, et al., Association of the muscarinic cholinergic 2 receptor (CHRM2) gene with major depression in women. *Am J Med Genet*, 2002, 114(5): 527–529.
188. Andersson, O., A. Stenqvist, A. Attersand, et al., Nucleotide sequence, genomic organization, and chromosomal localization of genes encoding the human NMDA receptor subunits NR3A and NR3B. *Genomics*, 2001, 78(3): 178–184.
189. Vengeliene, V., D. Bachteler, W. Danysz, et al., The role of the NMDA receptor in alcohol relapse: a pharmacological mapping study using the alcohol deprivation effect. *Neuropharmacology*, 2005, 48(6): 822–829.
190. Salter, M.W., and L.V. Kalia, Src kinases: a hub for NMDA receptor regulation. *Nat Rev Neurosci*, 2004, 5(4): 317–328.
191. Wernicke, C., J. Samochowiec, L.G. Schmidt, et al., Polymorphisms in the *N*-methyl-D-aspartate receptor 1 and 2B subunits are associated with alcoholism-related traits. *Biol Psychiatry*, 2003, 54(9): 922–928.
192. Schumann, G., D. Rujescu, A. Szegedi, et al., No association of alcohol dependence with a NMDA-receptor 2B gene variant. *Mol Psychiatry*, 2003, 8(1): 11–12.
193. Johnson, B.A., N. Ait-Daoud, C.L. Bowden, et al., Oral topiramate for treatment of alcohol dependence: a randomised controlled trial. *Lancet*, 2003, 361(9370): 1677–1685.
194. Feinn, R., M. Nellissery, and H.R. Kranzler, Meta-analysis of the association of a functional serotonin transporter promoter polymorphism with alcohol dependence. *Am J Med Genet B Neuropsychiatr Genet*, 2005, 133B(1): 79–84.
195. Halliday, G., J. Ellis, R. Heard, et al., Brainstem serotonergic neurons in chronic alcoholics with and without the memory impairment of Korsakoff's psychosis. *J Neuropathol Exp Neurol*, 1993, 52(6): 567–579.
196. Verheul, R., W. van den Brink, and P. Geerlings, A three-pathway psychobiological model of craving for alcohol. *Alcohol Alcohol*, 1999, 34(2): 197–222.
197. Heinz, A., D. Goldman, J. Gallinat, et al., Pharmacogenetic insights to monoaminergic dysfunction in alcohol dependence. *Psychopharmacology (Berl)*, 2004, 174(4): 561–570.
198. Pinto, E., J. Reggers, P. Gorwood, et al., The short allele of the serotonin transporter promoter polymorphism influences relapse in alcohol dependence. *Alcohol Alcohol*, 2008, 43(4): 398–400.
199. Pinto, D., A.T. Pagnamenta, L. Klei, et al., Functional impact of global rare copy number variation in autism spectrum disorders. *Nature*, 2010, 466(7304): 368–372.
200. Lappalainen, J., J.C. Long, M. Eggert, et al., Linkage of antisocial alcoholism to the serotonin 5-HT1B receptor gene in 2 populations. *Arch Gen Psychiatry*, 1998, 55(11): 989–994.

201. Vengeliene, V., A. Bilbao, A. Molander, et al., Neuropharmacology of alcohol addiction. *Br J Pharmacol*, 2008, 154(2): 299–315.
202. Wetherill, L., M.A. Schuckit, V. Hesselbrock, et al., Neuropeptide Y receptor genes are associated with alcohol dependence, alcohol withdrawal phenotypes, and cocaine dependence. *Alcohol Clin Exp Res*, 2008, 32(12): 2031–2040.
203. Dennis, R.A., and M.T. McCammon, Acn9 is a novel protein of gluconeogenesis that is located in the mitochondrial intermembrane space. *Eur J Biochem*, 1999, 261(1): 236–243.
204. Conner, B.T., E.P. Noble, S.M. Berman, et al., DRD2 genotypes and substance use in adolescent children of alcoholics. *Drug Alcohol Depend*, 2005, 79(3): 379–387.
205. McMillin, G.A., R. Mellis, and J. Bornhorst, Alcohol abuse and dependency genetics of susceptibility and pharmacogenetics of therapy, in *Critical Issues in Drug Abuse Testing*, A. Dasgupta, Editor. 2009, Humana Press, New York.
206. Haile, C.N., T.A. Kosten, and T.R. Kosten, Pharmacogenetic treatments for drug addiction: alcohol and opiates. *Am J Drug Alcohol Abuse*, 2008, 34(4): 355–381.
207. Fukasawa, T., A. Suzuki, and K. Otani, Effects of genetic polymorphism of cytochrome P450 enzymes on the pharmacokinetics of benzodiazepines. *J Clin Pharm Ther*, 2007, 32(4): 333–341.
208. Ooteman, W., M. Naassila, M.W. Koeter, et al., Predicting the effect of naltrexone and acamprosate in alcohol-dependent patients using genetic indicators. *Addict Biol*, 2009, 14(3): 328–337.
209. Bond, C., K.S. LaForge, M. Tian, et al., Single-nucleotide polymorphism in the human mu opioid receptor gene alters beta-endorphin binding and activity: possible implications for opiate addiction. *Proc Natl Acad Sci USA*, 1998, 95(16): 9608–9613.
210. van der Zwaluw, C.S., E. van den Wildenberg, R.W. Wiers, et al., Polymorphisms in the mu-opioid receptor gene (OPRM1) and the implications for alcohol dependence in humans. *Pharmacogenomics*, 2007, 8(10): 1427–1436.
211. Barr, T.L., S. Alexander, and Y. Conley, Gene expression profiling for discovery of novel targets in human traumatic brain injury. *Biol Res Nurs*, 2011, 13(2): 140–153.
212. Papapetropoulos, S., L. Shehadeh, and D. McCorquodale, Optimizing human post-mortem brain tissue gene expression profiling in Parkinson's disease and other neurodegenerative disorders: from target "fishing" to translational breakthroughs. *J Neurosci Res*, 2007, 85(14): 3013–3024.
213. Pasinetti, G.M., Use of cDNA microarray in the search for molecular markers involved in the onset of Alzheimer's disease dementia. *J Neurosci Res*, 2001, 65(6): 471–476.
214. Bahn, S., S.J. Augood, M. Ryan, et al., Gene expression profiling in the post-mortem human brain—no cause for dismay. *J Chem Neuroanat*, 2001, 22(1–2): 79–94.
215. Shelton, R.C., J. Claiborne, M. Sidoryk-Wegrzynowicz, et al., Altered expression of genes involved in inflammation and apoptosis in frontal cortex in major depression. *Mol Psychiatry*, 2011, 16: 751–762.
216. Kim, S., and M.J. Webster, Correlation analysis between genome-wide expression profiles and cytoarchitectural abnormalities in the prefrontal cortex of psychiatric disorders. *Mol Psychiatry*, 2010, 15(3): 326–336.
217. Mayfield, R.D., R.A. Harris, and M.A. Schuckit, Genetic factors influencing alcohol dependence. *Br J Pharmacol*, 2008, 154(2): 275–287.
218. Mortazavi, A., B.A. Williams, K. McCue, et al., Mapping and quantifying mammalian transcriptomes by RNA-Seq. *Nat Methods*, 2008, 5(7): 621–628.
219. Maher, C.A., C. Kumar-Sinha, X. Cao, et al., Transcriptome sequencing to detect gene fusions in cancer. *Nature*, 2009, 458(7234): 97–101.
220. Hawkins, R.D., G.C. Hon, and B. Ren, Next-generation genomics: an integrative approach. *Nat Rev Genet*, 2010, 11(7): 476–486.

221. Guttman, M., I. Amit, M. Garber, et al., Chromatin signature reveals over a thousand highly conserved large non-coding RNAs in mammals. *Nature*, 2009, 458(7235): 223–227.
222. Bierut, L.J., A. Agrawal, K.K. Bucholz, et al., A genome-wide association study of alcohol dependence. *Proc Natl Acad Sci USA*, 2010, 107(11): 5082–5087.
223. Edenberg, H.J., D.L. Koller, X. Xuei, et al., Genome-wide association study of alcohol dependence implicates a region on chromosome 11. *Alcohol Clin Exp Res*, 2010, 34(5): 840–852.
224. Agrawal, A., A.L. Hinrichs, G. Dunn, et al., Linkage scan for quantitative traits identifies new regions of interest for substance dependence in the Collaborative Study on the Genetics of Alcoholism (COGA) sample. *Drug Alcohol Depend*, 2008, 93(1–2): 12–20.
225. Lind, P.A., S. Macgregor, J.M. Vink, et al., A genomewide association study of nicotine and alcohol dependence in Australian and Dutch populations. *Twin Res Hum Genet*, 2010, 13(1): 10–29.
226. Treutlein, J., S. Cichon, M. Ridinger, et al., Genome-wide association study of alcohol dependence. *Arch Gen Psychiatry*, 2009, 66(7): 773–784.
227. Cook, E.H., Jr., and S.W. Scherer, Copy-number variations associated with neuropsychiatric conditions. *Nature*, 2008, 455(7215): 919–923.
228. Sebat, J., B. Lakshmi, D. Malhotra, et al., Strong association of de novo copy number mutations with autism. *Science*, 2007, 316(5823): 445–449.
229. St Clair, D., Copy number variation and schizophrenia. *Schizophr Bull*, 2009, 35(1): 9–12.
230. Ingason, A., I. Giegling, S. Cichon, et al., A large replication study and meta-analysis in European samples provides further support for association of AHI1 markers with schizophrenia. *Hum Mol Genet*, 2011, 19(7): 1379–1386.
231. Jordan, K.W., J. Nordenstam, G.Y. Lauwers, et al., Metabolomic characterization of human rectal adenocarcinoma with intact tissue magnetic resonance spectroscopy. *Dis Colon Rectum*, 2009, 52(3): 520–525.
232. Litten, R.Z., A.M. Bradley, and H.B. Moss, Alcohol biomarkers in applied settings: recent advances and future research opportunities. *Alcohol Clin Exp Res*, 2010, 34(6): 955–967.
232a. Lai, X., S. Liangpunsakul, D.W. Crabb, et al. A proteomic workflow for discovery of serum carrier protein-bound biomarker candidates of alcohol abuse using LC-MS/MS. *Electrophoresis* 2009, 30(12): 2207–2214.
233. Meyer, F., D. Paarmann, M. D'Souza, et al., The metagenomics RAST server—a public resource for the automatic phylogenetic and functional analysis of metagenomes. *BMC Bioinformatics*, 2008, 9: 386.
234. Turner, D.J., R. Tuytten, K.P. Janssen, et al., Toward clinical proteomics on a next-generation sequencing platform. *Anal Chem*, 2011, 83(3): 666–670.
235. Anni, H., and Y. Israel, Proteomics in alcohol research. *Alcohol Res Health*, 2002, 26(3): 219–232.
236. Feinberg, A.P., R.A. Irizarry, D. Fradin, et al., Personalized epigenomic signatures that are stable over time and covary with body mass index. *Sci Transl Med*, 2010b, 2(49): 49–67.
237. Satterlee, J.S., D. Schubeler, and H.H. Ng, Tackling the epigenome: challenges and opportunities for collaboration. *Nat Biotechnol*, 2010, 28(10): 1039–1044.
238. Portela, A., and M. Esteller, Epigenetic modifications and human disease. *Nat Biotechnol*, 2010, 28(10): 1057–1068.
239. Feinberg, A.P., Epigenomics reveals a functional genome anatomy and a new approach to common disease. *Nat Biotechnol*, 2010a, 28(10): 1049–1052.
240. Ishii, T., E. Hashimoto, W. Ukai, et al., [Epigenetic regulation in alcohol-related brain damage]. *Nihon Arukoru Yakubutsu Igakkai Zasshi*, 2008, 43(5): 705–713.

241. Huss, M., Introduction into the analysis of high-throughput-sequencing based epigenome data. *Brief Bioinform*, 2010, 11(5): 512–523.
242. de Magalhaes, J.P., C.E. Finch, and G. Janssens, Next-generation sequencing in aging research: emerging applications, problems, pitfalls and possible solutions. *Ageing Res Rev*, 2010, 9(3): 315–323.
243. Rivera, R.M., and L.B. Bennett, Epigenetics in humans: an overview. *Curr Opin Endocrinol Diabetes Obes*, 2010, 17(6): 493–499.
244. Pietrzykowski, A.Z., The role of microRNAs in drug addiction: a big lesson from tiny molecules. *Int Rev Neurobiol*, 2010, 91: 1–24.
245. Tsang, J.S., M.S. Ebert, and A. van Oudenaarden, Genome-wide dissection of microRNA functions and cotargeting networks using gene set signatures. *Mol Cell*, 2010, 38(1): 140–153.
246. Griffiths-Jones, S., The microRNA Registry. *Nucleic Acids Res*, 2004, 32(Database issue): D109–D111.
247. Griffiths-Jones, S., miRBase: the microRNA sequence database. *Methods Mol Biol*, 2006, 342: 129–138.
248. Guerrini, I., C.C. Cook, W. Kest, et al., Genetic linkage analysis supports the presence of two susceptibility loci for alcoholism and heavy drinking on chromosome 1p22.1-11.2 and 1q21.3-24.2. *BMC Genet*, 2005, 6: 11.
249. Joslyn, G., G. Brush, M. Robertson, et al., Chromosome 15q25.1 genetic markers associated with level of response to alcohol in humans. *Proc Natl Acad Sci USA*, 2008, 105(51): 20368–20373.
250. Windemuth, C., A. Hahn, K. Strauch, et al., Linkage analysis in alcohol dependence. *Genet Epidemiol*, 1999, 17 Suppl 1: S403–S407.
251. Haycock, P.C., Fetal alcohol spectrum disorders: the epigenetic perspective. *Biol Reprod*, 2009, 81(4): 607–617.
252. Dick, D.M., F. Aliev, J.C. Wang et al. A systematic single nucleotide polymorphism screen to fin-map alcohol dependence genes on chromosome 7 identifies association with a novel susceptibility gene ACN9. Biol Psychiatry 2008. 63(11): 1047–1053.
253. Edenberg H.J., J. Wang, H. Tian, et. al., A regulatory variation in OPRK1, the gene encoding the kappa-opioid receptor, is associated with alcohol dependence. *Hum Mol Genet*. 2008, 17(12):1783–1789.

4 Introduction to Drugs of Abuse

Larry Broussard and Catherine Hammett-Stabler

CONTENTS

4.1 Introduction 93
4.2 Testing for Drugs of Abuse..97
4.3 Sympathomimetic Amines Including Amphetamines and Ecstasy 100
4.4 Barbiturates.. 106
4.5 Benzodiazepines 107
4.6 Opiates and Opioids.. 109
4.7 Cocaine 113
4.8 Cannabinoids 114
4.9 Phencyclidine... 115
4.10 Club Drugs... 115
4.11 Inhalants 117
4.12 Specimen Validity Testing.. 118
References... 120

4.1 INTRODUCTION

A consensus definition of the term *drugs of abuse* is practically impossible to obtain because it is one of those terms that almost everyone defines differently. Any definition requires agreement on the meaning of the words, drug and abuse—a difficulty in itself as reflected by the finding that an Internet search of the term yields more than 2 million results. Furthermore, the context of use (medical, criminal justice, workplace, public health, public perception, etc.) contributes heavily to the definition. And finally, the term denotes a negative connotation (i.e., the misuse of a substance) to most people. For these reasons, any listing and classification of drugs of abuse depend on both the definition and context. Nevertheless, there are several drugs and classes of drugs that are usually included when compiling such a list (Table 4.1). Frequently such drugs are considered in groups such as illegal substances, prescription medications, over-the-counter medications, performance-enhancing substances, and so forth. The National Institute on Drug Abuse (NIDA) publishes several relevant lists including a chart of commonly abused drugs (1) and another of prescription drugs that are abused (2). The following substances are included in the NIDA commonly abused drugs list: nicotine, ethyl alcohol, cannabinoids, heroin, opium, cocaine, amphetamine, methamphetamine, methylenedioxymethamphetamine (MDMA),

TABLE 4.1

Drugs That May Be Considered Drugs of Abuse

Class	Drugs (Pharmaceutical or Chemical Names)	Street Names
	Opiates and Opioids	
	Heroin (diacetylmorphine)	China white, white horse, junk, H, dope, skag, brown sugar, smack
	Morphine (MS-Contin)	M, morph, Miss Emma, white stuff, monkey
	Codeine (Tylenol #3)	Cody, Captain Cody, schoolboy
	Hydrocodone (Lorcet, Lortab, Vicodin)	
	Hydromorphone (Dilaudid)	Juice
	Oxycodone (OxyContin, Percocet)	Oxy, oxy 80s, oxy c's, oc's, hillbilly heroin, percs
	Oxymorphone (Numorphan)	
	Propoxyphene (Darvon, Darvocet)	Pinks, footballs, 65's, D's, Dan's, dance
	Fentanyl (Duragesic, Sublimaze)	China white, dance fever, friend, goodfella, China girl
	Methadone (Methadose, Dolophine)	Amidone, chocolate chip cookies, wafer
	Meperidine (Demerol)	
	Buprenorphine (Buprenex, Subutex)	
	Barbiturates	
	Amobarbital (Amytal)	Barbs, reds, red birds, yellows, yellow jackets
	Butalbital (Fiorinal, Fioricet)	
	Butabarbital (Butisol)	
	Secobarbital (Seconal)	
	Phenobarbital	
	Antidepressants and Antianxiety Drugs	
Benzodiazepines	Alprazolam (Xanax)	Candy, downers, sleeping pills, tranks
	Chlordiazepoxide (Librium)	
	Clonazepam (Klonopin)	
	Clorazepate (Tranxene)	
	Diazepam (Valium)	
	Lorazepam (Ativan)	
	Oxazepam (Serax)	
	Flunitrazepam (Rohypnol)	
Tricyclic antidepressants	Amitriptyline (Elavil, Endep)	Downers, Blue angels, blue birds, TCA
	Amoxapine (Asendin)	
	Climipramine (Anafranil)	
	Desipramine (Norpramin, Pertofrane)	

TABLE 4.1 (continued)
Drugs That May Be Considered Drugs of Abuse

Class	Drugs (Pharmaceutical or Chemical Names)	Street Names
	Doxepin (Sinequan, Zonalon)	
	Imipramine (Tofranil)	
	Nortriptyline (Pamelor)	
	Protriptyline (Vivactil)	
	Trimipramine (Surmontil)	
Selective serotonin reuptake inhibitors	Citalopram (Celexa)	
	Fluvoxamine (Prozac)	
	Paroxetine (Paxil)	
	Sertraline (Zoloft)	
	Venlafaxine (Effexor)	
	Bupropion (Wellbutrin)	
	Duloxetine (Cymbalta)	
	Buspirone (BuSpar)	
Cannabinoids and Synthetics		
	Hashish	Hash, boom, gangster, hash oil, hemp
	Marijuana	Dope, ganja, grass, joint, Mary Jane, pot, reefer, weed, skunk
	JWH-018 series	Spice, gene, sence, K2,
	HU-210 series	
	CP-47, 497 series	
Club Drugs		
	Gamma-hydroxybutyrate (GHB)	Georgia home boy, grievous bodily harm, G, liquid ecstasy
	Flunitrazepam (also under benzodiazepines)	Roofies, roofinol, rophies, rope, Roche, roach, R2, Mexican valium, forget-me pill
	Methylenedioxymethamphetamine (MDMA)	Ecstasy, Adam, clarity, Eve, lover's speed, peace
Stimulants		
	Cocaine (cocaine hydrochloride)	C, coke, crack, blow, candy, rock, snow, Charlie
	Amphetamine (Adderall, Biphetamine, Dexedrine)	Speed, bennies, black beauties, LA turnaround, uppers
	Methamphetamine (Desoxyn)	Meth, ice, crank, crystal, fire, black beauties, biker's coffee, speed, glass, chalk, chicken feed, poor man's cocaine, trash, yaba

continued

TABLE 4.1 (continued)
Drugs That May Be Considered Drugs of Abuse

Class	Drugs (Pharmaceutical or Chemical Names)	Street Names
	Methylphenidate (Ritalin)	Skippy, vitamin R, JIF, MPH, R-ball
	Mephedrone	MCAT, drone, meow-meow
	Methcathione	Ephedrine, cat, MCAT
Dissociative Drugs and Hallucinogens		
	Dextromethorphan (DXM)	Robo, triple C
	Phencyclidine (PCP)	Angel dust, hog, love boat, peace pill
	Ketamine	K, special K, vitamin K, Cat valium
	Salvia divinorum	Magic mint, shepherdess's herb, maria pastora, magic mint
	Lysergic acid diethylamide (LSD)	Acid, sunshine
	Mescaline	Peyote, mesc, buttons, cactus
	Psilocybin	Magic mushrooms, shrooms, purple passion
Inhalants	Gases (butane, propane, aerosol propellants, nitrous oxide)	
	Nitrates (isoamyl, isobutyl, cyclohexyl)	
	Solvents (paint thinners, gasoline, glues)	
Phenylethylamines	2C-I (4-iodo-2,5-dimethoxyphenethylamine)	Nexus, cyber
	2C-T	
	2C-B-Fly (4-bromo-2,5-dimethyoxyphenethylamine)	
Tryptamines	DMT	Foxy
	5-MeO-DIPT	Foxy methoxy
	5-MeO-DMT	
	Bufotenine (5-OH-DMT)	

flunitrazepam, gamma-hydroxybutyrate (GHB), ketamine, phencyclidine (PCP), salvia divinorum, dextromethorphan, lysergic acid diethylamide (LSD), mescaline, psilocybin, anabolic steroids, and inhalants. Several of these substances are also included on the list of prescription drugs that are often abused. Other medications or classes of drugs on the prescription drug abuse chart include barbiturates, benzodiazepines, codeine, fentanyl, morphine, other opioids (such as oxycodone, meperidine, hydromorphone, hydrocodone, and propoxyphene), and methylphenidate. This table is not meant to be all inclusive but is given to illustrate that the range of drugs of abuse is extensive. This chapter will focus on the laboratory detection of many, but not all, of these substances. The genetic aspects of the metabolism and abuse of several of these substances are discussed in more detail in other chapters.

4.2   TESTING FOR DRUGS OF ABUSE

Drug screens or drugs of abuse panels are not equal. The drugs included in such a panel and the criteria for reporting positive or negative results vary depending on several factors including the purpose of testing, regulatory and contractual require-ments, and the capability of the laboratory performing the testing. The purpose of testing depends upon the setting and may have medical-legal implications. For example, in the emergency setting testing is usually requested to assist in determin-ing whether symptoms observed are caused by drugs versus trauma or disease (3). In the pain management setting, testing may be used to identify patients who are not taking their medications as prescribed and possibly diverting or selling them. In the workplace setting, the scope of testing may be mandated by a specific regulatory agency (such as the Department of Transportation, DOT, or the Nuclear Regulatory Commission, NRC), included in employment contracts (unions, etc.), or simply cho-sen by the employer. In the criminal justice setting, testing may be used as part of the sentencing or parole process. In the medical examiner setting, drug testing is routinely performed on autopsy specimens as part of postmortem investigation to determine cause and manner of death.

Testing for drugs of abuse can be performed using a variety of samples including serum/plasma/blood, urine, hair, oral fluid, and sweat, but urine is the sample most commonly used. Urine offers the advantages of noninvasive collection, relatively easy, cost-effective, proven technology, and a moderate window of detection for most of the drugs of interest (Table 4.2). When deciding to use urine as the specimen, one must weigh these advantages against the lack of correlation between concentration and impairment. Serum, plasma, or blood may be useful in selected situations as concentrations determined in these matrices may correlate with pharmacological effects or impairment in many cases. Unfortunately, the use of these samples in a therapeutic context is not recommended routinely as therapeutic ranges are simply not defined or have been found to lack correlation with efficacy. The use of oral fluid has gained a great deal of interest recently because of its ease of collection, but when interpreting results it should be remembered that the metabolic profile and window of detection for most of the drugs of interest differ from those reported in urine. Hair offers the advantage of the longest window of detection for many drugs reflecting their presence in the circulation at the follicle. Any drug present at this time becomes trapped in the growing hair shaft where it remains and can thus be quantified. Obviously, the presence of a drug in hair reflects use or exposure of sev-eral weeks, even months, past. Methods for analysis are fairly cumbersome, often requiring significant sample preparation. For these reasons, most testing begins with immunoassay screening for drugs or their metabolites in urine because the testing is relatively inexpensive, rapid, and can be automated.

TABLE 4.2

Approximate Windows of Detection of Drugs in Urine

Drug/Class	Detection Time (Days)
Amphetamines	2–3
Barbiturates	1–3 (short-acting)
	1–3 weeks for long-acting
Benzodiazepines	1–3 (short-term therapeutic use)
	1–3 weeks (long-term chronic use)
Cannabinoids	5–7 (single use)
	Up to 4 weeks (chronic use)
Cocaine metabolite	2–3
	Up to 16 days (chronic use)
Opiates	3–4
Phencyclidine (PCP)	3–8

Note: These times are approximate and can vary depending on many factors including amount and frequency of use, individual metabolism, coadministered drugs, illness, age, drug tolerance, fluid ingestion, sensitivity and specificity of assay, urine pH, and so forth.

Immunoassays utilize an antibody whose reaction with the drugs or metabolites is monitored using some type of measurable label such as an enzyme or chemiluminescent chemical. The primary analytical parameters affecting immunoassay use and interpretation of results are sensitivity and antibody specificity. Antigen–antibody reactions are not linear with respect to drug/metabolite concentration, so even if an immunoassay gives concentration values, the results are semiquantitative at best. Immunoassays are formulated to be most effective at cutoff concentrations such as those seen in Table 4.3 that are often established by workplace drug testing regulations rather than analytical or clinical considerations. A few instruments used in this testing allow the laboratory to set or establish their own cutoffs (usually lower than the workplace regulations). Laboratories that choose to do so must carefully validate and monitor subsequent performance. Values equal to or below the cutoff concentration are usually reported as negative even though there may be some drug or metabolite present. Because most clinicians interpret "negative" to mean "nothing is present," some clinical laboratories have adopted the practice of reporting results as < the cutoff or ≥ the cutoff. In the clinical setting it may be important to subject selected negative samples by screening to a more sensitive mass spectrometry based method. This has been particularly useful in pain management where, as will be discussed below, immunoassay cross-reactivity for many drugs is less than optimal for clinical use. It is also important to recognize that neither immunoassay nor mass spectrometry based methods can measure to zero.

TABLE 4.3
Cutoff Concentrations for Federal Workplace Drug Testing Programs

Drug/Class	Screening Cutoff (ng/mL)	Confirmation Cutoff (ng/mL)
Amphetamines		
Amphetamine/methamphetamine	500	250 (amp/meth)
MDMA	500	250
MDA		250
MDEA		250
Marijuana metabolites (cannabinoids)	50	15 (THCA)
Cocaine metabolite (benzoylecgonine)	150	100
Opiates		
Codeine/morphine	2000	2000 (cod/mor)
6-Acetyl-morphine	10 ng/mL	10 ng/mL
Phencyclidine (PCP)	25 ng/mL	25 ng/mL

Notes: MDMA, methylenedioxymethamphetamine; MDA, methylenedioxyamphetamine, MDEA, methylenedioxyethylamphetamine; THCA, delta-9-tetrahydrocannabinol carboxylic acid.

Specificity of an immunoassay is demonstrated by the amount (or lack) of cross-reactivity of the antibody with other similar, but unrelated compounds. The specificity of the assay may be dictated by its intended use. For example, in the clinical setting it may be desirable to detect all of the compounds of a given class, but in the workplace setting detection of only the compounds specified by law or contract is allowed.

As mentioned above, immunoassay cross-reactivity to other drugs is of particular concern when interpreting drug testing results. For example, opiate immunoassays do not detect all opioids, particularly oxycodone, and different immunoassays may have different reactivity. Up-to-date cross-reactivity data are typically listed in immunoassay package inserts or may be obtained from the manufacturer. Ideally, laboratories should perform cross-reactivity studies of the most common interfering substances on their reagent/instrument system, but at a minimum they should contact the manufacturer to verify that they have the most recent cross-reactivity information. When reviewing cross-reactivity data, the reader should realize that studies of potential interfering substances sometimes use concentrations that are lower than those encountered in the clinical setting and that manufacturers often do not test for cross-reactivity of endogenous metabolites. Unexpected immunoassay positive results should be verified using a confirmatory method such as gas chromatography mass spectrometry (GCMS) or liquid chromatography–tandem mass spectrometry (LCMSMS) because there have been several reports of false positives due to cross-reacting drugs (3–5). Confirmation testing may also be needed to identify which drug of a class is actually present.

Knowledge of the sensitivity and specificity of the assay is essential for interpretation of results, and implementation of follow-up action as false-positive (incorrectly reporting the presence of a drug) and false-negative (failure to detect the presence of a drug) results have serious consequences. The presence or absence of drug in a sample must be carefully interpreted with a strong knowledge of the characteristics of the method and of the drug of interest. Furthermore, detection of a drug or metabolite in urine does not automatically indicate that the drug is still in the circulation or causing biological activity, which may cause confusion for clinicians whose patient no longer shows clinical symptoms of intoxication at the time of sample collection.

The 1988 Mandatory Guidelines for Federal Workplace Drug Testing Programs and subsequent revisions mandate scientific and technical procedures for the drug-testing process including collection; transportation of specimens; testing procedures incorporating quality control, method evaluation, and results reporting; and standards for laboratory accreditation by the National Laboratory Certification Program (NLCP). Although private-sector industries are not required to follow the guidelines developed for the federal program, many choose to do so. Workplace drug testing performed under these regulations requires confirmation of all screen-positive samples using GCMS or another approved methodology. The cutoff concentrations for confirmation testing are often lower than those for screening because confirmation testing typically measures only the primary compound of interest (either drug or principal metabolite), whereas immunoassay screening may detect and measure multiple metabolites.

In the clinical setting, confirmation testing using a more sensitive and specific technique may not be necessary or even possible given the rapid turn-around times required. There are however situations, such as with the type of monitoring performed in pain management clinics, in which a "negative" drug screen warrants confirmation. Table 4.4 provides a listing of the various techniques used for confirmatory testing of the aforementioned specimen types. While it is hoped that the table is useful to the reader, it is by no means a complete listing of all published methods.

4.3 SYMPATHOMIMETIC AMINES INCLUDING AMPHETAMINES AND ECSTASY

Amphetamines, methamphetamine, and MDMA belong to the class of drugs known as sympathomimetic amines. These stimulants mimic the actions of naturally occurring neurotransmitters on the sympathetic nervous system both centrally and peripherally. Their structural similarity to the catecholamines, dopamine and norepinephrine, is such that the proteins involved in catecholamine transport fail to distinguish between them. Upon entry into the presynaptic terminal, these compounds stimulate the vesicles to release dopamine and norepinephrine. In addition, amphetamines inhibit the enzymatic action of monoamine oxidase preventing the deactivation of the free catecholamines released. The excess dopamine and norepinephrine are transported from the presynaptic terminal and into the synapse to produce euphoria and pleasure. Amphetamine and methamphetamine, the primary compounds in this class, are available as active ingredients in prescription medication

TABLE 4.4
Confirmation Procedures for Detection of Drugs in Biological Samples

Drug	Method	Specimen	References
Sympathomimetic Amines			
Amphetamine/methamphetamine	GCMS	Urine	6–12
	GCMS	Urine, blood, hair	13
	GCMS	Hair	14,15
	GCMS	Sweat	16
	GCMS	Oral fluid	17–19
	LCMS	Blood	20
	LCMS	Hair	21
	LCMSMS	Urine, blood, hair	22
	LCMSMS	Plasma, oral fluid	23
	LCMSMS	Oral fluid	24–29
	LCMSMS	Hair	31,32
MDA (methylenedioxyamphetamine)/	GCMS	Urine	8,10,11
MDMA (methylenedioxymethamphetamine)/	GCMS	Hair	14,15
MDEA (methylenedioxyethylamphetamine)	GCMS	Sweat	16
	GCMS	Oral fluid	17–19
	LCMS	Blood	20
	LCMSMS	Urine, blood, hair	22
	LCMSMS	Plasma, oral fluid	23
	LCMSMS	Oral fluid	24–30
	LCMSMS	Hair	32
Barbiturates			
Barbiturates	GCMS	Urine	35,36
	GCMS	Urine, serum, plasma	37
	LCMSMS	Oral fluid, urine, plasma	38
Benzodiazepines			
Benzodiazepines	GCMS	Urine	39,40
	GCMS	Oral fluid	17,18
	LCMS	Plasma, oral fluid	41
	LC DAD	Urine, plasma, saliva	42
	LCMSMS	Oral fluid	24–26,30,40
	LCMSMS	Hair	31,46
	UPLCMSMS	Blood, serum, urine	44
	LCMSMS	Blood, serum, urine	45,46
Opiates and Opioids			
Opiates/opioids including 6-acetylmorphine	GCMS	Urine	48–55
	GCMS	Hair	14

continued

TABLE 4.4 (continued)
Confirmation Procedures for Detection of Drugs in Biological Samples

Drug	Method	Specimen	References
	GCMS	Oral fluid	17,18
	LCMSMS	Plasma, oral fluid	23,56,57
	LCMSMS	Oral fluid	24,26–30
	LCMSMS	Hair	31,58
Methadone	GCMS	Urine	58–60
	GCMS	Hair	61
	GCMS	Oral fluid	62
Fentanyl	GCMS	Urine	58,63,64
	LCMSMS	Urine, blood	65–67
Cocaine			
Cocaine, benzoylecgonine	GCMS	Urine	71–74
	GCMS	Urine, blood	75–78
	GCMS	Oral fluid	17,18
	LCMSMS	Plasma, oral fluid	23
	LCMSMS	Oral fluid	26,27–30,79
	LCMSMS	Hair	31–81
Cannabinoids			
Cannabinoids	GCMS	Urine	82–92
	GCMS	Oral fluid	17,18
	GCMSMS	Hair	93
	GCMSMS	Oral fluid	94,95
	LCMS	Urine, blood, oral fluid	96
	LCMS	Oral fluid	97–99
	LCMSMS	Plasma, oral fluid	23
	LCMSMS	Oral fluid	24,25,100
	LCMSMS	Hair	101
Phencyclidine			
Phencyclidine	GCMS	Urine	102,103
	GCMS	Hair	104
	LCMSMS	Oral fluid	24,27–29, 105
Club Drugs			
LSD (lysergic acid diethylamide)	LCFD	Urine, blood	107
	LCFD	Serum, hair	108
	LCMSMS	Urine, blood, serum, plasma	109
	LCMSMS	Urine, blood	110
GHB (gamma-hydroxybutyrate)	GCMS	Urine	111

TABLE 4.4 (continued)

Confirmation Procedures for Detection of Drugs in Biological Samples

Drug	Method	Specimen	References
	GCMS	Blood	112
	GCMS	Urine, blood	113
	LCMS	Urine, serum	114
	LCMSMS	Urine	115
	LCMSMS	Urine, blood	116
	LCMSMS	Hair	117
Ketamine	GCMS	Urine	54
	GCMS	Oral fluid	30
	GCMS	Hair	118
	LCMS	Blood	19
	LCMSMS	Urine	109,119
	LCMSMS	Oral fluid	29
	LCMSMS	Hair	32
Inhalants			
Inhalants	HSGC	Blood, multiple body fluids	123
	HSGCMS	Blood	124,125
	HSGCMS	Blood, viscera	126

Notes: GCMS, gas chromatography mass spectrometry; LCMS, liquid chromatography–mass spectrometry; LCMSMS, liquid chromatography–tandem mass spectrometry; LCDAD, liquid chromatography diode array detection; UPLCMSMS, ultra-performance liquid chromatography-tandem mass spectrometry; LCFD, liquid chromatography fluorescence detection; HSGC, head space gas chromatography; HSGCMS, head space gas chromatography mass spectrometry.

(Adderall®, Dexedrine®, etc.) or as prodrugs that are subsequently metabolized to these compounds (Didrex®, Eldepryl®, etc.). Table 4.5 provides a listing of additional current pharmaceutical sources of amphetamine or methamphetamine. Therapeutic uses include treatment of narcolepsy, obesity, and attention deficit-hyperactivity disorders. Other sympathomimetic amines include pseudoephedrine, ephedrine, propylhexedrine, phenylephrine, phenmetrazine, and phentermine, which are available by prescription or as over-the-counter treatments for nasal congestion and appetite suppression. In 2005, the U.S. Food and Drug Administration (FDA) removed phenylpropanolamine, another sympathomimetic amine, from over-the-counter sale due to its association with hemorrhagic stroke. Methamphetamine recently became the most frequently encountered clandestinely produced controlled substance in the United States partly because of its ease of manufacture from ephedrine or pseudoephedrine. Purchase of ephedrine- and pseudoephedrine-containing products is now subject to reporting and record-keeping in an effort to limit the purchase of these products to an approximately 120-day supply.

TABLE 4.5

Pharmaceutical Sources of Amphetamine or Methamphetamine

Drug	
Adderall	
Dexedrine	
DextroStat	Contains d-amphetamine or d,l-amphetamine
Dexosyn	Contains d-methamphetamine
Vicks inhaler	Contains l-methamphetamine
Amphetaminil	
Clobenzorex	
Ethylamphetamine	
Fenethylline	
Fenproporex	
Mefenorex	Metabolized to amphetamine
Benzphetamine (Didrex)	
Dimethylamphetamine	
Famprofazone	
Fencamine	
Furfenorex	
Selegiline (Eldepryl)	Metabolized to methamphetamine and amphetamine

These drugs are rapidly absorbed after oral, intranasal, or intravenous administration. In addition to stimulation of the central nervous system, they stimulate adrenergic receptors and thus increase blood pressure. They are metabolized hepatically and excreted renally. Chapter 6 of this text discusses the pharmacogenomics of these compounds. Many sympathomimetic amines exist as stereoisomers. The pharmacological and toxicological activities observed depend on the chiral compound involved or dominating the product. Determining chirality is also critical in distinguishing use of a legitimate, legally used amphetamine versus an illicit form.

Ironically, several now illicit amphetamines were originally produced as a result of the practice employed by most pharmaceutical companies where slight modifications of the molecular structure of a given drug are made in an attempt to design additional drugs with similar or enhanced properties, or as a means of circumventing patent restrictions of competitors' products. MDMA (ecstasy; Adam) and methylenedioxyamphetamine (MDA) were the products of such investigations in the early 1900s, but were never marketed for any extended period of time. A third, similar amphetamine designer drug is methylenedioxyethylamphetamine (MDEA), sometimes known as Eve. In the 1960s and 1970s these illegally produced drugs became widely abused because of reported hallucinogenic and psychoactive properties. MDMA, MDA, and MDEA gained notoriety as club drugs because of their association with dance parties known as raves. All three of these drugs produce perception distortions, a desire to communicate, and euphoria. Undesirable and dangerous side effects include tachycardia, hyperthermia, jaw clenching (sometimes leading to the use of pacifiers to prevent grinding of teeth), nystagmus, panic attacks, and neurotoxicity.

Federally regulated workplace drug testing has always included testing for amphetamine and methamphetamine, and in October 2010 screening for MDMA with confirmation testing for MDMA, MDA, and MDEA was added. The current screening cutoff for amphetamines (amphetamine and methamphetamine) mandated for federal workplace testing is 500 ng/mL and 250 ng/mL for MDMA (Table 4.3). Because of the structural similarity of the other sympathomimetic amines, cross-reactivity of antibodies is a primary concern when interpreting results of immunoassay screening for amphetamines. Different amphetamines assays are designed to target methamphetamine, amphetamine, both, and other sympathomimetic amines. For regulated workplace drug testing an immunoassay specific for only amphetamine and methamphetamine is desirable, but in the clinical setting it is desirable to have an assay that will detect all sympathomimetic amines including amphetamine and methamphetamine (3). Immunoassays specifically designed to detect MDMA (ecstasy) and related compounds are also available. Investigation of positive amphetamines screening results should include inquiry about ingestion of diet and herbal preparations because sympathomimetic amines are known appetite suppressants, and some of these products work simply because they contain active pharmaceutical agents. For example, the FDA warned consumers in January 2006 that the Brazilian dietary supplements Emagrece Sim® and Herbathin® contained several active drug ingredients including fenproporex that is metabolized to amphetamine. As previously mentioned, several prescription medications contain amphetamine, methamphetamine, or prodrugs to these compounds, and ingestion of these drugs would be expected to cause a positive drug screen (Table 4.5).

Confirmation using GCMS or other approved methodologies (Table 4.4) is required before a positive amphetamine, methamphetamine, or MDMA is reported in the regulated workplace drug setting. Because methamphetamine is metabolized to amphetamine, a positive methamphetamine result is reported only if the specimen also contains amphetamine at a concentration greater than 100 ng/mL, in addition to methamphetamine that must be present at a concentration of greater than the 250 ng/mL cutoff. This is not the case in the clinical setting; nevertheless, confirmation using GCMS or LCMSMS is very useful because many amphetamine immunoassays exhibit the cross-reactivity problems noted earlier. Numerous methods using both technologies are described for urine, plasma/serum, blood, hair, sweat, and oral fluid (6–32).

Because GCMS procedures using nonchiral derivatives and nonchiral columns do not differentiate the d (+)- or l(–)- isomers of amphetamine and methamphetamine, it is necessary to perform isomer resolution to determine that a positive result is due to the presence of the d-(illicit) isomer (12,15,21). In most instances the laboratory does not automatically perform isomer resolution analysis so the physician must order the analysis after receiving the positive amphetamine or methamphetamine results. Primary examples are the presence of l-methamphetamine in Vicks® inhaler, which cannot be distinguished from use of illicit methamphetamine that is typically found as the d-isomer or a racemic mixture, depending on method of production, and the excretion of l-methamphetamine and l-amphetamine by patients taking selegiline (Eldepryl®) for Parkinson's. The generally accepted interpretation of isomer resolution results is that greater than 80% of the l-isomer is considered consistent with use of legitimate medication, or conversely, greater than 20% of the d-isomer (and total concentration above the cutoff) is considered evidence of illicit use. Confirmation testing for MDMA screen-positive samples includes testing for MDMA, MDA, and MDEA with a cutoff of 250 ng/mL. MDA is both a separate drug and a metabolite of MDMA, so a positive MDA can indicate ingestion of either MDMA or MDA or both.

All GCMS confirmation procedures for amphetamines and related compounds should include preventative measures to avoid loss of these volatile compounds during the evaporation step of extraction or during analysis. Procedures to reduce/eliminate loss of amphetamines during evaporation include lowering the temperature for evaporation, performing incomplete evaporation, or adding methanolic HCl prior to evaporation in order to produce more stable hydrochloride salts. Use of derivatizing agents decreases the volatility of amphetamines, improves chromatography and quantitation, and reduces the effect of potentially interfering compounds by forming higher molecular weight fragments yielding different mass ions and ion ratios. Laboratories that are Substance Abuse and Mental Health Services Administration (SAMHSA)-certified must perform interference studies for amphetamines confirmation assays by analyzing samples containing interferents (phentermine at 50,000 ng/mL and phenylpropanolamine, ephedrine, pseudoephedrine at 1 mg/mL, and MDMA, MDEA, and MDA at 5000 ng/mL) in the presence of and without amphetamine and methamphetamine at 40% of the cutoff (33). Interference studies for MDMA, MDEA, and MDA confirmation assays must include the same interferents listed above with the substitution of amphetamine and methamphetamine at 5000 ng/mL for the MDMA, MDEA, and MDA (33). Hydroxynorephedrine, norephedrine, norpseudoephedrine, phenylephrine, and propylhexedrine are other compounds with structures similar to amphetamine and methamphetamine that may be included in interference studies.

When using blood, plasma, or serum, the collection tubes should be carefully validated to test for drug loss or interfering substances. This is particularly true when testing samples collected using gel-barrier tubes.

4.4 BARBITURATES

Barbiturates are general central nervous system (CNS) depressants that are typically classified by their duration of action. The therapeutic uses of the drugs found in this group correlate to this classification system. Ultrashort-acting barbiturates, such as thiopental, are used as anesthetics, whereas the short- (pentobarbital and secobarbital) and intermediate-acting barbiturates (amobarbital, butalbital, and butabarbital) are prescribed as sedative-hypnotic agents. Phenobarbital, a long-acting barbiturate, is used to control seizures. When used as a sedative-hypnotic or anticonvulsant, the route of administration is usually oral and absorption takes place primarily in the intestine. All barbiturates are strong inducers of hepatic microsomal enzymes; for this reason, they are often found to influence the metabolism of coadministered drugs. For example, phenobarbital induces uridine diphosphate glucuronosyl transferase (UGT) enzymes and the CYP2C and CYP3A subfamilies of cytochrome P450. Other drugs metabolized by these enzymes would be metabolized more rapidly when coadministered with phenobarbital. Similarly secobarbital is a strong CYP2A6 and CYP2C8/9 inducer (34).

Barbiturates became popular as treatment for anxiety, insomnia, and seizure disorders in the 1960s and 1970s. With this increased use, abuse of barbiturates for reduction of anxiety, decrease of inhibition, and treatment of unwanted side effects of illicit drugs also increased. Not surprisingly, abuse tends to most commonly occur with short- to intermediate-acting barbiturates and is rare with the long-acting barbiturates such as phenobarbital. The clinical presentation of barbiturate abuse and toxicity ranges from ataxia, slurred speech, lethargy, confusion, and nystagmus to respiratory depression, hypotension, hypothermia, loss of deep tendon reflexes, cardiovascular collapse, and coma. Toxicity is usually apparent within 15 to 30 minutes after ingestion of short-acting barbiturates. For the longer-acting drugs, symptoms of toxicity may appear within 1 to 2 hours following an overdose but may be delayed due to slowed gastric emptying. Since the 1970s barbiturate abuse and use has declined dramatically as the use of benzodiazepines, a safer group of sedative-hypnotics, grew.

The immunoassays for barbiturates are designed to detect secobarbital, usually with a screening cutoff of 200 or 300 ng/mL. Cross-reactivity to other barbiturates varies but is generally sufficient to allow their detection at therapeutic levels. Acceptable confirmation methodologies include GC, high-performance liquid chromatography (HPLC), GCMS, or LCMSMS as noted in Table 4.4 (35–38). The most common barbiturates included in a confirmation panel are amobarbital, butalbital, pentobarbital, phenobarbital, secobarbital, and less often, butabarbital.

Although GCMS analysis can be performed without derivatization, derivatization of the drugs is preferred to improve the chromatography. Several techniques are used, including flash methylation in which the sample extract is mixed with a methylation reagent, such as trimethylanilinium hydroxide, and injected into a hot injection port where the reaction takes place.

4.5 BENZODIAZEPINES

Benzodiazepines are among the most frequently prescribed drugs in the United States today. The name of this class of drugs is derived from the common general structure consisting of a benzene ring fused to a seven-membered diazepine ring. Benzodiazepines are selective CNS depressants prescribed for their anxiolytic, sedative, and anticonvulsant activities. They are also used as preanesthetic and intraoperative medications. The benzodiazepine class is extensive and includes alprazolam (Xanax®), chlordiazepoxide (Librium®), clonazepam (Klonopin®), diazepam (Valium®), estazolam (ProSom®), flunitrazepam (Rohypnol®), flurazepam (Dalmane®), lorazepam (Ativan®), midazolam (Versed®), oxazepam (Serax®), prazepam (Centrax®), quazepam (Doral®), temazepam (Restoril®), and triazolam (Halcion®). With the exception of flunitrazepam, those listed are available for use in the United States. It should be noted that a number of others are available in other countries and occasionally make their way into this country. There are at least two classification systems for benzodiazepines based on either the time of action, like the barbiturates, or structure, but neither system is used extensively. The structure classification system includes three groups: 1,4-benzodiazepines (e.g., diazepam), diazolobenzodiazepines (midazolam), and triazolobenzodiazepines (alprazolam).

After oral administration, benzodiazepines are well absorbed and distributed throughout the body. They are extensively metabolized in the liver via dealkylation, reduction, and hydroxylation, followed by conjugation. The major urinary products of benzodiazepines are the glucuronide metabolite conjugates. Interpretation of urinary metabolite results is difficult because the metabolism of different benzodiazepines often results in common metabolites such as oxazepam, temazepam, and nordiazepam. The cytochrome P450s 3A3 and 3A4 mediate the hydroxylation and dealkylation reactions, so coadministration of drugs that inhibit or induce these enzymes affects benzodiazepine metabolism (44). Similarly any drugs that alter glucuronyl transferase activity may affect benzodiazepine metabolism.

The widespread use of benzodiazepines has led to abuse of these drugs, typically in conjunction with other drugs (including alcohol or other CNS depressants) as part of polydrug abuse. Some abusers take benzodiazepines as a means to mitigate the stimulant effects of ecstasy or cocaine. Long-term use of benzodiazepines can lead to development of a tolerance, requiring larger doses to achieve the desired effects. In turn, physical and psychological dependence can develop, and a withdrawal syndrome may occur following abrupt cessation of use. The clinical presentation of benzodiazepine toxicity includes initial lethargy, slurred speech, and may progress to coma. Respiratory depression is the most serious consequence of an overdose, and the degree varies with the specific compound ingested and any co-ingestions.

Immunoassays for benzodiazepines target oxazepam or nordiazepam at cutoff concentrations of 200 or 300 ng/mL. Cross-reactivity to other benzodiazepines varies substantially. Those benzodiazepines that are not metabolized to oxazepam or nordiazepam, such as flurazepam, alprazolam, lorazepam, clonazepam, and triazolam, may not be detected. To date, no immunoassay detects every benzodiazepine or all of the metabolites, and as a result, false-negatives occur. Because metabolites are excreted as glucuronide conjugates, the sensitivity of immunoassays may be improved by using a hydrolysis step prior to immunoassay testing. This is cumbersome and not realistic in many settings. It is thus important that the laboratory clearly understand which benzodiazepines and metabolites are detected.

Methods for confirmation testing of benzodiazepines include GC and HPLC, and the better methods utilize detection using mass spectrometry as seen in Table 4.4 (39–46). Confirmation methods using GCMS generally report cutoff concentrations of 50 ng/mL. Most procedures detect oxazepam, nordiazepam, temazepam, α-hydroxylprazolam, and diazepam at a minimum. The first step in confirmation testing is hydrolysis of the conjugated metabolites using either glucuronidase or HCl. HCl hydrolysis cleaves the benzodiazepine ring to form benzophenones that can be separated and identified. Disadvantages of benzophenone-producing methods include inability to identify specific benzodiazepines present and the inability to detect triazolobenzodiazepines such as alprazolam. For this reason, some laboratories have turned to LCMSMS, and some of these methods permit the identification of multiple benzodiazepines and metabolites at low concentrations (43–46). The use of deuterated internal standards (for each benzodiazepine measured) facilitates quantification and minimizes issues such as ion suppression. An advantage of LCMSMS is that many of these methods can be adapted for the testing of other matrices such as blood, serum, and meconium. As mentioned previously interpretation of results to determine benzodiazepines ingested is challenging because several of these drugs are metabolized to the same metabolites.

Flunitrazepam (Rohypnol®) is a potent sedative hypnotic that is banned in the United States but is prescribed in other countries for the treatment of insomnia. Initial effects are rapid, occurring within 15 to 30 minutes of ingestion. The individual also experiences disinhibition and anterograde amnesia (forgetting events that occurred during intoxication). Many individuals report residual "hangover" effects persisting into the day following ingestion. In the early 1990s these properties led to the use of flunitrazepam (often with alcohol) as a "date rape" drug. The colorless, odorless drug could be easily added to a drink without means of visual detection. In response, the manufacturer reformulated the tablets to impart an easily identifiable blue color to clear beverages and haziness to colored beverages. In the United States flunitrazepam use as a date rape drug has declined primarily in response to the reformulation and implementation of stricter legal penalties for drug-facilitated sexual assault (DFSA). Unfortunately, illicitly manufactured and procured tablets continue to surface and obviously do not have the coloring safeguards. It is important to remember that benzodiazepines immunoassays vary considerably in their ability to detect flunitrazepam and many have very little cross-reactivity with the drug or its metabolites. There are, however, specific immunoassays designed specifically for flunitrazepam detection. Confirmation testing using GCMS or LCMSMS typically targets detection of flunitrazepam and its principal metabolite, 7-aminoflunitrazepam.

4.6 OPIATES AND OPIOIDS

The term *opiates* refers to naturally occurring alkaloids of the poppy plant, *Papaver somniferum*, notably morphine and codeine, and semisynthetic alkaloids with similar structures. The semisynthetic alkaloids of this class include heroin along with the prescriptive medications buprenorphine (Buprenex®), dihydrocodeine, heroin, hydrocodone (Vicodin®), hydromorphone (Dilaudid®), oxycodone (Percodan®, Oxycontin®), and oxymorphone (Opana®). The term *opioids* refers to compounds that have affinity toward opioid receptors and exhibit pharmacological properties similar to morphine but may be structurally unrelated to morphine. These include fentanyl (Duragesic®, Sublimaze®), meperidine (Demerol®), methadone, pentazocine (Talwin®), propoxyphene (Darvon®), and tramadol (Ultram®). Thus, it can be said that all opiates are opioids, but all opioids are not opiates.

These drugs play an important role in the treatment of chronic pain. Chronic pain is an extremely complex condition and afflicts many patients worldwide. By many estimates, ~30% of the U.S. population is currently suffering from chronic pain. Current pharmacological therapy for pain management focuses on opioids, so much so, that up to 90% of chronic pain patients presenting to pain centers are prescribed one or more of these drugs. Opioids have important therapeutic uses beyond pain management. They are also used for their antitussive and antidiarrheal effects, for relief of acute pulmonary edema, and in the treatment of heroin addiction.

Use and abuse of opioids dates to ancient times, and the addictive properties of this class of drugs were recognized in the United States during the Civil War. The genetic aspects of opiate addiction, as well as metabolism, are discussed in detail in Chapter 9. The clinical presentation of opiate toxicity includes CNS depression, pulmonary edema, hypotension, decreased bowel sounds, hyporeflexia, miosis, and respiratory depression in which breathing is shallow. The presence of miosis is rarely seen with other clinical causes of CNS depression so that patients with this finding are usually given the narcotic antagonist naloxone (Narcan®) as part of the initial evaluation and treatment (47).

Opiate immunoassays are designed to detect morphine and codeine at defined cut-off levels. In the clinical and nonregulated workplace settings, cutoffs of 300 ng/mL are often used. Unfortunately, there is considerable variability between immunoassays regarding their ability to detect other opiates such as hydromorphone, hydrocodone, and oxycodone. This is very much dependent upon the specificity of the antibody and also varies between lots of a given assay. When abuse of oxycodone increased, a separate immunoassay was developed. Similarly an assay for buprenorphine was developed when it was introduced as a treatment option for heroin dependency. Laboratories using these assays should include the cutoff used in reporting results. Generally, the opiate assays do not detect the nonopiate opioids, such as methadone, propoxyphene (withdrawn from the U.S. market in November 2010), or fentanyl. Separate immunoassays are available for these compounds, but many clinicians are unaware that the opiate assays they order are not able to detect these clinically relevant compounds. Numerous methods are described for confirmation of opiates and opioids in urine and other matrices using GCMS or LCMSMS as seen in Table 4.4 (14,17,18,23,24,26–31,48–56). Because many of these drugs are extensively metabolized by conjugation with glucuronic acid, samples are hydrolyzed when it is desired to determine total drug, as is the case when using urine.

Interpretation of opiate results requires knowledge of natural sources of morphine as well as an understanding of the pharmacology of the various drugs. For example, poppy seeds contain morphine naturally, and their ingestion may cause a positive opiates result regardless of the testing methods (immunoassay or chromatography). In an effort to eliminate or reduce the number of positive results due to poppy seed ingestion, the Department of Defense raised the cutoff levels from 300 ng/mL to 2000 ng/mL in 1994. In December 1998, federally mandated workplace programs followed suit. Lower cutoffs continue to be used in clinical and nonregulated settings.

Heroin (diacetylmorphine) is rapidly metabolized to 6-acetylmorphine (6-AM; 6-MAM) and subsequently to morphine. Testing for 6-AM in addition to morphine and codeine is now required in the federally regulated drug testing system. Detection of morphine and 6-AM, a metabolite unique to heroin, confirms heroin use, but because it has a shorter half-life than morphine the absence of 6-AM does not confirm nonuse. In addition, there have been cases reported in which 6-AM was detected in the absence of detectable morphine. The recent identification of hydromorphone in urine samples with very high morphine concentrations has now been determined to reflect a minor metabolic pathway for morphine metabolism found in some patients. In these cases, the percentage of hydromorphone present is less than 6% of the total morphine.

SAMHSA-certified laboratories must perform interference studies for opiates confirmation assays by analyzing samples containing interferents (hydrocodone, hydromorphone, oxycodone, oxymorphone, and norcodeine at 5000 ng/mL) in the presence and absence of morphine and codeine at 40% of the cutoff (33). Similarly interference studies for 6-acetylmorphine assays must test free morphine, codeine, hydrocodone, hydromorphone, oxycodone, oxymorphone, and norcodeine at 5000 ng/mL as potential interferents (33).

Methadone, a synthetic opioid originally developed by German scientists prior to World War II, is used as a treatment for heroin (and other opioids) dependence and for chronic pain. Methadone is relatively inexpensive when compared to other opioids, and by 2008 the number of patients taking methadone for pain was three times the number of patients taking it for addiction. Methadone abuse can be dangerous as evidenced by the fact that even though only 5% of the nation's opioid prescriptions are for methadone, 30% of opioid-related deaths involve the drug. This disproportionate increase in methadone use, misuse, and mortality has led to an increased emphasis on monitoring patients being treated with methadone regardless of purpose.

Methadone is administered orally and is subject to first-pass metabolism by the liver. Interindividual differences in bioavailability range from 80% to 95%. Additionally, wide variations in inter- and intraindividual pharmacokinetics are encountered. The primary methadone metabolite is 2-ethylidene-1,5-dimethyl-3,3-diphenylpyrrolidine (EDDP), which is inactive. Both methadone and EDDP are excreted primarily through the kidneys. Approximately 5% to 50% of a dose is eliminated as methadone and 3% to 25% as EDDP. Excretion of unmodified methadone in urine is pH dependent, with increased excretion observed when the urine is acidic.

Although methadone's structure is different from that of morphine or heroin, it interacts with opioid receptors similarly. When used in opioid treatment programs, daily doses of methadone reduce cravings and prevent symptoms of opiate withdrawal while blocking their euphoric effects. When used for addiction treatment, methadone must be dispensed by an opioid treatment program (OTP) that is SAMHSA certified. The treatment process is usually long term and typically includes three phases including an extended maintenance phase with the ultimate goal of gradually tapering the dosage to the point of complete withdrawal from methadone. Urine drug testing is the most objective technique for patient monitoring in an OTP and is used for several purposes. SAMHSA requires the patient undergo a minimum of eight drug tests per year, but some states require more frequent testing. Initial urine drug testing is used to determine the primary opioid along with any other drugs that may be abused. Subsequent testing is used to verify abstinence and compliance (methadone) and to detect abuse of drugs other than methadone.

Unlike methadone's use in addiction treatment where it must be dispensed by OTPs, methadone when used to treat pain may be prescribed by any appropriately licensed and registered practitioner and dispensed by licensed and registered retail pharmacies. For several reasons standardized dosing of methadone is difficult and treatment should be individualized. The use of urine drug testing as part of a comprehensive patient monitoring program for pain management has been shown to be an effective method of detecting misuse of opioids. There is no mandated requirement for periodic drug testing when methadone is used for pain management, but the *Clinical Guidelines for the Use of Chronic Opioid Therapy in Chronic Noncancer*

Pain recommends periodic drug screens in patients at high risk of misuse, abuse, and diversion (sale) (34).

Detection of both methadone and its primary metabolite, EDDP, should be included when using urine drug testing to monitor methadone treatment for addiction or pain management. When immunoassay is used as the testing technique, two separate assays are required because the immunoassays for methadone and EDDP demonstrate little or no cross-reactivity to each other. The most frequently used cutoff concentration for methadone immunoassays is 300 ng/mL and for EDDP is 100 ng/mL. Methadone immunoassays with cutoff concentrations of 200 and 250 ng/mL are also available. Chromatography-based methods (Table 4.4) provide a means to quantify both parent and metabolite simultaneously (58–62).

The variable pharmacokinetics and potential drug–drug interactions associated with methadone contribute to a range of methadone and EDDP concentrations in the urine of patients being monitored. Interpretation of methadone and EDDP results is shown in Table 4.6. Compliant patients are expected to be positive for both methadone and EDDP, but a sample that is negative for methadone and positive for EDDP could indicate the patient is a rapid metabolizer. Negative results for both methadone and EDDP typically indicate that methadone was not ingested and could result from bingeing. These results could also occur if levels of both methadone and EDDP are below the cutoff concentrations as would occur in a dilute urine sample. A technique known as spiking has been used by patients to mimic compliance when diverting methadone for sale. Because this involves the addition of a small amount of the dose to the urine after collection (before testing), this yields a strongly positive result for methadone but a negative EDDP. It should be noted that this is also attempted with other drugs used in other rehabilitation programs (such as buprenorphine).

Fentanyl, a fast-acting and potent synthetic narcotic analgesic and anesthetic, is approximately 75 to 100 times more potent than morphine. It is available as lozenges (Actiq®) and transdermal patches (Duragesic®) for management of severe pain including breakthrough pain in cancer patients. A common mode of fentanyl abuse is injection, but the patches can be abused via transdermal application or ingestion of the fentanyl-containing gel in the patch. Several analogs of fentanyl synthesized by pharmaceutical companies are used clinically. These drugs and their uses include sufentanil for cardiac surgery, alfentanil for minor surgeries, lofentanil for trauma

---

TABLE 4.6

Interpretation of Urine Test Results for Methadone and Its Metabolite 2-Ethylidene-1,5-Dimethyl-3,3-Diphenylpyrrolidine (EDDP)

Methadone	EDDP	Interpretation
Positive	Positive	Methadone ingested, normal situation
Negative	Negative	No methadone ingested, bingeing, or levels below detection limit
Negative	Positive	Fast metabolism, interaction with other drugs
Positive	Negative	Methadone added to sample ("spiker")

Notes: EDDP, 2-ethylidene-1,5-dimethyl-3,3-diphenylpyrrolidine.

patients, and remifentanil that can be used in conjunction with hypnotic drugs such as propofol. Availability has led to abuse of fentanyl and its analogs by health-care professionals. Signs and symptoms of abuse and intoxication resemble those described for opiates. Other fentanyl derivatives have been illicitly produced and gained popularity as street or designer drugs. Abuse of two of these designer fentanyl derivatives, α-methylfentanyl (China white) and 3-methylfentanyl, has resulted in epidemics and multiple deaths. Opiate immunoassays do not detect fentanyl or its analogs, but immunoassays specific for these drugs have been developed. Confirmation procedures (Table 4.4) include GCMS- and LCMSMS-based methods (63–67).

Meperidine, another synthetic narcotic agonist that is less potent than morphine, is another opioid commonly abused by health-care professionals. It is available in injectable and oral forms. An illicit meperidine designer drug derivative that is also abused is 1-methyl-4-phenyl-4-propionoxypiperdine (MPPP). Clandestine production of MPPP can produce a highly neurotoxic by-product, 1-methyl-4-phenyl-1,2,3,6-tetrahydropyridine (MPTP), that can cause parkinsonism. Detection of meperidine and its analogs requires chromatographic methods such as GC, GCMS, and LCMSMS (Table 4.4) because opiate immunoassays do not cross-react with these compounds (68–70).

4.7 COCAINE

Cocaine, a naturally occurring alkaloid obtained from the South American shrub *Erythroxylum coca*, is one of the most common illicit drugs of abuse. This potent CNS stimulant also has local anesthetic and vasoconstrictive properties and is used clinically as a local anesthetic in otolaryngological procedures (as a 10% to 20% solution) and in ophthalmologic procedures (as a 1% to 4% solution). The illicit forms of cocaine are the hydrochloride salt and the freebase form known as crack. The hydrochloride salt is typically cut with agents such as mannitol, lactose, and sucrose to dilute the product or agents such as caffeine, pseudoephedrine, and lidocaine to enhance the effects of the actual drug. Crack cocaine is prepared by adding baking soda to the hydrochloride salt in an aqueous solution and then heating to remove water. The precipitated product is usually smoked for rapid absorption. Leaves from the coca are used to prepare teas that while commonly used in South America are illegal in the United States. The teas prepared from these products contain enough cocaine to cause a positive urine drug screen, but the slow absorption via the gastrointestinal (GI) tract diminishes any stimulatory effects. Unfortunately, some individuals who are unaware of the illegal status of these teas face serious consequences after ingestion, from loss of insurance to even loss of their jobs. Two of the herbal teas that have been encountered are Health Inca and Mate De Coca.

Cocaine is rapidly metabolized to several metabolites including ecgonine methyl ester and benzoylecgonine. A unique cocaine metabolite, cocaethylene, is produced when co-ingested with ethanol. This metabolite has significant clinical implications due to its toxicity. The genetic aspects of cocaine metabolism and abuse are discussed in Chapter 7. The clinical manifestations of acute cocaine toxicity include CNS stimulation, psychosis, convulsions, small-muscle twitching, mydriasis, ventricular arrhythmias, and ultimately possible respiratory paralysis, coma, or myocardial infarction leading to death. Some symptoms of chronic cocaine use include psychiatric disturbances, rhinitis, possible nasal septum perforation, distorted perception, tachycardia, and tachypnea.

The primary compound measured by cocaine (metabolite) immunoassays is the urinary metabolite benzoylecgonine. The antibody specificity of current assays for cocaine metabolites is excellent although the cross-reactivity for cocaine metabolites other than benzoylecgonine may vary between assays. Many assays, for example, do not readily detect parent cocaine, and the cross-reactivity for cocaethylene varies considerably. In October 2010, the cutoff concentrations used in federally regulated workplace drug testing programs were lowered to 150 ng/mL for immunoassays and to 100 ng/mL for GCMS confirmation methods (Table 4.4) (17,18,23,71–78). Much lower cutoffs are often found in clinical and nonregulated workplace settings. Several LCMSMS methods (Table 4.4) have been described including a method for detection in meconium in which sample pretreatment is minimized to a simple deproteinization step in which the specimen is mixed with an acetonitrile–internal standard mixture prior to injection (79).

4.8 CANNABINOIDS

Cannabinoids are a group of compounds present in marijuana, a recreational drug that originates from the dried leaves and flowers of *Cannabis sativa*. The principal psychoactive cannabinoid is D^9-tetrahydrocannabinol (THC). The drug produces a range of behavioral effects that prevent its classification as a stimulant, sedative, tranquilizer, or hallucinogen. These effects include euphoria, relaxation, lack of concentration, altered time perception, mood changes, and impaired learning and memory. Physiological effects include tachycardia, dry mouth and throat, redness of the conjunctiva, and in some subjects, hypotension. Smoking, the principal means of administration, provides rapid and efficient delivery. Oral and sublingual administration results in lower THC concentrations than smoking. More than 100 THC metabolites have been identified, and more of the dose is excreted in feces than in urine. The major urinary metabolites are 11-nor-tetrahydrocannabinol-9-carboxylic acid (THCA) and its glucuronide conjugate. The genetic aspects of marijuana metabolism and abuse are discussed in Chapter 8.

Screening immunoassays for cannabinoids detect multiple metabolites of marijuana with a cutoff concentration of 50 ng/mL used for regulated drug testing and as low as 20 ng/mL in the clinical and nonregulated setting. GCMS confirmation identifies and quantifies THCA in urine (Table 4.4) using a cutoff of 15 ng/mL in regulated drug testing (82–92). Hair and oral fluid are often used for testing as well (93–95). GCMS confirmation procedures include a hydrolysis step of the glucuronide conjugate. LCMSMS also usually includes a hydrolysis step and limits of quantification (LOQ) as low as 3 ng/mL are reported (96–101).

Chronic marijuana users may produce positive results for longer periods of time because of accumulation of cannabinoid metabolites in fatty tissue followed by slow release. This variable release from tissue and differences in hydration status can lead to a false assumption of new marijuana use if a negative result is followed by a positive result. Creatinine normalization (dividing the THCA result by the creatinine result) has been used to address this problem. For less-than-daily users, an increase of greater than 50% in the normalized THCA concentration is considered indicative of new marijuana use. For chronic users, the excretion pattern is more complicated as they have smaller decreases in excretion later in the elimination phase. Increases of greater than 150% for these users have been suggested as an indicator of new use. Marijuana smoke can be inhaled by a nonsmoking individual, but studies have shown that it is virtually impossible that this passive inhalation can cause a positive result when using a screening cutoff of 50 ng/mL.

4.9 PHENCYCLIDINE

Pharmacologically, phencyclidine (1-phenylcyclohexylpiperidine, PCP) is classified as a dissociative anesthetic but it also exhibits hallucinogenic, stimulant, depressant, and analgesic properties. Originally marketed as a short-acting analgesic or general anesthetic in humans and later as a large-animal veterinary tranquilizer, it has had no legal use since 1979. Illegal use of PCP has fluctuated and tends to be highly regionalized in the United States. PCP can be self-administered by oral, intravenous, or smoked routes. Crystalline PCP may be sprinkled on marijuana, and liquid PCP can be soaked into marijuana or tobacco cigarettes. In addition to hallucinations, other effects of PCP may include staggering gait, slurred speech, nystagmus, mood elevation, agitation, increased blood pressure, violent behavior, and coma. PCP is metabolized by hepatic CYP3A4 with at least three hydroxylated metabolites excreted in urine in conjugated and unconjugated forms.

Although PCP immunoassays target the parent drug at a cutoff concentration of 25 ng/mL, most cross-react with metabolites. In addition, unrelated drugs such as dextromethorphan, diphendyramine, thioridazine, and zolpidem have been reported to cause false-positive PCP screening results for some immunoassays. It has even been suggested that some PCP assays could possibly be used to detect dextromethorphan abuse. Chromatography-based methods are described for testing urine, hair, and oral fluid (Table 4.4) (24,27–29,102–105). The cutoff concentration for GCMS confirmation of PCP in regulated drug testing is also 25 ng/mL.

4.10 CLUB DRUGS

Club drugs are so named because of the popularity and use of these drugs in bars, nightclubs, concerts, and parties. Which of these is found depends on local availability, costs, and local lore. These drugs are used for their psychoactive effects or in some cases surreptitiously used to facilitate sexual assault (date rape drugs). Drugs common to the club drug scene include MDMA (see 4.3), flunitrazepam (see 4.5), g-hydroxybutyrate (GHB), g-butyrolactone (GBL), 1.4-butanediol (1,4-BD), lysergic acid diethylamide (LSD), nitrates, ketamine, diphenhydramine, and dextromethorphan. Hallucinogenic tryptamines and phenylalkylamines have recently been rediscovered by this population of users. The compounds are similar to psilocybin, psilocin, and bufotenine, and include *N,N*-dipropyltryptamine (DPT), *N,N*-dimethyltryptamine (DMT), *N,N*-α-methyltryptamine (AMT), 5-methoxy-*N,N*-α-methyltryptamine (5-MeO-AMT), 5-methoxy-N, N-dimethyltryptamine (5-MeO-DMT), 4-iodo-2,5-dimethoxy-phenethylamine (2C-I), and 4-iodo-2,5-dimethoxyamphet-amine (DOI). Also of growing concern is the appearance of piperazine derivatives in the club scene. These drugs, originally used for their antihelminthic properties in animals, are also hallucinogenic. N-Benzylpiperazine (BZP) and 1-(3-trifluoro-methylphenyl) piperazine (TFMPP) are two of the most popular piperazines and have been referred to as "Legal E" and "Legal X" because of their similarities to MDMA (106). Of the drugs listed above only LSD, MDMA, and flunitrazepam are detected by specific immunoassays. Detection using other methods such as GC, HPLC, GCMS, or LCMSMS (Table 4.4) often requires suspicion of use and utiliza-tion of assays targeted to suspected drugs.

LSD, a very powerful hallucinogen, is a colorless, odorless, tasteless liquid that is available in several forms including tablets and blotter paper impregnated with the drug. Oral ingestion is the most common route, but it may also be adminis-tered nasally or by injection. In addition to hallucinations characterized by visual illusions and alterations in sound and intensity of colors, LSD causes physiological changes including mydriasis (pupil dilation of 3 to 5 mm), lacrimation, tachycardia, and hyperpyrexia. Metabolism of LSD, which is not completely understood, includes production of hydroxylated metabolites excreted in urine as glucuronide metabolites. Excretion of unchanged LSD in urine is low, representing from 1% to ~20% of the original dose. The combination of ingestion of small doses of the drug and rapid extensive metabolism results in low urinary concentrations; thus LSD immunoas-says target a common cutoff of 0.5 ng/mL (107–110). The rapid metabolism of LSD results in a detection window as short as 12 to 24 hours. Confirmation methods include GCMS and LCMSMS with cutoff concentrations as low as 0.2 ng/mL. The reported low rate of positive results could be due to low usage, low dosage, rapid metabolism, or a combination of these factors.

Gamma hydroxybutyrate (GHB, Xyrem®) is a CNS depressant used in the treat-ment of narcolepsy. The compound exists at low concentrations naturally in the brain as a metabolite of the neurotransmitter gamma-aminobutyric acid (GABA). It is now a controlled substance but was previously sold as a food supplement. In this form, GHB was a popular alternative to steroids for building muscle mass without exercise because it was believed to cause an increased release of growth hormone. Several GHB analogs, including GBL and 1,4-BD, are available as industrial solvents and are abused because they are converted endogenously into GHB. The drug is also abused for its CNS depressant properties. Depending on the dosage, GHB can cause a range of effects from wakefulness and euphoria to deep sleep or coma. Like flunitrazepam, GHB is used as a date rape drug because of its sedative and anterograde amnesia-producing properties.

GHB is rapidly and extensively metabolized with an elimination half-life of less than 1 hour and excretion of less than 5% of the ingested dose as the unchanged drug. In order to detect GHB, ingestion samples must be collected within 6 to 12 hours; and even then, it is difficult to detect the drug. The short detection window and the effects after ingestion may lead to delayed collection of samples and negative test-ing results. It is often recommended that both urine and blood samples be collected when a drug-facilitated sexual assault (DFSA) involving GHB is suspected. There are no immunoassays available for testing of GHB and its analogs. Procedures available for GHB testing include GC, GCMS, and LCMSMS (111–117).

Interpretation of GHB levels in blood and urine samples can be complicated because it exists as an endogenous compound. GHB concentrations exceeding 10 ng/mL in urine and 2 ng/mL in blood are considered indicative of exogenous GHB exposure. Blood should not be collected in citrate-buffered collection tubes because of reports that this may cause falsely elevated GHB levels (118).

Ketamine, 2-2-chlorophenyl-2-(methylamino)cyclohexanone, is a dissociative anesthetic that is abused as an hallucinogen and as a date rape drug because of its CNS depression properties including amnesia. It is structurally and pharmacologically related to PCP. When used as an anesthetic it is administered intravenously (IV) and when used illicitly routes of administration include IV, oral, intramuscular, smoking, and subcutaneous. Ketamine is rapidly metabolized with a half-life of 2 to 3 hours and conjugated metabolites are excreted in urine. There are no immunoassays available for testing of ketamine and its metabolites. Procedures available for ketamine testing include GC, GCMS, and LCMSMS (19,29,30,32,54,109,119–121).

4.11 INHALANTS

The terms *inhalants* or *volatile organic compounds* (VOCs) are used to describe a wide range of volatile chemicals that may be inhaled accidentally or intentionally. The common feature is volatility, the property of existing in or being able to be converted to a form that may be inhaled. Volatile compounds are present in many commercial products—solvents, contact adhesives, typewriter correction fluid, gasoline, lighter fluid, refrigerants, fire extinguishers—and are the propellants for virtually all aerosol products. Extensive availability and low cost of inhalants have contributed to an increased incidence of intentional inhalation of volatile substances (inhalant abuse, volatile substance abuse, glue sniffing) in younger adolescents even though legislation has been enacted to limit accessibility and to make their use by adolescents illegal. Inhalants continue to be one of the most dangerous classes of abused substances worldwide because of their high prevalence in underdeveloped countries (122).

Products preferred by inhalant users include airplane glue, hair spray or aerosols, gasoline, paint or solvents, marker pens or correction fluid, and amyl or butyl nitrates (poppers). These products are generally relatively inexpensive, easy to obtain, and contain volatile substances free of secondary toxic components. Some of the volatile substances in these products include toluene, chloroform, butane, propane, acetone, and many halogenated hydrocarbons. The development of replacements of the ozone-depleting CFCs has led to the introduction of new compounds (Freon replacements) as propellants. The abuse of many of these substances has also been reported.

Depending on the product, volatile substances may be inhaled directly from the container (snorting or sniffing), from a plastic bag (bagging) particularly if the product is an aerosol or a viscous liquid such as glue, or from a saturated cloth (huffing). Of these routes of administration bagging usually results in the highest concentration, snorting the lowest, and huffing an in-between concentration. Clues to inhalant abuse include chronic sore throat, cough, and runny nose; unexplained listlessness; moodiness; weight loss; bloodshot eyes or blurred vision; and chemical odors on breath, hair, bed linen, and clothes. Oral and nasal ulceration or a rash around the mouth ("glue sniffer's or huffer's rash") may be observed. Sometimes the products may be discovered in the room of the abuser.

Some pharmacokinetic properties and principles apply to all inhalants. The metabolism of volatile substances includes elimination unchanged in exhaled air and elimination as metabolites in exhaled air and urine. The primary site of metabolism is the liver, with the metabolism often including oxidation or reduction followed by conjugation leading to a more polar and water-soluble compound. The metabolites of some volatiles are more toxic than the parent compounds.

The abuse appeal of these inhaled substances is that they produce effects similar to those caused by ethanol (i.e., euphoria and loss of inhibition, which may be followed by hallucinations, confusion, nausea, vomiting, and ataxia). Convulsions, coma, or death may result from larger doses. Causes of death associated with inhalant abuse include asphyxiation, suffocation, and dangerous high-risk behavior. Cardiac arrhythmias leading to cardiac arrest are reported when an intoxicated subject becomes alarmed or frightened ("sudden sniffing death syndrome"). Problems caused by chronic inhalant abuse include central nervous system (CNS) damage characterized by loss of cognitive and other higher functions, gait disturbance, and loss of coordination. Other features of chronic abuse include nosebleed and rhinitis, halitosis, oral and nasal ulceration, conjunctivitis and bloodshot eyes, anorexia, thirst, lethargy, weight loss, and fatigue (122).

Routine urine drug screens do not detect the commonly abused inhalants. Blood is the specimen of choice, although analysis of urine for metabolites sometimes extends the time frame for detection of exposure. Proper collection techniques include use of a glass tube with minimal headspace remaining after collection, an anticoagulant such as lithium heparin or EDTA, and a cap that ensures a tight seal. The most common method for the detection of volatile substances in blood and other samples is headspace gas chromatography with flame ionization, electron capture, or mass spectrometer detection devices (123–126).

4.12   SPECIMEN VALIDITY TESTING

Drug abusers try to avoid detection by drug testing in many ways including use of products designed to prevent detection of drugs present in a urine sample. An Internet search of the topic "pass a drug test" yields more than 2 million results. Some products require ingestion (usually in conjunction with large amounts of water) prior to submission of a urine sample while other products are to be added to or substituted for the urine sample. Collection procedures have been designed to prevent such substitution, dilution, or adulteration of the specimen. These procedures include collection site preparation (i.e., no hot water, addition of a coloring agent to the toilet, no coats or purses, etc.) and use of collection cup temperature monitoring devices. Tests of specimen validity can be performed in an attempt to detect tampering with the samples.

In addition to testing for the drugs in Table 4.3, SAMHSA-certified laboratories are now required to perform specimen validity testing to identify dilute, substituted, adulterated, or invalid samples submitted for regulated drug testing. This testing

TABLE 4.7
Reporting Specimen Validity Test Results

Test(s)	Result(s)	Reported As
pH	<3.0 or >11.0	Adulterated
	>3.0 and <4.5	Invalid
	>9.0 and <11.0	Invalid
Nitrite	>500 mcg/mL	Adulterated
	>200 and <500 mcg/mL	Invalid
Creatinine and specific gravity	>2.0 and <20 mg/dL	
	>1.0010 and <1.0030	Dilute
Creatinine and specific gravity	<2.0 mg/dL	
	<1.0010 or >1.0200	Substituted
Creatinine and specific gravity	< 2.0 mg/dL	
	Acceptable (1.0011–1.0199)	Invalid
Creatinine and specific gravity	>2.0 mg/dL	
	<1.0010	Invalid

Source: National Laboratory Certification Program Manual for Laboratories and Inspectors, October 1, 2010.

includes creatinine, specific gravity (when the creatinine is below 20 ng/mL), pH, and nitrites measurements. Testing for known adulterants such as glutaraldehyde, pyridinium chlorochromate, or oxidizing chemicals as a class is optional. Criteria for the testing process are similar to those of testing for drugs—that is, the results must be obtained using approved initial and confirmatory tests on two separate aliquots of urine. Guidelines for the interpretation of specimen validity testing are listed in Table 4.7. A sample is reported as adulterated when test results show the presence of a substance that is not a normal constituent or an endogenous substance present in an abnormal concentration. A sample is reported as substituted when creatinine and specific gravity values are outside the physiologically producible ranges of human urine indicating submission of a nonurine specimen. A result of "substituted" or "adulterated" indicates intention to circumvent testing for drugs and, in some settings, is treated the same as a positive drug result. A sample is reported dilute when the creatinine and specific gravity values are lower than expected but are still within the physiologically producible ranges of human urine. A dilute specimen does not necessarily indicate specimen tampering because it can result from consumption of large amounts of fluid or by addition of liquid to the urine specimen.

If the laboratory is unable to complete testing or obtain a valid drug test result due to the specimen containing an unidentified adulterant or interfering substance, or having an abnormal physical characteristic or endogenous substance at an abnormal concentration, the result is reported as invalid. Reporting the specimen validity test results does not preclude reporting the drug test results. Thus a result of dilute, adulterated, substituted, or invalid can be reported in conjunction with positive results for specific drugs if the criteria for identification and reporting of the positive result are met. Regulated samples reported as positive for any drug, substituted, adulterated, or invalid must be retained in secured frozen storage for a minimum of 1 year.

REFERENCES

1. NIDA chart of drugs of abuse. Available online at: http://www.nida.nih.gov/DrugPages/ DrugsofAbuse.html Accessed on 2/24/2011.
2. NIDA chart of drugs of abuse. Available online at: http://www.nida.nih.gov/DrugPages/ PrescripDrugsChart.html. Accessed on 2/24/2011.
3. Hammett-Stabler CA, Pesce A, Cannon D. The medical use of urine drug screening. *Clin Chim Acta* 2002;315:125–135.
4. Paul BD, Past MR. (2009). Confirmation methods in drug testing: an overview. In A. Dasgupta (Ed.), *Critical Issues in Alcohol and Drugs of Abuse Testing*. Washington, DC: AACC Press, Table 9-2, page 134.
5. Bosker WM, Huestis MA. Oral fluid testing for drugs of abuse. *Clin Chem* 2009;55(11):1910–1931.
6. Czarny R, Hornbeck C. Quantitation of methamphetamine and amphetamine in urine by capillary GC/MS. II. Derivatization with 4-carboethoxyhexafluorobutyryl chloride. *J Anal Toxicol* 1989;13:257–262.
7. Paul BD, Past MR, McKinley RM, Foreman JD, McWhorter LK, Snyder JJ. Amphetamine as an artifact of methamphetamine during periodate degradation of interfering ephedrine, pseudoephedrine, and phenylpropanolamine: an improved procedure for accurate quantitation of amphetamine in urine. *J Anal Toxicol* 1994;18:331–336.
8. Klette KL, Jamerson MH, Morris-Kukoski CL, Kettle AR, Snyder JJ. Rapid simultaneous determination of amphetamine, methamphetamine, 3,4-methylenedioxyamphetamine, 3,4-methylenedisoxymethamphetamine, and 3,4-methylenedioxyethylamphetamine in urine by fast gas chromatography-mass spectrometry. *J Anal Toxicol* 2005;29:669–674.
9. Valentine JL, Middleton R. GC-MS identification of sympathomimetic amine in urine: rapid methodology applicable for emergency clinical toxicology. *J Anal Toxicol* 2000;24:211–222.
10. Stout PR, Horn CK, Klette KL. Rapid simultaneous determination of amphetamine, methamphetamine, 3,4-methylenedioxyamphetamine, 3,4-methylenedioxymethamphetamine, and 3,4-methylenedioxyethylamphetamine in urine by solid-phase extraction and GC-MS: a method optimized for high-volume laboratories. *J Anal Toxicol* 2002;26:253–261.
11. Gan BK, Baugh D, Liu RH, Walia AS. Simultaneous analysis of amphetamine, methamphetamine, and 3,4-methylenedioxymethamphetamine (MDMA) in urine samples by solid-phase extraction, derivatization, and gas chromatography/mass spectrometry. *J Forensic Sci* 1991;36:1331–1341.
12. Paul BD, Jemionek J, Jacobs A, Searles DA. Enantiomeric separation and quantitation of (+/–)-amphetamine, (+/–)-methamphetamine, (+/–)-MDA, (+/–)-MDMA, and (+/–)-MDEA in urine specimens by GC-EI-MS after derivatization with (R)-(–)- or (S)-(+)-alpha-methoxy-alpha-trifluoromethylphenylacetic acid chloride derivatization. *J Anal Toxicol* 2005;29:652–657.
13. Miki A, Katagi M, Zaitsu K, Nishioka H, Tsuchihashi H. Development of a two-step injector for GC-MS with on-column derivatization, and its application to the determination of amphetamine-type stimulants (ATS) in biological specimens. *J Chromatogr B* 2008;865(1–2):25–32.

14. Wu YH, Lin KL, Chen SC, Chang YZ. Integration of GC/EI-MS and GC/NCI-MS for simultaneous quantitative determination of opiates, amphetamines, MDMA, ketamine, and metabolites in human hair. *J Chrom B: Anal Tech in Biomed Life Sci* 2008;870(2):192–202.

15. Strano-Rossi S, Botre F, Bermejo AM, Tabernero MJ. A rapid method for the extraction, enantiomeric separation and quantification of amphetamines in hair. *Forensic Sci Int* 2009;193(1–3):95–100.

16. DeMartinis BS, Barnes AJ, Scheidweiler KB, Huestis MA. Development and validation of a disk solid phase extraction and gas chromatography-mass spectrometry method for MDMA, MDA, HMMA, HMA, MDEA, methamphetamine and amphetamine in sweat. *J Chromatogr B* 2007;852(1–2):450–458.

17. Gunnar T, Ariniemi K, Lillsunde P. Validated toxicological determination of 30 drugs of abuse as optimized derivatives in oral fluid by long column fast gas chromatography/ electron impact mass spectrometry. *J Mass Spectrom* 2005;40:739–753.

18. Pujadas M, Pichini S, Civit E, Santamarifia E, Perez K, de la Torre R. A simple and reliable procedure for the determination of psychoactive drugs in oral fluid by gas chromatography-mass spectrometry. *J Pharm Biomed Anal* 2007;44:594–601.

19. Scheidweiler KB, Huestis MA. A validated gas chromatographic-electron impact ionization mass spectrometric method for methylenedioxymethamphetamine (MDMA), methamphetamine and metabolites in oral fluid. *J Chromatogr B Analyt Technol Biomed Life Sci* 2006;835:90–99.

20. Apollonio LG, Pianca DJ, Whittall IR, Maher WA, Kyd JM. A demonstration of the use of ultra-performance liquid chromatography-mass spectrometry (UPLC/MS) in the determination of amphetamine-type substances and ketamine for forensic and toxicological analysis. *J Chromatogr B* 2006;836(1–2):111–115.

21. Nishida K, Itah S, Inoue N, Kudo K, Ikeda N. High-performance liquid chromatographic-mass spectrometric determination of methamphetamine and amphetamine enantiomers, desmethylselegiline and selegiline, in hair samples of long-term methamphetamine abusers or selegiline users. *J Anal Toxicol* 2006;30(4):232–237.

22. Cheze M, Deveaux M, Martin C, Lhermitte M, Pepin G. *Forensic Sci Int* 2007;170(2–3):100–104.

23. Sergi M, Bafile E, Compagnone D, Curini R, D'Ascenzo G, Romolo FS. Multiclass analysis of illicit drugs in plasma and oral fluids by LC-MS/MS. *Anal Bioanal Chem* 2009;393:709–718.

24. Badawi N, Simonsen KW, Steentoft A, Bernhoft IM, Linnet K. Simultaneous screening and quantification of 29 drugs of abuse in oral fluid by solid-phase extraction and ultra-performance LC-MS/MS. *Clin Chem* 2009;55:2004–2018.

25. Oiestad EL, Johansen U, Christophersen AS. Drug screening of preserved oral fluid by liquid chromatography-tandem mass spectrometry. *Clin Chem* 2007;53:300–309.

26. Concheiro M, Gray TR, Shakleya DM, Huestis MA. High-throughput simultaneous analysis of buprenorphine, methadone, cocaine, opiates, nicotine, and metabolites in oral fluid by liquid chromatography tandem mass spectrometry. *Anal Bioanal Chem* 2010;398(2):915–924.

27. Fritch D, Blum K, Nonnemacher S, Haggerty BJ, Sullivan MP, Cone EJ. Identification and quantitation of amphetamines, cocaine, opiates, and phencyclidine in oral fluid by liquid chromatography-tandem mass spectrometry. *J Anal Toxicol* 2009;33(9):569–577.

28. Kala SV, Harris SE, Freijo TD, Gerlich S. Validation of analysis of amphetamines, opiates, phencyclidine, cocaine, and benzoylecgonine in oral fluids by liquid chromatography-tandem mass spectrometry. *J Anal Toxicol* 2008;32(8):605–611.

29. Sergi M, Compagnone D, Curini R, D'Ascenzo G, Del Carlo M, Napotetano S, Risoluti R. Micro-solid phase extraction coupled with high-performance liquid chromatography-tandem mass spectrometry for the determination of stimulants, hallucinogens, ketamine and phencyclidine in oral fluids. *Anal Chim Acta* 2010;675(2):132–137.

30. Wylie FM, Torrance H, Anderson RA, Oliver JS. Drugs in oral fluid, part I: validation of an analytical procedure for licit and illicit drugs in oral fluid. *Forensic Sci Int* 2005;150:191–198.

31. Miller EI, Wylie FM, Oliver JS. Simultaneous detection and quantification of amphetamines, diazepam and its metabolites, cocaine and its metabolites, and opiates in hair by LC-ESI-MS-MS using a single extraction method. *J Anal Toxicol* 2008;32(7):457–469.

32. Tabernero MJ, Felli ML, Bermejo AM, Chiarotti M. Determination of ketamine and amphetamines in hair by LC/MS/MS. *Anal Bioanal Chem* 2009;395(8):2547–2557.

33. National Laboratory Certification Program Manual for laboratories and inspectors. October 1, 2010.

34. Chou R, Fanciullo GJ, Fine PG, et al. Clinical guidelines for the use of chronic opioid therapy in chronic noncancer pain. *J Pain* 2009; 10(2): 113–130. Available online at: http://www.jpain.org/article/S1526-5900(08)00831-6/fulltext Accessed February 1, 2012.

35. Pocci R, Dixit V, Dixit VM. Solid-phase extraction and GC/MS confirmation of barbiturates from human urine. *J Anal Toxicol* 1992;16:45–47.

36. Lui RH, McKeehan AM, Edwards C, Foster G, Bensley WE, Langner JG, Walla AS. Improved gas chromatography/mass spectrometry analysis of barbiturates in urine using centrifuge-based solid-phase extraction, methylation, with d5-pentobarbital as internal standard. *J Forensic Sci* 1994;39(6):1504–1514.

37. Johnson LL, Garg U. Quantitation of amobarbital, butalbital, pentobarbital, phenobarbital, and secobarbital in urine, serum, and plasma using gas chromatography-mass spectrometry (GC-MS). *Methods Mol Biol* 2010;603:65–74.

38. Fritch D, Blum K, Nonnemacher S, Kardos K, Buchhalter A, Cone E. Barbiturate detection in oral fluid, plasma, and urine. *Ther Drug Monit* 2011;33(1):72–79.

39. Dickson PH, Markus W, McKernan J, Nipper HC. Urinalysis of α-hydroxyalprazolam, α-hydroxytriazolam, and other benzodiazepine compounds by GC/EIMS. *J Anal Toxicol* 1992;16:67–71.

40. Meatherall R. GC-MS confirmation of urinary benzodiazepine metabolites. *J Anal Toxicol* 1994;18:369–381.

41. Quintela O, Cruz A, Castro A, Concheiro M, Lopez-Rivadulla M. Liquid chromatography-electrospray ionization mass spectrometry for the determination of nine selected benzodiazepines in human plasma and oral fluid. *J Chromatogr B Analyt Technol Biomed Life Sci* 2005;825:63–71.

42. Uddin MN, Samanidou VF, Papadoyannis IN. Validation of SPE-HPLC determination of 1,4-benzodiazepines and metabolites in blood plasma, urine, and saliva. *J Sep Sci* 2008;31:3704–3717.

43. Ngwa G, Fritch D, Blum K, Newland G. Simultaneous analysis of 14 benzodiazepines in oral fluid by solid-phase extraction and LC-MS-MS. *J Anal Toxicol* 2007;31(7):369–377.

44. Marin SJ, McMillin GA. LC-MS/MS analysis of 13 benzodiazepines and metabolites in urine, serum, plasma, and meconium. *Methods Mol Biol* 2010;603:89–105.

45. Gunn J, Kriger S, Terrell AR. Simultaneous determination and quantification of 12 benzodiazepines in serum or whole blood using UPLC/MS/MS. *Methods Mol Biol* 2010;603:107–119.

46. Laloup M, del Mar Ramirex Fernandez M, De Boeck G, Wood M, Maes V, Samyn N. Validation of a liquid chromatography-tandem mass spectrometry method for the simultaneous determination of 26 benzodiazepines and metabolites, zolpidem, and zopiclone, in blood, urine, and hair. *J Anal Toxicol* 2005;29:616–626.

47. Wu AHB, ed. *Tietz Clinical Guide to Laboratory Tests*, 4th ed. St. Louis, MO: Saunders Elsevier, 2006.
48. Paul BD, Mell LD, Mitchell JM, Irving J, Novak AJ. Simultaneous identification and quantitation of codeine and morphine in urine by capillary gas chromatography and mass spectroscopy. *J Anal Toxicol* 1985;9:222–226.
49. Paul BD, Shimomura ET, Smith, ML. A practical approach to determine cutoff concentrations for opiate testing with simultaneous detection of codeine, morphine, and 6-acetylmorphine in urine. *Clin Chm* 1999;45:510–519.
50. Paul BD, Mitchell JM, Mell LD, Irving J. Gas chromatography/electron impact mass fragmentometric determination of urinary 6-acetylmorphine, a metabolite of heroin. *J Anal Toxicol* 1989;13:2–7.
51. McKinley S, Snyder JJ, Welsh E, Kazarian CM, Jamerson MH, Klette KL. Rapid quantification of urinary oxycodone and oxymorphone using gas chromatography-mass spectrometry. *J Anal Toxicol* 2007;31:434–441.
52. Broussard L, Presley LC, Tanous M, Queen C. Improved gas chromatography-mass spectrometry method for simultaneous identification and quantification of opiates in urine as propionyl and oxime derivatives. *Clin Chem* 2001;47(1):127–129.
53. Meatherall R. GC-MS confirmation of codeine, morphine, 6-acetylmorphine, hydrocodone, hydromorphone, oxycodone, and oxymorphone in urine. *J Anal Toxicol* 1999;23:177–186.
54. Nowatzke W, Zeng J, Saunders A, Bohrer A, Koenig J, Turk J. Distinction among eight opiate drugs in urine by gas chromatography-mass spectrometry. *J Pharm Biomed Anal* 1999;20:815–828.
55. Chen BG, Wang SM, Liu RH. GC-MS analysis of multiply derivatized opiods in urine. *J Mass Spectrom* 2007;42:1012–1023.
56. Dahn T, Gunn J, Kriger S, Terrell AR. Quantitation of morphine, codeine, hydrocodone, hydromorphone, oxycodone, oxymorphone, and 6-monoacetylmorphine (6-MAM) in urine, blood, serum, or plasma using liquid chromatography with tandem mass spectrometry detection. *Methods Mol Biol* 2010;603:411–422.
57. Coles R, Kushnir MM, Nelson GJ, McMillin GA, Urry FM. Simultaneous determination of codeine, morphine, hydrocodone, hydromorphone, oxycodone, and 6-acetylmorphine in urine, serum, plasma, whole blood, and meconium by LC-MS-MS. *J Anal Toxicol* 2007 Jan–Feb;31(1):1–14.
58. Strano-Rossi S, Bermejo AM, de la Torre X, Botrè F. Fast GC-MS method for the simultaneous screening of THC-COOH, cocaine, opiates and analogues including buprenorphine and fentanyl, and their metabolites in urine. *Anal Bioanal Chem* 2011;399(4):1623–1630.
59. Snozek CL, Bjergum MW, Langman LJ.Gas chromatography-mass spectrometry method for the determination of methadone and 2-ethylidene-1,5-dimethyl-3, 3-diphenylpyrrolidine (EDDP). *Methods Mol Biol* 2010;603:351–358.
60. El-Beqqali A, Abdel-Rehim M. Quantitative analysis of methadone in human urine samples by microextraction in packed syringe-gas chromatography-mass spectrometry (MEPS-GC-MS). *J Sep Sci* 2007;Oct 30(15):2501–2505.
61. Sporkert F, Pragst F. Determination of methadone and its metabolites EDDP and EMDP in human hair by headspace solid-phase microextraction and gas chromatography-mass spectrometry. *J Chromatogr B Biomed Sci Appl* 2000 Sep 15;746(2):255–264.
62. Vindenes V, Yttredal B, Oiestad EL, Waal H, Bernard JP, Mørland JG, Christophersen AS. Oral fluid is a viable alternative for monitoring drug abuse: detection of drugs in oral fluid by liquid chromatography-tandem mass spectrometry and comparison to the results from urine samples from patients treated with Methadone or Buprenorphine. *J Anal Toxicol* 2011 Jan 35(1):32–39.

63. Goldberger BA, Chronister CW, Merves ML. Quantitation of fentanyl in blood and urine using gas chromatography-mass spectrometry (GC-MS). *Methods Mol Biol* 2010;603:245–252.
64. Van Nimmen NF, Poels KL, Veulemans HA. Highly sensitive gas chromatographic-mass spectrometric screening method for the determination of picogram levels of fentanyl, sufentanil and alfentanil and their major metabolites in urine of opioid exposed workers. *J Chromatogr B Analyt Technol Biomed Life Sci* 2004 May 25;804(2):375–387.
65. Wang L, Bernert JT. Analysis of 13 fentanils, including sufentanil and carfentanil, in human urine by liquid chromatography-atmospheric-pressure ionization-tandem mass spectrometry. *J Anal Toxicol* 2006 Jun;30(5):335–341.
66. Verplaetse R, Tytgat J. Development and validation of a sensitive ultra performance liquid chromatography tandem mass spectrometry method for the analysis of fentanyl and its major metabolite norfentanyl in urine and whole blood in forensic context. *J Chromatogr B Analyt Technol Biomed Life Sci* 2010 Jul 15;878(22):1987–1996.
67. Huynh NH, Tyrefors N, Ekman L, Johansson M. Determination of fentanyl in human plasma and fentanyl and norfentanyl in human urine using LC-MS/MS. *J Pharm Biomed Anal* 2005 Apr 29;37(5):1095–1100.
68. Moore C, Rana S, Coulter C. Determination of meperidine, tramadol and oxycodone in human oral fluid using solid phase extraction and gas chromatography-mass spectrometry. *J Chromatogr B Analyt Technol Biomed Life Sci* 2007 May 1;850(1–2):370–375.
69. Ishii A, Tanaka M, Kurihara R, Watanabe-Suzuki K, Kumazawa T, Seno H, Suzuki O, Katsumata Y. Sensitive determination of pethidine in body fluids by gas chromatography-tandem mass spectrometry. *J Chromatogr B Analyt Technol Biomed Life Sci* 2003 Jul 15;792(1):117–121.
70. Springer D, Peters FT, Fritschi G, Maurer HH. Studies on the metabolism and toxicological detection of the new designer drug 4′-methyl-α-pyrrolidinopropiophenone in urine using gas chromatography-mass spectrometry. *J Chromatogr B Analyt Technol Biomed Life Sci* 2002 Jun 15;773(1):25–33.
71. Paul BD, McKinley RM, Walsh JW, Jamir TS, Past MR. Effect of freezing on concentration of drugs of abuse in urine. *J Anal Toxicol* 1993;17:378–380.
72. Ramacharitrar V, Levine B, Smialek JE. Benzoylecgonine and ecgonine methyl ester concentrations in urine specimens. *J Forensic Sci* 1995;40:99–101.
73. Paul BD, McWhorter LK, Smith ML. Electron ionization mass fragmentometric detection of urinary ecgonidine, a hydrolytic product of methylecgonidine, as an indicator of smoking cocaine. *J Mass Spectrom* 1999;34:651–660.
74. Gerlits J. GC/MS quantitation of benzoylecgonine following liquid-liquid extraction of urine. *J Forensic Sci* 1993;1210–1214.
75. Paul BD, Lalani S, Bosy T, Jacobs AJ, Huestis MA. Concentration profiles of cocaine, pyrolytic methylecgoninine, and thirteen metabolites in human blood and urine: determination by gas chromatography-mass spectrometry. *Biomed Chromatogr* 2005;19:677–688.
76. Williams RH, Maggiore JA, Shah SM, Erickson TB, Negrusz A. Cocaine and its metabolites in plasma and urine samples from patients in an urban emergency medicine setting. *J Anal Toxicol* 2000;24:478–481.
77. Jenkins AJ, Goldberger B. Identification of unique cocaine metabolites and smoking by-products in postmortem blood and urine specimens. *J Forensic Sci* 1997;42:824–827.
78. Aderjan RE, Schmitt G, Wu M, Meyer C. Determination of cocaine and benzoylecgonine by derivatization with iodomethane-D_3 or PFPA/HFIP in human blood and urine using GC/MS (EI or PCI mode). *J Anal Toxicol* 1993;17:51–55.
79. Gunn J, Kriger S, Terrell AR. Detection and quantification of cocaine and benzoylecgonine in meconium using solid phase extraction and UPLC/MS/MS. *Methods Mol Biol* 2010;603:165–174.

80. Quintela O, Lendoiro E, Cruz A, deCastro A, Quevedo A, Jurado C, Lopez-Rivadulla M. Hydrophilic interaction liquid chromatography-tandem mass spectrometry (HILIC-MS/MS) determination of cocaine and its metabolites benzoylecgonine, ecgonine methyl ester, and cocaethylene in hair samples. *Anal Bioanal Chem* 2010;396(5):1703–1712.

81. Lopez P, Martello S, Bermejo AM, DeVincenzi E, Tabernero MJ, Chiarotti M. Validation of ELISA screening and LC-MS/MS confirmation methods for cocaine in hair after simple extraction. *Anal Bioanal Chem* 2010;397(4):1539–1548.

82. Whiting JD, Manders WW. Confirmation of 9-carboxy-THC in urine by gas chromatography/mass spectrometry. *Aviat Space Environ Med* 1983;54:1031–1033.

83. Paul BD, Mell LD, Mitchell JM, Mckinley RM, Irving J. Detection and quantitation of urinary 11-nor-delta-9-tetrahydrocannabinol-9-carboxylic acid, a metabolite of tetrahydrocannabinol, by capillary gas chromatography and electron impact mass fragmentometry. *J Anal Toxicol* 1987;11:1–5.

84. Paul BD, Jacobs A. Effects of oxidizing adulterants on detection of 11-nor-delta-9-THC-9-carboxylic acid in urine. *J Anal Toxicol* 2002;26:460–463.

85. Jamerson MH, Welton RM, Morris-Kukoski CL, Klette KL. Rapid quantification of urinary 11-nor-delta-9-tetrahydrocannabinol-9-carboxylic acid using fast gas chromatography-mass spectrometry. *J Anal Toxicol* 2005;29:664–668.

86. Baker TS, Harry JV, Russell JW, Myers RL. Rapid method for the GC/MS confirmation of 11-nor-9-carboxy-delta-9-tetrahydrocannabinol in urine. *J Anal Toxicol* 1984;8:255–259.

87. Dixit V, Dixit VM. Solid-phase extraction of 11-nor-delta-9-tetrahydrocannabinol-9-carboxylic acid from human urine with gas chromatography-mass spectrometry. *J Chromatogr B Biomed Appl* 1991;567:81–91.

88. Singh J, Johnson L. Solid-phase extraction of THC metabolite from urine using Empore disk cartridge prior to analysis by GC-MS. *J Anal Toxicol* 1997;21:384–387.

89. O'Dell L, Rymut K, Chaney G, Darpino T, Telepchak M. Evaluation of reduced solvent volume solid-phase extraction cartridges with analysis by gas chromatography-mass spectrometry for determination of 11-nor-9-carboxy-delta-9-THC in urine. *J Anal Toxicol* 1997;21:433–437.

90. De Cock KJS, Delbeke FT, De Boer D, Van Eenoo P, Roels K. Quantitation of 11-nor-delta-9-tetrahydrocannabinol-9-carboxylic acid with GC-MS in urine collected for doping analysis. *J Anal Toxicol* 2003;27:106–109.

91. Abraham TT, Lowe RH, Pirnay SO, Darwin WD, Huestis MA. Simultaneous GC-EI-MS determination of delta-9-tetrahydrocannabinol, 11-hydroxy-delta-9-tetrahydrocannabinol, and 11-nor-9-carboxy-delta-9-tetrahydrocannabinol in human urine following tandem enzyme-alkaline hydrolysis. *J Anal Toxicol* 2007;31:477–485.

92. Stout PR, Klette KL. Solid-phase extraction and GC-MS analysis of THC-COOH method optimization for high-throughput forensic drug-testing laboratory. *J Anal Toxicol* 2001;25:550–554.

93. Emidio ES, Prata VM, Dorea HS. Validation of an analytical method for analysis of cannabinoids in hair by headspace solid-phase microextraction and gas chromatography-ion trap tandem mass spectrometry. *Anal Chim Acta* 2010;670(1–2):63–71.

94. Niedbala S, Kardos K, Salamone S, Fritch D, Bronsgeest M, Cone EJ. Passive cannabis smoke exposure and oral fluid testing. *J Anal Toxicol* 2004;28:546–552.

95. Day D, Kuntz DJ, Feldman M, Presley L. Detection of THCA in oral fluid by GC-MS-MS. *J Anal Toxicol* 2006;30:645–650.

96. Teixeira H, Verstraete A, Proenca P, Corte-Real F, Monsanto P, Vieira DN. Validated method for the simultaneous determination of Delta-9-THC and Delta-9-THC-COOH in oral fluid, urine and whole blood using solid-phase extraction and liquid chromatography-mass spectrometry with electrospray ionization. *Forensic Sci Int* 2007;170(2–3):148–155.

97. Quintela O, Andrenyak DM, Hoggan AM, Crouch DJ. A validated method for the detection of Delta-9-tetrahydrocannabinol and 11-nor-9-carboxy-Delta-9-tetrahydro-cannabinol in oral fluid samples by liquid chromatography coupled with quadrupole-time-of-flight mass spectrometry. *J Anal Toxicol* 2007;31(3):157–164.

98. Concheiro M, de Castro A, Quintela O, Cruz A, Lopez-Rivadulla M. Development and validation of a method for the quantitation of Delta-9-tetrahydrocannabinol in oral fluid by liquid chromatography electrospray-mass-spectrometry. *J Chromatogra B Analyt Technol Biomed Life Sci* 2004;810:319–324.

99. Teixeira H, Proenca P, Verstraete A, Corte-Real F, Vieira DN. Analysis of delta-9-tetra-hydrocannabinol in oral fluid samples using solid-phase extraction and high-performance liquid chromatography-electrospray ionization mass spectrometry. *Forensic Sci Int* 2005;150:205–211.

100. Laloup M, Ramirez Fernandez M, Wood M, De Boeck G, Henquet C, Maes V, Samyn N. Quantitative analysis of delta-9-tetrahydrocannabinol in preserved oral fluid by liquid chromatography-tandem mass spectrometry. *J Chromatogr A* 2005;1082(1):15–24.

101. Coulter C, Taruc M, Tuyay J, Moore C. Quantitation of tetrahydrocannabinol in hair using immunoassay and liquid chromatography with tandem mass spectrometric detection. *Drug Testing Anal* 2009;1(5):234–239.

102. ElSohly M, Little TL, Mitchell JM, Paul BD, Mell LD, Irving J. GC/MS analysis of phencyclidine acid metabolite in human urine. *J Anal Toxicol* 1988;12:180–182.

103. Stevenson CC, Cibull DL, Platoff GE, Bush DM, Gere JA. Solid phase extraction of phencyclidine from urine followed by capillary gas chromatography/mass spectrometry. *J Anal Toxicol* 1992;16:337–339.

104. Nakahara Y, Takahashi K, Sakamoto T, Tanaka A, Hill VA, Baumgartner WA. Hair analysis for drugs of abuse. XVII. Simultaneous detection of PCP, PCHP, and PCP diol in human hair for confirmation of PCP use. *J Anal Toxicol* 1997;21(5):356–362.

105. Coulter C, Crompton K, Moore C. Detection of phencyclidine in human oral fluid using solid-phase extraction and liquid chromatography with tandem mass spectrometric detection. *J Chrom B: Anal Tech Biomed Life Sci* 2008;863(1):123–128.

106. Hammett-Stabler C, Broussard L. (2011). Toxicology and the clinical laboratory. In W. Clarke (Ed.), *Contemporary Practice in Clinical Chemistry*, 2nd ed. Washington, DC: AACC Press.

107. Bergemann D, Geier A, von Meyer L. Determination of lysergic acid diethylamide in body fluids by high-performance liquid chromatography and fluorescence detection—a more sensitive method suitable for routine use. *J Forensic Sci* 1999;44(2):372–374.

108. Rohrich J, Zorntiein S, Becker J. Analysis of LSD in human body fluids and hair samples applying ImmunElute columns. *Forensic Sci Int* 2000;107(1):181–190.

109. De Kanel J, Vicery WE, Waldner B, Monahan RM, Diamond FX. Automated extraction of lysergic acid diethylamide (LSD) and *N*-demethyl-LSD from blood, serum, plasma, and urine samples using the Zymark Rapid Trace with LC/MS/MS confirmation. *J Forensic Sci* 1998;43(3):622–625.

110. Chung A, Hudson J, McKay G. Validated ultra-performance liquid chromatography-tandem mass spectrometry method for analyzing LSD, iso-LSD, Nor-LSD, and O-H-LSD in blood and urine. *J Anal Toxicol* 2009;33(5):253–259.

111. McCusker RR, Paget-Wilkes H, Chronister CW, Goldberger BA, ElSohly MA. Analysis of gamma-hydroxybutyrate (GHB) in urine by gas chromatography-mass spectrometry. *J Anal Toxicol* 1999;23:301–305.

112. Elian AA. GC-MS determination of gamma-hydroxybutyric acid (GHB) in blood. *Forensic Sci Int* 2001;122:43–47.

113. Couper FJ, Logan BK. Determination of γ-hydroxybutyrate (GHB) in biological specimens by gas chromatography-mass spectrometry. *J Anal Toxicol* 2000;24:1–7.

114. Kaufmann E, Alt A. Determination of GHB in urine and serum by LC/MS using a simple one-step derivative. *Forensic Sci Int* 2007;168:133–177.

115. Wood M, Laloup M, Samyn N, Morris MR, deBruijn EA, Maes RA, Young MS, Maes V, DeBoeck G. Simultaneous analysis of gamma-hydroxybutyric acid and its precursors in urine using liquid chromatography-tandem mass spectrometry. *J Chromatogr A* 2004;1056:83–90.

116. Johansen SS, Windberg CN. Simultaneous determination of γ-hydroxybutyrate (GHB) and its analogues (GBL, 1,4-BD, GVL) in whole blood and urine by liquid chromatography coupled to tandem mass spectrometry. *J Anal Toxicol* 2011;35(1):8–14.

117. Stout PA, Simons KD, Kerrigan S. Quantitative analysis of gamma-hydroxybutyrate at endogenous concentrations in hair using liquid chromatography tandem mass spectrometry. *J Forensic Sci* 2010;55(2)531–537.

118. LeBeau M. (2010). Gamma-hydroxybutyric acid (GHB). In B. Levine (Ed.), *Principles of Forensic Toxicology* 3rd ed., pp. 207–214. Washington, DC: AACC Press.

119. Xiang P, Shen M, Zhuo X. Hair analysis for ketamine and its metabolites. *Forensic Sci Int* 2006;162(1–3):131–134.

120. Parkin MC, Turfus SC, Smith NW, Halket JM, Braithwaite RA, Eliott SP, Osselton MD, Cowan DA, Kicman AT. Detection of ketamine and its metabolites in urine by ultra high pressure liquid chromatography-tandem mass spectrometry. *J Chrom B* 2008;876(1):137–142.

121. Harun N, Anderson RA, Cormack PA. Analysis of ketamine and norketamine in hair samples using molecularly imprinted solid-phase extraction (MISPE) and liquid chromatography-tandem mass spectrometry (LC-MS/MS). *Anal Bioanal Chem* 2010;396(7):2449–2459.

122. Broussard L. (2010). Inhalants. In B. Levine (Ed.), *Principles of Forensic Toxicology,* 3rd ed., pp. 413–421. Washington, DC: AACC Press.

123. Sharp M-E. A comprehensive screen for volatile organic compounds in biological fluids. *J Anal Toxicol* 2001;25:631–635.

124. Wasfi IA, Al-Awadhi AH, Al-Hatali ZN, Al-Rayami FJ, Al Katheeri NA. Rapid and sensitive static headspace gas chromatography-mass spectrometry method for the analysis of ethanol and abused inhalants in blood. *J Chromatogr B* 2004;799:331–336.

125. Liu J, Kenji H, Kashimura S, Kashiwagi M, Hamanaka T, Miyoshi A, Kageura M. Headspace solid-phase microextraction and gas chromatographic-mass spectrometric screening for volatile hydrocarbons in blood. *J Chromatogr B* 1997;748:1–9.

126. Tranthim-Fryer DJ, Hansson RC, Norman KW. Headspace/solid-phase microextraction/ gas chromatography-mass spectrometry: a screening technique for the recovery and identification of volatile organic compounds (VOC's) in postmortem blood and viscera samples. *J. Forensic Sci* 2001;46(4):934–946.

5 Pharmacogenomics of Amphetamine and Related Drugs

Steven C. Kazmierczak

CONTENTS

5.1 Introduction 129
    5.1.1 Methylenedioxymethamphetamine (MDMA, or Ecstacy)............... 130
    5.1.2 Methamphetamine 131
5.2 Pharmacokinetics of Amphetamine and Related Compounds..................... 132
5.3 Heritability of Methamphetamine Use... 132
5.4 Molecular Mechanism of Amphetamine Abuse... 133
    5.4.1 Identification of Genes Leading to Addiction Vulnerability............ 133
5.5 Conclusions.. 134
References.. 135

5.1 INTRODUCTION

Illicit psychostimulant drugs such as amphetamines and derivatives represent a highly addictive class of compounds. Included in this group are L-amphetamine, ephedrine, methamphetamine, methylphenidate and pemoline, the latter used previously to treat attention-deficit hyperactivity disorder and narcolepsy. An additional compound also frequently included in this group is cathinone which is the active ingredient found in the leaves of the Khat shrub (*Catha edulis*) found primarily in East Africa and southern Arabia. Cathinone is easily extracted from fresh leaves of the Khat shrub upon chewing or following brewing and produces effects that are similar to that of amphetamine.

Amphetamine was first synthesized in Germany in 1887, while methamphetamine was synthesized in Japan in 1919. The medical utility of amphetamines includes their previous use as an appetite suppressant, and in the treatment of narcolepsy and the symptoms of attention deficit-hyperactivity disorder (ADHD) in children. With repeated use, individuals develop tolerance to many of the effects of amphetamines such as appetite suppression, euphoria, and insomnia. However, tolerance does not develop in children taking amphetamines for ADHD or individuals taking amphetamines for narcolepsy.

The nonmedical use of amphetamines has seen a dramatic increase in recent years due to the relative ease by which these drugs can be synthesized, and both

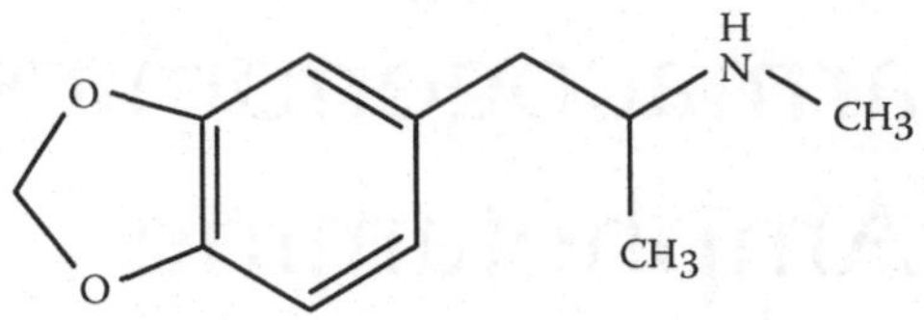

MDMA

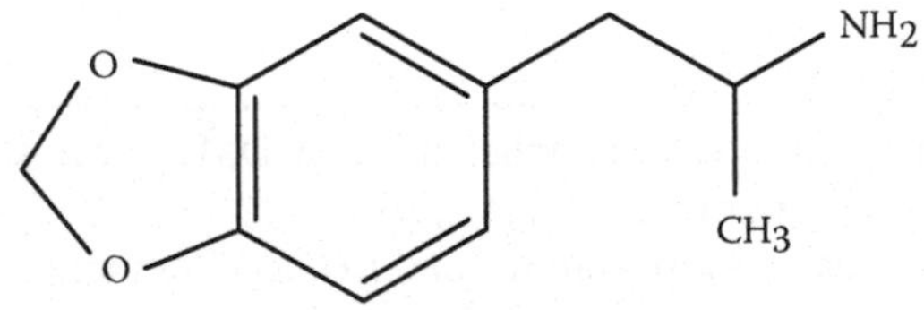

MDA

FIGURE 5.1 Chemical structure of 3,4-methylenedioxymethamphetamine (MDMA) and 3,4-methylenedioxyamphetamine (MDA).

methamphetamine and 3,4-methylenedioxymethamphetamine (MDMA, or "ecstasy") have seen a significant increase in usage. The chemical structures of MDMA and MDA are given in Figure 5.1.

5.1.1 METHYLENEDIOXYMETHAMPHETAMINE (MDMA, OR ECSTACY)

Ecstacy or MDMA is a synthetic amphetamine that is classified as a hallucinogen due to its potential to induce hallucinations when used in extremely high doses. The use of MDMA is often associated with dance parties or "raves." Users of MDMA experience increased self-confidence and empathy and intimacy with other people along with enhanced sensation of proximity. Also reported are euphoria and increased physical energy. However, negative psychological effects such as anxiety, paranoia, and depression have also been reported.

Following ingestion of a moderate dose, the effects of MDMA are observed within 1 hour and last up to 4 hours. Blood MDMA concentrations peak approximately 2 hours following ingestion and can remain detectable for up to 24 hours. Ingestion of large amounts of the drug can result in disproportional increases in blood concentrations. Individuals with severe intoxication usually have plasma drug concentrations of 8 mg/L or more, while concentrations of 1 mg/L or less are usually associated with lesser clinical effects. With a half-life of close to 8 hours, patients with toxic concentrations of MDMA may show effects of the drug for up to 24 hours postingestion. MDMA has a large volume of distribution and readily crosses the blood-brain barrier. Less than 10% of ingested MDMA is metabolized by the liver to produce methylenedioxyamphetamine (MDA), and approximately 65% of MDMA is eliminated unchanged via the kidney.

Both physical and psychological effects are typically encountered for 48 to 72 hours following use of MDMA, and these effects are typical of the "crash" often seen following the use of amphetamines. Physical effects associated with acute withdrawal of the drug include muscle stiffness and muscle pain, blurred vision, loss of appetite, nausea, dry mouth, and insomnia. The psychological effects associated with acute withdrawal include difficulty in concentration, anxiety, fatigue, and depression.

Chronic users of MDMA quickly develop tolerance and require progressively greater amounts of drug to achieve the same effects. Chronic use of the drug results in neurotoxicity and the effect is related to the dose and the frequency of use. Long-term use is associated with decreased brain serotonin concentrations and loss of neurons, transporters, and terminals.

5.1.2 METHAMPHETAMINE

Methamphetamine is the second most popular drug of abuse worldwide, with an annual global prevalence estimated to be 0.4%, and an annual prevalence among adults in the United States estimated to be 0.8% (1,2). Methamphetamine is a potent synthetic psychostimulant that can be administered via a variety of ways including injection, smoking, snorting, ingestion, or transrectally. Although methamphetamine use is a global problem, the vast majority of users of methamphetamine can be found in East and Southeast Asia as well as in North America. Most of the methamphetamine found in the North America is produced in Mexico due to restrictions on precursor chemicals in the United States and Canada.

In 2005 there were close to 100,000 admissions to emergency departments in the United States associated with recent methamphetamine use (7). Both the acute and chronic effects of methamphetamine use are highly dependent on factors such as purity of the drug, the amount of drug consumed, duration of consumption, and route of administration. The most common features associated with methamphetamine overdose are tachycardia, hypertension, rapid respiration, dilated pupils, and agitation. Rhabdomyolysis may be seen in a small proportion of patients and cardiac, hepatic, and renal failure can also occur. Fatalities associated with methamphetamine use are usually the result of hyperpyrexia, aspiration asphyxia, acute cardiac failure, cerebrovascular hemorrhage as a result of hypertension, and pulmonary edema or pulmonary congestion.

Low to moderate doses of methamphetamine stimulate the central nervous system resulting in increased self-confidence and an overall feeling of well-being, decreased appetite and increased alertness, increased blood pressure and heart rate, pupil dilation, and increased temperature. All of these effects are intensified at higher doses, and violent behavior is typically present. Plasma methamphetamine concentrations greater than 100 ug/L were found to typically be associated with rapid and confused speech, hyperthermia, agitation, paranoia, and restlessness (3).

There are two different genetic aspects of response to amphetamine, methamphetamine, and related designer drugs (such as MDMA) on individuals: a genetic aspect of metabolism through polymorphic CYP2 D6 liver enzymes and differences in addiction response due to genetic variations in the dopamine transporter, dopamine receptors, catechol-O-methyltransferase receptor and norepinephrine transporter protein. People with the met/met catechol-O-methyltransferase genotype appear to be at increased risk of an adverse effect from amphetamines and related compounds.

5.2 PHARMACOKINETICS OF AMPHETAMINE AND RELATED COMPOUNDS

Methamphetamine is metabolized primarily in the liver, and the metabolites that are produced do not contribute significantly to the clinical effects of the parent drug. The major metabolic pathways include N-demethylation to produce amphetamine. This step is catalyzed by cytochrome P450 2D6 (*CYP2D6*). Another major metabolite produced by *CYP2D6* is 4-hydroxymethamphetamine through hydroxylation and β-hydroxylation to produce norephedrine. Genetic polymorphism in this enzyme may contribute to the interindividual variability in metabolism (4).

Following ingestion approximately 70% of a dose of methamphetamine is excreted in the urine within 24 hours. The primary compounds detected in the urine include methamphetamine (30% to 50%), up to 10% amphetamine, and up to 15% 4-hydroxymethamphetamine. Because methamphetamine has a urinary half-life of approximately 24 hours, repeated use of methamphetamine will result in the drug being detectable for up to 7 days following use (5). The half-life of methamphetamine in plasma is approximately 10 hours. Thus, an intravenous dose of 10 mg of methamphetamine can be detected in plasma for up to 48 hours afterwards (6). It should be noted that the route of administration does not appreciably alter the plasma half-life of the drug. The effects of methamphetamine infusion are fairly rapid with effects on the cardiovascular system detectable within 2 minutes following infusion.

5.3 HERITABILITY OF METHAMPHETAMINE USE

Addiction to methamphetamine is a highly heritable disorder. Estimates of heritability for stimulants such as methamphetamine have been found to range from 0.33 to 0.44 for male twins and as high as 0.79 for female twins. Stated differently, depending on gender, 33% to 79% of the variance in stimulant abuse can be attributed to genetic factors, and 21% to 67% can be attributed to environmental factors (9).

Genome scans have been conducted in an attempt to identify chromosomal loci that may be associated with vulnerability to methamphetamine addiction. Linkage and genome-wide association studies have identified candidate genes and polymorphisms conferring susceptibility to methamphetamine addiction. Close to 40 different genes have been studied with respect to methamphetamine use disorders (9). The majority of the genes identified in methamphetamine use disorders are involved in neurotransmitter reception, signaling, or metabolism. Of those genes studied, approximately half have been shown to have a significant association with

TABLE 5.1

Genes Found to Be Associated with Significant Risk of a Methamphetamine Use Disorder

Gene	Gene Name	Function	Reference
AKT1	v-akt murine thymoma viral oncogene homolog 1	Mediates dopamine-associated behavior	11
ARRB2	Arrestin, beta 2	Mediates dopamine signaling pathways	12
DTNBP1	Dystrobevin-binding protein 1	Mediates glutamate/dopamine systems	13
GSTP1	Glutathione S-transferase P1	Detoxification of xenobiotics	14
OPRM1	μ-Opioid receptor 1	Mediates opiate response	15,16
PDYN	Prodynorpin	Regulates dopamine release	17
SNCA	α-Synuclein	Involved in dopamine uptake	18

methamphetamine use disorders. Minor alleles for markers in seven genes conferred significant risk for a methamphetamine use disorder. Table 5.1 lists the genes associated with significant risk of a methamphetamine use disorder and the function that the gene is assumed to regulate.

5.4 MOLECULAR MECHANISM OF AMPHETAMINE ABUSE

Amphetamines exert several neurotoxic effects that can be permanent. The drug shows structural similarity to monoamines and can substitute for this compound at the dopamine transporter, noradrenaline transporter, serotonin transporter, and vesicular monoamine transporter-2 (10). The result of this insult includes decreases in dopamine concentrations, decreased activity of tyrosine hydroxylase, and decreases in the concentration of dopamine uptake sites and vesicular monoamine transporters. Methamphetamine causes the redistribution of monoamines from storage vesicles into the cytosol by reversing the function of the vesicular monoamine transporter-2 and reversing the pH gradient that normally serves to aid in the accumulation of monoamine in the vesicles. Also, there is a reversal in function of the dopamine transporter, noradrenaline transporter, and serotonin transporter. This reversal in function results, respectively, in the release of dopamine, noradrenaline, and serotonin from the cytosol into the synapses. These monoamines that are released into the synapse are then able to continually stimulate postsynaptic monoamine receptors. Methamphetamine also inhibits monoamine oxidase, thereby attenuating the metabolism of monoamine. Methamphetamine has been found to be twice as effective as dopamine in causing the release of noradrenaline. In addition, the effect of methamphetamine on noradrenaline is 60-fold greater that its effect on serotonin release.

5.4.1 IDENTIFICATION OF GENES LEADING TO ADDICTION VULNERABILITY

A number of factors can influence the response of an individual to a certain drug or medication. Poor compliance and drug–drug interactions are well-known causes of adverse drug reactions and poor therapeutic outcome. Another factor that has been recognized to have a significant impact on an individual's response to drugs is the genetics of the person using the drug. Pharmacogenetics is the area of study that considers the genetics of each individual and how these genetic differences can be used to assess one's response to a particular drug. Differences in response to a particular drug can occur at the level of drug transporters, drug metabolizing enzymes, and drug targets. It is well appreciated that interindividual differences in drug response are significantly impacted by variability in the genes that encode for the enzymes involved in drug metabolism.

With respect to drug metabolism, the cytochrome P450 genes show significant polymorphism. As a result, the response of individuals to a particular drug can be highly variable. The variations that have been categorized include copy number variants, missense mutations, insertions and deletions, and mutations affecting gene expression and activity of the cytochrome P450 genes (4).

Cytochrome P450 enzymes play a key role in drug metabolism and are suggested to be responsible for approximately 80% of all phase I drug metabolism (8). In particular, *CYP2D6* is the most important polymorphic enzyme in drug metabolism. Of particular interest is the fact that this enzyme is the only drug metabolizing cytochrome enzymes which is not inducible. Thus, genetic variation in *CYP2D6* is largely responsible for the interindividual variation in the metabolism of methamphetamine. Polymorphisms in cytochrome P450 genes include gene deletions, gene duplications, and mutations resulting in inactive gene products.

More than 100 different functional *CYP2D6* alleles or gene variants that affect the function or activity of the enzyme have been described. These variants are divided into alleles resulting in ultrarapid, normal, decreased, and no or null enzyme activity. *CYP2D6*4* and *CYP2D6*5* are the most important null alleles resulting from, respectively, a splice defect and gene deletion. Alleles with severely decreased enzyme activity include *CYP2D6*10*, *CYP2D6*17*, and *CYP2D6*41* as a result of a splicing defect. Ultrarapid enzyme activity is caused by duplication or multiduplication of active *CYP2D6* genes. Table 5.1 shows those genes that confer significant risk for a methamphetamine use disorder (11–18).

Some minor alleles have been found to confer a significant protective effect against methamphetamine use disorders. Those genes that confer a protective effect are encoded for proteins that modulate dopaminergic functions, function in the metabolism of catecholamine transmitters, are involved in metabolism of methamphetamine, function as major inhibitory neurotransmitter receptors, and interact with dopamine transporters.

5.5 CONCLUSIONS

The fields of genomics and proteomics provide very powerful tools for understanding and identifying genes and proteins that control behaviors such as addiction. Although significant progress has been made in understanding the complex nature of abuse and dependence on amphetamines, much work still needs to be done in elucidating the genetics of this disorder. A number of candidate genes have been identified that can impart a greater or lesser vulnerability to addiction. The hope is that a better understanding of the genetics of this disorder will enable improved treatment and prevention of amphetamine addiction.

REFERENCES

1. United Nations Office on Drugs and Crime. *2007World Drug Report.* Vienna, Austria: United Nations Office on Drugs and Crime; 2007.
2. Substance Abuse and Mental Health Services Administration. *Results from the 2006 National Survey on Drug Use and Health: National Findings.* NSDUH Series H-32, DHHS Publication no. SMA 07-4293. Rockville, MD: Office of Applied Studies; 2007.
3. Logan BK. Methamphetamine and driving impairment. *J Forensic Sci* 1996;41:457–464.
4. Olyer JM, Cone EJ, Joseph RE, Moolchan ET, Huestis MA. Duration of detectable methamphetamine and amphetamine excretion in urine after controlled oral administration of methamphetamine to humans. *Clin Chem* 2002;48:1703–1714.
5. Mendelson J, Uemura N, Harris D, Nath RP, Fernandez E, Jacob P, et al. Human pharmacology of the methamphetamine stereoisomers. *Clin Pharmacol Ther* 2006;80:403–420.
6. Substance Abuse and Mental Health Services Administration. *Drug Abuse Warning Network, 2005: National Estimates of Drug-Related Emergency Department Visits.* DAWN Series D-29, DHHS Publication no. (SMA) 07-4256. Rockville, MD: Substance Abuse and Mental Health Services Administration, Office of Applied Studies; 2007.
7. Ingelman-Sundberg M, Sim SC, Gomez A, Rodriguez-Antona C. Influence of cytochrome P450 polymorphisms on drug therapies: Pharmacogenetic, pharmacoepigenetic and clinical aspects. *Pharmacol Ther* 2007;116:496–526.
8. Eichelbaum M, Ingelman-Sundberg M, Evans WE. Pharmacogenomics and individualized drug therapy. *Annu Rev Med* 2006;57:119–137.
9. Bousman CA, Glatt SJ, Everall IP, Tsuang MT. Genetic association studies of methamphetamine use disorders: a systematic review and synthesis. *Am J Med Genet Part B* 2009;150B:1025–1049.
10. Cruickshank CC, Dyer KR. A review of the clinical pharmacology of methamphetamine. *Addiction* 2009;104:1085–1099.
11. Ikeda M, Iwata N, Suzuki T, Kitajima T, Yamanouchi, Kinoshiya Y, et al. Positive association of AKT1 haplotype to Japanese methamphetamine use disorder. *Int J Neuropsychopharmacol* 2006;9:77–81.
12. Ikeda M, Ozaki N, Suzuki T, Kitajima T, Yamanouchi Y, Kinoshiya Y, et al. Possible association of beta-arrestin 2 gene with methamphetamine use disorder, but not schizophrenia. *Genes Brain Behav* 2007;6:107–112.
13. Kishimoto M, Ujike H, Motohashi Y, Tanaka Y, Okahisa Y Kotaka T, et al. The dysbindin gene (DTNBP1) is associated with methamphetamine psychosis. *Biol Psychiatry* 2008;63:191–196.
14. Hashimoto T, Hashimoto K, Matsuzawa D, Shimizu E, Sekine Y, Inada T, et al. A functional glutathione S-transferase P1 gene polymorphism is associated with methamphetamine-induced psychosis in Japanese population. *Am J Med Genet B* 2005;135B:5–9.
15. Ide S, Kobayashi H, Tanaka K, Ujike H, Sekine Y, Ozaki N, et al. Gene polymorphisms of the mu opioid receptor in methamphetamine abusers. *Ann NY Acad Sci* 2004;1025:316–324.
16. Ide S, Kobayashi H, Ujike H, Ozaki N, Sekine Y, Inada T, et al. Linkage disequilibrium and association with methamphetamine dependence/psychosis of mu-opioid receptor gene polymorphisms. *Pharmacogenomics J* 2006;6:179–188.

17. Nomura A, Ujike H, Tanaka Y, Otani K, Morita Y, Kishimoto M, et al. Genetic variant of prodynorphin gene is a risk factor for methamphetamine dependence. *Neurosci Lett* 2006;400:158–162.
18. Kobayashi H, Ide S, Hasegawa J, Ujike H, Sekine Y, Ozaki N, et al. Study of association between alpha-synuclein gene polymorphism and methamphetamine psychosis/dependence. *Ann NY Acad Sci* 2004;1025:325–334.

6 Pharmacogenomics of Cocaine

Loralie J. Langman and Christine L.H. Snozek

CONTENTS

6.1 Introduction 137
6.2 Pharmacology of Cocaine... 138
6.3 Genetics of Cocaine Pharmacodynamics 139
    6.3.1 Dopamine Receptors (*DRD1–DRD5*)... 139
    6.3.2 Dopamine Transporter (*DAT1/SLC6A3*)... 140
    6.3.3 Other Monoaminergic-Related Genes... 141
    6.3.4 Serotonin Transporter (*SLC6A4*)... 141
    6.3.5 Norepinephrine Transporter (*SLC6A2*) 142
    6.3.6 Dopamine β-Hydroxylase (DBH).. 142
6.4 Opioid Receptors 142
    6.4.1 μ Opioid Receptor (*OPRM1*) 143
    6.4.2 κ Opioid Receptor (*OPRK1*)... 143
6.5 Genetics of Cocaine Metabolism.. 143
6.6 Metabolic Enzymes 145
    6.6.1 Human Carboxylesterase 1 (CES1).. 145
    6.6.2 Human Butyrylcholinesterase (BChE).. 145
6.7 Conclusions.. 147
References.. 147

6.1 INTRODUCTION

Cocaine is an alkaloid found in *Erythroxylon coca*, which grows principally in the northern Andes in South America, and to a lesser extent in India, Africa, and Java (1). The drug has been used for its stimulant properties for over 2000 years. In clinical medicine, cocaine is used mainly for local anesthesia and vasoconstriction in nasal surgery, and to dilate pupils in ophthalmology. Sigmund Freud famously proposed its use to treat depression and alcohol dependence, but the realities of cocaine addiction quickly brought this idea to an end. Cocaine abuse is firmly entrenched in the drug culture in the United States and remains one of the most commonly used illicit drugs (2). According to the National Survey on Drug Use and Health, the rate of past-year use for cocaine (powder and crack combined) among individuals aged 12 and older has remained relatively stable since 2002; there were 1.6 million current users of cocaine in 2009 (3).

Cocaine is sold on the street in two forms. The hydrochloride salt (powder) form of cocaine is administered by nasal insufflation (i.e., snorting) or, less frequently, by intravenous injection. The other form, commonly called crack or crack cocaine, is a free-base form that has not been neutralized by an acid to make the hydrochloride salt. It comes as a rock crystal that is heated and its vapors smoked. The term refers to the crackling sound heard when it is heated (2). The use of crack cocaine should not be confused with "free-basing." The latter is a process in which the user purifies cocaine hydrochloride by mixing an aqueous solution of cocaine with baking soda or ammonia, then adding diethyl ether (ethyl acetate) to extract the free form of the drug. The organic solvent can then be evaporated to dryness and the purified drug smoked. However, because of the extremely flammable nature of diethyl ether, and therefore the risk of igniting any remaining solvent, free-basing is no longer commonly practiced (2).

As will be discussed below, cocaine is a powerful stimulant that results in a sensation of euphoria. The feeling of well-being combined with potent activation of neural reward pathways make cocaine an extremely addictive substance. There is a great deal of interest in understanding the addictive process, in order to effectively treat or prevent drug dependence. Much research has therefore been devoted to the study of genetic polymorphisms that are associated with variability in responses to illicit drugs and development of addiction. This chapter will focus on pharmacogenetic data relevant to cocaine use and abuse, with emphasis on individual differences in pharmacokinetics (PK), especially metabolism, and pharmacodynamics (PD).

6.2 PHARMACOLOGY OF COCAINE

Cocaine is a potent central nervous system (CNS) stimulant that elicits a state of increased alertness and euphoria (2), with actions similar to those of amphetamine but of shorter duration (4). Cocaine blocks the reuptake of multiple monoamine neurotransmitters at nerve synapses, but the majority of its CNS effects are thought to be related to inhibition of dopamine reuptake, which prolongs the action of dopamine in the synapse. The euphoria associated with prolonged dopaminergic signaling is what leads to recreational abuse of cocaine. Cocaine also blocks reuptake of serotonin and norepinephrine at presynaptic nerve terminals (Figure 6.1); this produces a sympathomimetic response (including cardiovascular effects such as increased blood pressure, heart rate, and body temperature). Cocaine is effective as a local anesthetic and vasoconstrictor; therefore it is used clinically for nasal surgery, rhinoplasty, and emergency nasotracheal intubation.

The CNS and cardiovascular responses to cocaine exhibit acute tolerance—that is, the effects at a given blood concentration of cocaine are more pronounced if the drug level is increasing (during absorption) than at similar concentrations during the clearance phase (a phenomenon referred to in pharmacology as clockwise hysteresis) (5,6). Thus, a single, nonpeak concentration can be associated with greater or lesser physiological responses depending on the time since cocaine administration. This phenomenon makes attempts to correlate single-time point blood concentrations with psychomotor effects difficult. In addition, because both the rate and direction of change in drug concentration significantly influence the intensity of response to

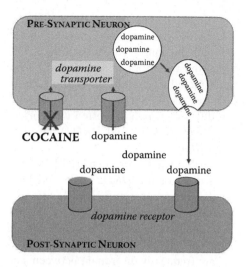

FIGURE 6.1   The effect of cocaine on synaptic dopamine regulation.

cocaine, the drug's stimulant effects are dependent both on dose and on route of administration, with intravenous administration and smoking resulting in the most rapid increases in concentration and therefore the strongest "high."

Acute cocaine toxicity produces a sympathomimetic response that may result in mydriasis, diaphoresis, hyperactive bowel sounds, tachycardia, hypertension, hyperthermia, hyperactivity, agitation, seizures, or coma. Excited delirium and extreme physical activity may also lead to rhabdomyolysis, acute renal failure, and disseminated intravascular coagulopathy. Sudden death due to cardiotoxicity can occur with cocaine use. Death may also follow the sequential development of hyperthermia, agitated delirium (cocaine-induced psychosis), and respiratory arrest.

6.3   GENETICS OF COCAINE PHARMACODYNAMICS

The major neurotransmitters involved in the addictive process are the monoamines: dopamine, serotonin, and norepinephrine (7,8). The monoaminergic neurotransmitters act as messengers throughout various regions of the nervous system and function in a variety of aspects pertinent to addiction, including impulse control, behavior modulation, reward response, and positive motivation. The monoamines are of particular interest with respect to psychostimulants, most of which have a substantial effect on these neurotransmitter systems. It has been estimated that the genetic contribution to variability in response to stimulants is approximately 60% (9,10), thus there is a great deal of interest in polymorphisms of genes related to monoaminergic signaling.

6.3.1   DOPAMINE RECEPTORS (*DRD1–DRD5*)

Drug-induced dopamine elevations appear to mediate reinforcement independently of conscious pleasure perception, increasing the desire to acquire more of the drug regardless of whether its use is enjoyable (11). Dopamine receptors therefore play a large role in substance abuse, functioning in both short-term rewarding effects and the long-term development of dependence (12,13). Signaling through dopamine receptors is implicated in physiological changes following drug exposure, including regulation of transcription and alterations in gene expression (14). These receptors are encoded by at least five separate genes (*DRD1–DRD5*), the products of which are often categorized as D1-like (*DRD1* and *DRD5*) or D2-like (*DRD2*, *DRD3*, and *DRD4*) based on function. The best studied of these genes is *DRD2*.

A restriction fragment length polymorphism (RFLP), TaqIA, is present some 10,000 base pairs downstream of the *DRD2* gene and has been variably associated with abuse and dependence on numerous substances, including cocaine (15) and other psychostimulants (16). Another RFLP, TaqIB, affects intron 2 of *DRD2* and was significantly associated with cocaine dependence (15) and polysubstance abuse in Caucasians (17). However, for both polymorphisms the findings remain inconclusive and are frequently discrepant between studies. Intriguingly, the TaqIA and TaqIB polymorphisms are in linkage disequilibrium, suggesting that haplotype analysis might provide more definitive results (18).

Other *DRD2* polymorphisms have been linked to drug dependence as well. One example is a promoter region variant (−141C Ins/Del) where the −141C *Del* allele is thought to reduce expression of DRD2 (19). Several other *DRD2* variants can also cause decreased gene expression, including the C957T polymorphism. This sequence change, though silent at the amino acid level, alters mRNA folding to cause decreased stability and translational efficiency. In addition, the ability of dopamine to stabilize *DRD2* mRNA was greatly attenuated with the 957T allele (20). Variants leading to reduced expression of *DRD2* are of particular interest, because it has been suggested that low dopamine receptor levels may support use of dopaminergic stimulants that compensate for the inherent deficiency (21).

Other dopamine receptor genes (*DRD1*, *DRD3,* and *DRD4*) have been studied as well, with conflicting outcomes. A variable number of tandem repeats (VNTR) polymorphism in exon 3 of *DRD4* occurred more frequently in methamphetamine abusers than in the controls; however, *DRD2* and *DRD3* showed no association (22). This variant has been linked to other addiction-related phenomena (23–25), but these results have been challenged by other studies (26). The existence of numerous *DRD4* alleles and the low prevalence of some variants are undoubtedly confounding factors. Nevertheless, some larger studies (23) have indicated that these genes may exhibit a legitimate association with the addictive process.

6.3.2 DOPAMINE TRANSPORTER (*DAT1/SLC6A3*)

The dopamine transporter gene (*SLC6A3* or *DAT1*) encodes a protein involved in two key functions: release of dopamine into the synapse to activate neurotransmission, and reuptake of dopamine into presynaptic neurons to terminate the signal. The former process is stimulated by amphetamines, while the latter is inhibited by cocaine, resulting in similar dopaminergic responses and activation of neural reward pathways. Several studies indicate that *DAT1* polymorphisms are important in determining susceptibility to addiction and interindividual variability in response to psychostimulants and other drugs. The most commonly studied genetic polymorphism in *DAT1* is a variable number of tandem repeats (VNTR) in the 3'-untranslated region of exon 15; the repeated 40-nucleotide sequence is most commonly present 10 times in the gene. It is thought that the 9-repeat variant shows different transcriptional activity compared to the 10-repeat gene, although studies have yet to conclusively determine the relative *in vivo* expression of these alleles (27,28).

The *DAT1* VNTR polymorphism has been implicated in individual responses to stimulants, particularly cocaine and amphetamines. An early study showed increased likelihood of cocaine-induced paranoia in individuals with at least one copy of the 9-repeat allele, although no association between the variant allele and cocaine dependence could be determined (29). Cocaine-induced paranoia is a common idiosyncratic reaction, the development of which is thought to be more dependent upon the susceptibility of individual users than the amount of drug used, although frequent high-dose exposure to cocaine does increase the likelihood of mood disturbances and paranoia.

Relatively few studies have examined non-VNTR *DAT1* polymorphisms in addiction, though several such variants have been described. Two nonsynonymous single nucleotide polymorphisms (SNPs), T265C and T1246C, encode valine to alanine substitutions at amino acid residues 55 and 382, respectively (30). Both represent conservative changes but appear to alter the uptake of dopamine and affinity of cocaine binding with the transporter (31,32). Numerous studies (33–35) suggest that there is a great deal of potential for a role for *DAT1* variants in addiction, though assessment of haplotypes is likely to be more informative than study of individual polymorphisms in isolation.

6.3.3 OTHER MONOAMINERGIC-RELATED GENES

Several other genes involved in neurotransmission by monoamines have been studied in the context of cocaine. Most have received limited attention to date, and thus they will be discussed only briefly.

6.3.4 SEROTONIN TRANSPORTER (*SLC6A4*)

Analogous to the role of the dopamine transporter, the function of the serotonin transporter (encoded by *SLC6A4*) is to regulate synaptic serotonin levels, both during physiological processes and in response to exogenous substances including cocaine (13). The *SLC6A4* polymorphism most commonly studied in addiction is the presence or absence of a 44-nucleotide sequence in the promoter region, resulting in a long (L) or short (S) allele, respectively (36). This polymorphism is referred to as the 5-HTTLPR (serotonin transporter linked polymorphic region). The S variant produces lower serotonin transporter levels and activity compared to the L allele, in a dominant fashion (36,37). Although some reports link the S allele to substance abuse (38), several other studies show no correlation between an S genotype and addiction to various substances, including cocaine (39).

A second *SLC6A4* polymorphism is an intron 2 VNTR, with three known alleles consisting of 9, 10, or 12 copies of a 16–17 nucleotide repeated sequence. In a case-control study, an association of the 10-repeat allele with heroin addiction was found (40). A trend toward diminished amphetamine response was seen for combined analysis of the 5-HTTLPR and intron 2 VNTR polymorphisms (41), suggesting that variants of this gene may prove more useful in combination with other genetic information. Given the interconnected nature of the monoamine neurotransmitter systems, it is not surprising that combined analysis of multiple genes in the pathway provides greater pharmacogenetic information than single genes in isolation.

6.3.5 Norepinephrine Transporter (*SLC6A2*)

As is the case with other monoamine transporters, the norepinephrine transporter is involved in neurotransmitter reuptake, particularly with norepinephrine and dopamine (42,43). Several polymorphisms have been identified, including variants affecting the promoter, nonsynonymous SNPs, as well as synonymous and intronic variants (44). The 1369C allele encodes for a proline substitution at residue 457, which appears to result in differential response to cocaine, suggesting a possible role in addiction to that drug (45). Norepinephrine appears to contribute less to substance abuse and dependence compared with dopamine and serotonin, thus *SLC6A2* variants may have only minor influences on the pharmacogenetics of drug use.

6.3.6 Dopamine B-Hydroxylase (DBH)

The dopamine β-hydroxylase (DβH, DBH) enzyme converts dopamine into norepinephrine (46); thus, altered expression of *DBH* could affect both the dopaminergic and noradrenergic systems. Although several studies have confirmed the ability of various polymorphisms to affect DBH production (47–49), less is known about the functional significance of this differential gene expression. One study did link a *DBH* haplotype with low DBH levels and cocaine-induced paranoia (47). Low DBH per se does not necessarily correlate with risk for addiction; for example, the C-1021T promoter polymorphism strongly affects DBH expression but did not show a link to cocaine dependence (50), suggesting that any association between DBH and substance abuse is more complex than a simple gene dosage model.

6.4 OPIOID RECEPTORS

Compounds derived from the opium poppy have been used since ancient times as analgesics, antitussives, and soporifics. Endogenous receptors for opiates were reported in 1973 (51–53) and consist of the μ (MOR), Δ (DOR), and κ (KOR) receptors, encoded by *OPRM1*, *OPRD1*, and *OPRK1*, respectively. The opioid receptors mediate both analgesic and rewarding properties of opioid compounds, as well as other physiological effects. Opioid receptors are also important in modulating responses to a variety of drugs, including alcohol and psychostimulants such as cocaine (54).

6.4.1   μ Opioid Receptor (*OPRM1*)

Although the direct pharmacological targets of cocaine and other stimulants are the monoamine transporters, cocaine use also affects expression and function of the μ opioid receptor (MOR), particularly during long-term abuse. Chronic experimental administration of cocaine increases MOR density and mRNA levels in several regions of rat brain (55–57), and similar results have been seen in cocaine-dependent humans (58,59). Many SNPs have been identified in *OPRM1* and evaluated for association with addiction to cocaine and other drugs of abuse.

To date, the best-studied *OPRM1* polymorphisms are A118G and C17T (60,61). There is evidence to suggest that A118G is associated with the polysubstance dependence (ethanol, cocaine, and opiates) but not cocaine in isolation (62,63). However, a meta-analysis suggests that A118G does not appear to affect risk for substance dependence (64). Additional carefully designed studies are required to elucidate what role this polymorphism has in drug abuse. Similarly, evaluation of the role of C17T in addictions again shows mixed results. In one study, the 17T allele was more prevalent in opiate- or cocaine-dependent subjects of various ethnic groups; similar findings have been reported in opioid-dependent individuals stratified by ethnic/cultural group (60,61). However, studies in other populations found no association of this SNP with alcohol or mixed (opiate or cocaine) dependence (65).

6.4.2   κ Opioid Receptor (*OPRK1*)

Similar to the MOR, the KOR is also involved in response to addictive drugs, most notably cocaine and opiates. The role of the KOR in cocaine addiction is thought to be a consequence of dopamine regulation: signaling through the KOR is associated with reduction of dopamine levels, while its major endogenous ligand, dynorphin, can attenuate cocaine-mediated blockade of dopamine reuptake (54). Elevation of synaptic dopamine levels provides reward and reinforcement, thus modulation of this neurotransmitter is a key component of addiction to numerous substances including cocaine (56,66,67). For this reason, the role of the KOR in modulating dopamine levels and response to substances such as cocaine implicates *OPRK1* as a likely target for understanding hereditary predisposition toward addiction.

At least seven SNPs in the human *OPRK1* gene have been reported (68,69), of which only G36T in exon 2 has been extensively studied. Animal studies have suggested a role for the G36T polymorphism in drug addiction and alcoholism. Most *OPRK1* polymorphisms described to date, including G36T, are silent (i.e., do not affect the amino acid sequence). However, synonymous SNPs can affect mRNA stability and folding, thereby influencing the eventual expression of the gene product.

6.5   GENETICS OF COCAINE METABOLISM

Cocaine has a complex metabolic pathway (Figure 6.2) that occurs via both non-enzymatic hydrolysis and enzymatic transformation in the plasma and liver. It is rapidly metabolized to benzoylecgonine (BE) and ecgonine methyl ester, both of

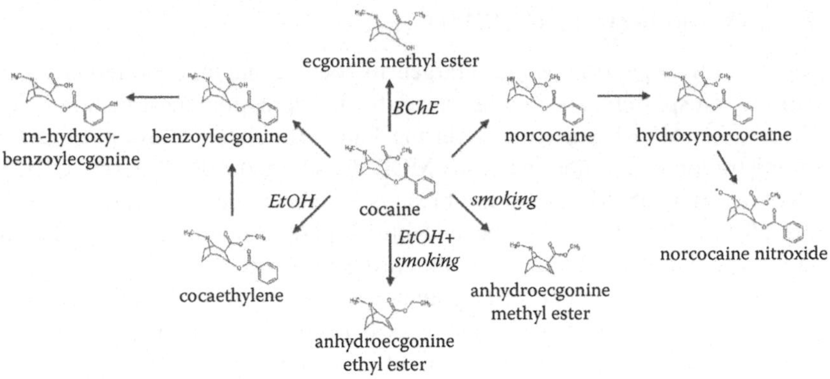

FIGURE 6.2   Cocaine and select metabolites.

which are inactive (2,70). Despite lacking pharmacological activity, BE is extremely relevant to clinical testing because its half-life is longer than that of cocaine; thus it is the most commonly monitored analyte in urine for determination of cocaine use.

BE is further metabolized to minor metabolites such as m-hydroxybenzoylecgonine (m-HOBE) (1,71), which is an important metabolite in the meconium of cocaine-exposed babies (72,73). In adults, m-HOBE has a longer half-life than BE and thus has the potential to be detected for longer periods of time (71,74). Further, m-HOBE is believed to arise exclusively via *in vivo* metabolism (75), therefore, its presence confirms cocaine use if positive drug test results are challenged on the grounds that BE can form spontaneously *in vitro* in urine contaminated with cocaine.

Cocaine is frequently used with other drugs, most commonly ethanol. In simultaneous cocaine and ethanol use, liver methylesterase catalyzes the transesterification of cocaine and ethanol to cocaethylene (2,76,77). This reaction occurs about 3.5 times faster than cocaine hydrolysis to BE (78). Cocaethylene appears to posses the same CNS stimulatory activity as cocaine (79,80) and has a longer half-life (81,82); thus coadministration of ethanol with cocaine produces greater euphoria and enhanced perception of well-being relative to cocaine use alone (2,83). Simultaneous use of even small amounts of cocaine and ethanol creates greater risk for toxicity than either drug alone; it has been suggested that simultaneous use carries an 18- to 25-fold increase in risk for immediate death over use of cocaine alone (2,83–85).

Other cocaine metabolites are also of toxicological interest. Norcocaine is an *N*-demethylated metabolite that can be converted into the hepatotoxic metabolites hydroxyl-norcocaine and norcocaine-nitroxide (86,87). In animals, these metabolites have been reported to inhibit mitochondrial respiration leading to ATP depletion and subsequent cell death (88). Norcocaine concentrations have been shown to be present in greater concentrations in cholinesterase-deficient subjects (89) and in simultaneous cocaine and ethanol users (90). Anhydroecgonine methyl ester (AEME, methyl ecgonidine) and anhydroecgonine ethyl ester (AEEE, ethyl ecgonidine) are pyrolysis products formed after cocaine is smoked (crack cocaine); the latter forms in the presence of coadministered ethanol (91,92).

6.6 METABOLIC ENZYMES

6.6.1 HUMAN CARBOXYLESTERASE 1 (CES1)

Human carboxylesterase 1 (hCE1 or CES1) is a serine esterase involved in hydrolysis of various xenobiotics and endogenous substrates with ester, thioester, or amide bonds, and is thought to have a role in drug metabolism and detoxification (93). CES1 is primarily expressed in the liver, with lesser amounts in the intestine, kidney, lung, testes, heart, monocytes, and macrophages (94,95). CES1 hydrolyzes the methyl ester linkage to generate benzoylecgonine, the primary and inactive metabolite of cocaine (94,96); CES1 is also responsible for the formation of the toxic cocaine metabolite cocaethylene (94,96,97).

There are few reports of genetic mutations that result in complete loss of hydrolytic activity (98). The predicted clinical effects of such a mutation would be significant alterations in the pharmacokinetics and drug responses of CES1 substrates such as cocaine. Therefore, a deficiency of this enzyme would lead to increased cocaine concentration, increased area under the concentration-time curve (AUC), and possible prolongation of cocaine's effects. However, these mutations are rare, and there is no literature available on the actual effect of deficiency of this enzyme in relation to cocaine metabolism.

6.6.2 HUMAN BUTYRYLCHOLINESTERASE (BChE)

Human butyrylcholinesterase (BChE), also called serum cholinesterase or pseudo-cholinesterase, hydrolyzes the larger benzoyl ester linkage on cocaine (96). BChE is also responsible for the hydrolysis of succinylcholine, a muscle relaxant given to patients undergoing surgery to facilitate tracheal intubation during anaesthesia. The effect of succinylcholine usually disappears within 3 to 5 minutes; however, some patients exhibit prolonged apnea after standard doses. In the early 1950s it was suggested that deficiencies of BChE could prolong apnea after succinylcholine administration (99). Kalow and Genest first associated BChE with this phenomenon and categorized different enzyme forms based on combinations of catalytic activity measurement in serum and inhibition tests (100).

Serum BChE concentrations, and measures of catalytic activity such as the dibucaine number (percentage of activity inhibited by dibucaine) and the fluoride number (percentage of activity inhibited by sodium fluoride), are conventionally used to identify BChE phenotypes associated with succinylcholine sensitivity. Mutant BChE is less sensitive to enzyme inhibitors than the wild-type form, and thus the dibucaine and fluoride numbers can define the three main BChE types:

1. The U form with normal dibucaine number and fluoride number
2. The A form, less inhibited by dibucaine than U phenotype
3. The F form with lower dibucaine and fluoride inhibition

There are additional subtypes of BChE characterized by altered enzyme production, stability, and half-life. The quantitative variants display decreased numbers (H-, J-,

TABLE 6.1
Butyrylcholinesterase Variants

	Phenotypic Description	Amino Acid Alterations	Formal Name for Genotype	References
Usual	Normal	None	BCHE	(120,121)
Atypical	Dibucaine resistant	Asp70Gly	BCHE*70G	(105,119,122)
Fluoride 1	Fluoride resistant	Thr243et	BCHE*243M	(103)
Fluoride 2	Fluoride resistant	Gly390Val	BCHE*390V	(123)
K variant	30% reduction in activity	Ala539Thr	BCHE*539T	(124)
J variant	66% reduction in activity	Glu467Val	BCHE*497V	(124)
Silent	No activity	117 Gly-Frameshift stop codon, 12 amino acids further along, at position 129	BCHE*FS117	(125,126)

Note: The table omits a number of very rare variants that have been seen in only a few isolated families: http://www.ncbi.nlm.nih.gov/gene?Db=omim&DbFrom=gene&Cmd=Link&LinkName=gene_omim&LinkReadableName=OMIM&IdsFromResult=590

and K variants) or absence (S form, silent form) of effective circulating molecules (101). Inhibition tests are unable to discriminate between these quantitative variants and the U form, except when they are simultaneously present with the A form (102–104). Numerous mutations (Table 6.1) in the coding region of the gene have been described and result in the variations detailed above (102,105).

Studies evaluating the relationship between blood cocaine concentrations and their effects have demonstrated that peak plasma cocaine concentrations tend to correlate with peak pharmacological and behavioral effects (2,106). It would be anticipated that reduced levels of plasma BChE would increase the plasma half-life of cocaine, resulting in increased levels of cocaine in the CNS and potentiation of dopaminergic transmission. This hypothesis was supported by a study using knockout mice where animals deficient in BChE demonstrated behavioral and physiological responses to cocaine for a significantly longer period of time than did wild-type mice (107).

Even though there is some evidence that peak cocaine correlates with peak pharmacological effect, deaths due to cocaine do not appear to be dose related, and blood levels do not accurately predict toxicity (108). One explanation for the lack of consistent concentration-dependence may be the existence of genetic mutations found in the *BCHE* gene. It was suggested that BChE levels in humans may be predictive of complications from cocaine use. BChE hydrolyzes cocaine to ecgonine methyl ester, a pharmacologically inert substance (109). Reductions in the activity of BChE would increase the amount of cocaine available to undergo alternate metabolism, including

N-demethylation by hepatic carboxylesterase into the metabolically active, vasoconstrictive and toxic metabolite norcocaine (107,110,111). Studies in a knockout mouse model (112) suggest that mice with atypical or silent BChE may be more susceptible to the pathophysiological effects of cocaine at typical doses (113). It would be anticipated that a reduction in the level of endogenous BChE in humans would increase the risk of cocaine toxicity, with potentially fatal consequences (114).

Because of its vital role in cocaine metabolism and toxicity, BChE holds promise as a therapeutic agent. Increasing the available metabolic capacity of BChE would shorten the duration of response to cocaine and may be able to reduce and finally eliminate the rewarding effects of cocaine, thereby weaning a user from cocaine addiction (115). Exogenously delivered BChE prevents cocaine seizures in rats (116) and has been proposed as a treatment for cocaine overdose and addiction in humans (117–119).

6.7 CONCLUSIONS

Cocaine is a powerfully addictive substance that modulates monoaminergic signaling and other neurological reward pathways, yet the genetic variants controlling individual responses and susceptibility to cocaine abuse remain largely unknown. The drug exerts its pharmacological effects through increasing synaptic levels of dopamine, serotonin, and to a lesser extent norepinephrine. Genes encoding proteins involved in synaptic monoamine signaling are therefore of great interest to cocaine pharmacogenetics. Dopamine in particular is thought to play a major role in the development of addiction to numerous drugs; in addition, the opioidergic system has been associated with abuse of various substances including cocaine.

However, despite several promising targets for scientific inquiry, most studies show only weak or contradictory associations between genetic polymorphisms and cocaine use or abuse. Addiction in general is thought to have a large hereditary component, yet the current understanding of which genes and specific variants control an individual's risk remains poor. This is true for both cocaine and most other drugs of abuse, yet the study of cocaine dependence is in some ways even more challenging: for instance, it is notably more difficult to locate nonaddicted cocaine users for a control group than it is to recruit nonalcoholic drinkers. Regardless, the knowledge of cocaine pharmacogenetics has profited greatly from the findings of related studies, for example, the delineation of numerous *BCHE* variants as they affect response to succinylcholine. Increased availability of tools such as genome-wide association studies may build upon this foundation to expand the existing knowledge of individual responses to cocaine.

REFERENCES

1. Isenschmid DS. Cocaine. In: Levine B, ed. *Principles of Forensic Toxicology*, 2nd ed. Washington, DC: AACC Press, 2003:207–228.
2. Isenschmid DS. Cocaine—Effects on human performance and behavior. *Forensic Sci Rev* 2002;14:61.

3. Substance Abuse and Mental Health Services Administration. (2010). *Results from the 2009 National Survey on Drug Use and Health: Volume I. Summary of National Findings* (Office of Applied Studies, NSDUH Series H-38A, HHS Publication No. SMA 10-4586 Findings). Rockville, MD.

4. Catterall WA, Mackie K. Chapter 14. Local anesthetics. In: Brunton LL, Lazo JS, Parker KL, eds. *Goodman and Gilman's The Pharmacological Basis of Therapeutics*, 11th ed: New York: McGraw-Hill, 2006: http://www.accessmedicine.com/content. aspx?aID=938035.

5. Ambre J, Ruo TI, Nelson J, Belknap S. Urinary excretion of cocaine, benzoylecgonine, and ecgonine methyl ester in humans. *J Analyt Toxicol* 1988;12:301–306.

6. Jatlow PI. Drug of abuse profile: Cocaine. *Clin Chem* 1987;33:66B–71B.

7. Axelrod J. Biochemical pharmacology of catecholamines and its clinical implications. *Trans Am Neurol Assoc* 1971;96:179–186.

8. Molinoff PB, Axelrod J. Biochemistry of catecholamines. *Ann Rev Biochem* 1971;40:465–500.

9. Kendler KS, Karkowski LM, Neale MC, Prescott CA. Illicit psychoactive substance use, heavy use, abuse, and dependence in a US population-based sample of male twins. *Arch Gen Psychiatry* 2000;57:261–269.

10. Tsuang MT, Lyons MJ, Eisen SA, Goldberg J, True W, Lin N, et al. Genetic influences on DSM-III-R drug abuse and dependence: a study of 3,372 twin pairs. *Am J Med Genet* 1996;67:473–477.

11. Rutter JL. Symbiotic relationship of pharmacogenetics and drugs of abuse. *AAPS J* 2006;8:E174–E184.

12. Boeck AT, Fry DL, Sastre A, Lockridge O. Naturally occurring mutation, Asp70his, in human butyrylcholinesterase. *Ann Clin Biochem* 2002;39:154–156.

13. Jensen FS, Bartels CF, La Du BN. Structural basis of the butyrylcholinesterase H-variant segregating in two Danish families. *Pharmacogenetics* 1992;2:234–240.

14. Goodman A. Neurobiology of addiction. An integrative review. *Biochem Pharmacol* 2008;75:266–322.

15. Noble EP, Blum K, Khalsa ME, Ritchie T, Montgomery A, Wood RC, et al. Allelic association of the D2 dopamine receptor gene with cocaine dependence. *Drug Alcohol Dependence* 1993;33:271–285.

16. Persico AM, Bird G, Gabbay FH, Uhl GR. D2 dopamine receptor gene TaqI A1 and B1 restriction fragment length polymorphisms: enhanced frequencies in psychostimulant-preferring polysubstance abusers. *Biol Psychiatry* 1996;40:776–84.

17. O'Hara BF, Smith SS, Bird G, Persico AM, Suarez BK, Cutting GR, Uhl GR. Dopamine D2 receptor RFLPs, haplotypes and their association with substance use in black and Caucasian research volunteers. *Hum Heredity* 1993;43:209–218.

18. Xu K, Lichtermann D, Lipsky RH, Franke P, Liu X, Hu Y, et al. Association of specific haplotypes of D2 dopamine receptor gene with vulnerability to heroin dependence in 2 distinct populations. *Arch Gen Psychiatry* 2004;61:597–606.

19. Arinami T, Gao M, Hamaguchi H, Toru M. A functional polymorphism in the promoter region of the dopamine D2 receptor gene is associated with schizophrenia. *Hum Mol Genet* 1997;6:577–82.

20. Duan J, Wainwright MS, Comeron JM, Saitou N, Sanders AR, Gelernter J, Gejman PV. Synonymous mutations in the human dopamine receptor D2 (DRD2) affect mRNA stability and synthesis of the receptor. *Hum Mol Genet* 2003;12:205–216.

21. Noble EP, Zhang X, Ritchie TL, Sparkes RS. Haplotypes at the DRD2 locus and severe alcoholism. *Am J Med Genet* 2000;96:622–631.

22. Chen CK, Hu X, Lin SK, Sham PC, Loh el W, Li T, et al. Association analysis of dopamine D2-like receptor genes and methamphetamine abuse. *Psychiatr Genet* 2004;14:223–226.

23. Ekelund J, Lichtermann D, Jarvelin MR, Peltonen L. Association between novelty seeking and the type 4 dopamine receptor gene in a large Finnish cohort sample. *Am J Psychiatr* 1999;156:1453–1455.

24. Laucht M, Becker K, Blomeyer D, Schmidt MH. Novelty seeking involved in mediating the association between the dopamine D4 receptor gene exon III polymorphism and heavy drinking in male adolescents: results from a high-risk community sample. *Biol Psychiatry* 2007;61:87–92.

25. Skowronek MH, Laucht M, Hohm E, Becker K, Schmidt MH. Interaction between the dopamine D4 receptor and the serotonin transporter promoter polymorphisms in alcohol and tobacco use among 15-year-olds. *Neurogenetics* 2006;7:239–246.

26. Luciano M, Zhu G, Kirk KM, Whitfield JB, Butler R, Heath AC, et al. Effects of dopamine receptor D4 variation on alcohol and tobacco use and on novelty seeking: multivariate linkage and association analysis. *Am J Med Genet B Neuropsychiatr Genet* 2004;124B:113–123.

27. Fuke S, Suo S, Takahashi N, Koike H, Sasagawa N, Ishiura S. The VNTR polymorphism of the human dopamine transporter (DAT1) gene affects gene expression. *Pharmacogenomics J* 2001;1:152–156.

28. Heinz A, Goldman D, Jones DW, Palmour R, Hommer D, Gorey JG, et al. Genotype influences in vivo dopamine transporter availability in human striatum. *Neuropsychopharmacology* 2000;22:133–139.

29. Gelernter J, Kranzler HR, Satel SL, Rao PA. Genetic association between dopamine transporter protein alleles and cocaine-induced paranoia. *Neuropsychopharmacology* 1994;11:195–200.

30. Vandenbergh DJ, Rodriguez LA, Hivert E, Schiller JH, Villareal G, Pugh EW, et al. Long forms of the dopamine receptor (DRD4) gene VNTR are more prevalent in substance abusers: no interaction with functional alleles of the catechol-o-methyltransferase (COMT) gene. *Am J Med Genet* 2000;96:678–683.

31. Lin Z, Uhl GR. Human dopamine transporter gene variation: effects of protein coding variants V55A and V382A on expression and uptake activities. *Pharmacogenomics J* 2003;3:159–168.

32. Uhl GR, Lin Z. The top 20 dopamine transporter mutants: structure-function relationships and cocaine actions. *Eur J Pharmacol* 2003;479:71–82.

33. Le Strat Y, Ramoz N, Pickering P, Burger V, Boni C, Aubin HJ, et al. The 3′ part of the dopamine transporter gene DAT1/SLC6A3 is associated with withdrawal seizures in patients with alcohol dependence. *Alcoholism, Clin Exp Res* 2008;32:27–35.

34. Segman RH, Kanyas K, Karni O, Lerer E, Goltser-Dubner T, Pavlov V, Lerer B. Why do young women smoke? IV. Role of genetic variation in the dopamine transporter and lifetime traumatic experience. *Am J Med Genet B Neuropsychiatr Genet* 2007;144B:533–540.

35. Stapleton JA, Sutherland G, O'Gara C. Association between dopamine transporter genotypes and smoking cessation: a meta-analysis. *Addiction Biol* 2007;12:221–226.

36. Lesch KP, Bengel D, Heils A, Sabol SZ, Greenberg BD, Petri S, et al. Association of anxiety-related traits with a polymorphism in the serotonin transporter gene regulatory region. *Science* (New York, 1996;274:1527–1531).

37. Little KY, McLaughlin DP, Zhang L, Livermore CS, Dalack GW, McFinton PR, et al. Cocaine, ethanol, and genotype effects on human midbrain serotonin transporter binding sites and mRNA levels. *Am Journal of Psychiatry* 1998;155:207–213.

38. Gerra G, Garofano L, Santoro G, Bosari S, Pellegrini C, Zaimovic A, et al. Association between low-activity serotonin transporter genotype and heroin dependence: behavioral and personality correlates. *Am J Med Genet B Neuropsychiatr Genet* 2004;126B:37–42.

39. Patkar AA, Berrettini WH, Hoehe M, Hill KP, Sterling RC, Gottheil E, Weinstein SP. Serotonin transporter (5-HTT) gene polymorphisms and susceptibility to cocaine dependence among African-American individuals. *Addiction Biol* 2001;6:337–345.

40. Tan EC, Yeo BK, Ho BK, Tay AH, Tan CH. Evidence for an association between heroin dependence and a VNTR polymorphism at the serotonin transporter locus. *Mol Psychiatry* 1999;4:215–217.

41. Loscher W, Fassbender CP, Gram L, Gramer M, Horstermann D, Zahner B, Stefan H. Determination of GABA and vigabatrin in human plasma by a rapid and simple HPLC method: correlation between clinical response to vigabatrin and increase in plasma GABA. *Epilepsy Res* 1993;14:245–255.

42. Horn AS. Structure-activity relations for the inhibition of catecholamine uptake into synaptosomes from noradrenaline and dopaminergic neurones in rat brain homogenates. *Brit J Pharmacol* 1973;47:332–338.

43. Raiteri M, Del Carmine R, Bertollini A, Levi G. Effect of sympathomimetic amines on the synaptosomal transport of noradrenaline, dopamine and 5-hydroxytryptamine. *European J Pharmacol* 1977;41:133–143.

44. Stober G, Hebebrand J, Cichon S, Bruss M, Bonisch H, Lehmkuhl G, et al. Tourette syndrome and the norepinephrine transporter gene: results of a systematic mutation screening. *Am J Medical Genetics* 1999;88:158–63.

45. Paczkowski FA, Bonisch H, Bryan-Lluka LJ. Pharmacological properties of the naturally occurring Ala(457)Pro variant of the human norepinephrine transporter. *Pharmacogenetics* 2002;12:165–173.

46. Rosano TG, Eisenhofer G, Whitley RJ. Catecholamines and serotonin. In: Carl A. Burtis ERA, and David E. Bruns, ed. *Tietz Textbook of Clinical Chemistry*, 4th ed. St. Louis, MO: Elsevier Saunders, 2006:2448.

47. Cubells JF, Kranzler HR, McCance-Katz E, Anderson GM, Malison RT, Price LH, Gelernter J. A haplotype at the DBH locus, associated with low plasma dopamine beta-hydroxylase activity, also associates with cocaine-induced paranoia. *Mol Psychiatry* 2000;5:56–63.

48. Wei J, Ramchand CN, Hemmings GP. Possible control of dopamine beta-hydroxylase via a codominant mechanism associated with the polymorphic (GT)n repeat at its gene locus in healthy individuals. *Hum Genet* 1997;99:52–55.

49. Zabetian CP, Anderson GM, Buxbaum SG, Elston RC, Ichinose H, Nagatsu T, et al. A quantitative-trait analysis of human plasma-dopamine beta-hydroxylase activity: evidence for a major functional polymorphism at the DBH locus. *Am J Hum Genet* 2001;68:515–522.

50. Guindalini C, Laranjeira R, Collier D, Messas G, Vallada H, Breen G. Dopamine-beta hydroxylase polymorphism and cocaine addiction. *Behav Brain Funct* 2008;4:1.

51. Pert CB, Snyder SH. Opiate receptor: demonstration in nervous tissue. *Science* (New York, 1973;179:1011–1014.

52. Simon EJ, Hiller JM, Edelman I. Stereospecific binding of the potent narcotic analgesic (3H) Etorphine to rat-brain homogenate. *Natl Acad Sci U.S.* 1973;70:1947–1949.

53. Terenius L. Stereospecific interaction between narcotic analgesics and a synaptic plasm a membrane fraction of rat cerebral cortex. *Acta Pharmacologica et Toxicologica* 1973;32:317–320.

54. Kreek MJ, LaForge KS, Butelman E. Pharmacotherapy of addictions. *Nature Rev* 2002;1:710–726.

55. Azaryan AV, Coughlin LJ, Buzas B, Clock BJ, Cox BM. Effect of chronic cocaine treatment on mu- and delta-opioid receptor mRNA levels in dopaminergically innervated brain regions. *J Neurochem* 1996;66:443–448.

56. Unterwald EM, Rubenfeld JM, Kreek MJ. Repeated cocaine administration upregulates kappa and mu, but not delta, opioid receptors. *Neuroreport* 1994;5:1613–1616.

57. Yuferov V, Zhou Y, Spangler R, Maggos CE, Ho A, Kreek MJ. Acute "binge" cocaine increases mu-opioid receptor mRNA levels in areas of the rat mesolimbic mesocortical dopamine system. *Brain Res Bull* 1999;48:109–112.

58. Gorelick DA, Kim YK, Bencherif B, Boyd SJ, Nelson R, Copersino M, et al. Imaging brain mu-opioid receptors in abstinent cocaine users: time course and relation to cocaine craving. *Biol Psychiatry* 2005;57:1573–1582.

59. Zubieta JK, Gorelick DA, Stauffer R, Ravert HT, Dannals RF, Frost JJ. Increased mu opioid receptor binding detected by PET in cocaine-dependent men is associated with cocaine craving. *Nature Med* 1996;2:1225–1229.

60. Berrettini WH, Hoehe MR, Ferraro TN, Demaria PA, Gottheil E. Human mu opioid receptor gene polymorphisms and vulnerability to substance abuse. *Addict Biol* 1997;2:303–308.

61. Bond C, LaForge KS, Tian M, Melia D, Zhang S, Borg L, et al. Single-nucleotide polymorphism in the human mu opioid receptor gene alters beta-endorphin binding and activity: possible implications for opiate addiction. *Proc Natl Acad Sci USA* 1998;95:9608–9613.

62. Gelernter J, Kranzler H, Cubells J. Genetics of two mu opioid receptor gene (OPRM1) exon I polymorphisms: population studies, and allele frequencies in alcohol- and drug-dependent subjects. *Mol Psychiatry* 1999;4:476–483.

63. Luo X, Kranzler HR, Zhao H, Gelernter J. Haplotypes at the OPRM1 locus are associated with susceptibility to substance dependence in European-Americans. *Am J Med Genet B Neuropsychiatr Genet* 2003;120B:97–108.

64. Arias A, Feinn R, Kranzler HR. Association of an Asn40Asp (A118G) polymorphism in the mu-opioid receptor gene with substance dependence: A meta-analysis. *Drug Alcohol Dependence* 2006;83:262–268.

65. Gelernter J, Kranzler H, Satel SL. No association between D2 dopamine receptor (DRD2) alleles or haplotypes and cocaine dependence or severity of cocaine dependence in European- and African-Americans. *Biol Psychiatry* 1999;45:340–345.

66. Maisonneuve IM, Ho A, Kreek MJ. Chronic administration of a cocaine "binge" alters basal extracellular levels in male rats: an in vivo microdialysis study. *J Pharmacol Exp Ther* 1995;272:652–657.

67. Unterwald EM, Ho A, Rubenfeld JM, Kreek MJ. Time course of the development of behavioral sensitization and dopamine receptor up-regulation during binge cocaine administration. *J Pharmacol Exp Ther* 1994;270:1387–1396.

68. LaForge K, Kreek MJ, Uhl GR, Sora I, Yu L, Befort K, et al. Symposium XIII: allelic polymorphism of human opioid receptors: functional studies: genetic contributions to protection from, or vulnerability to, addictive diseases. 1999 Proceedings of the 61st Annual Scientific Meeting of the College on Problems of Drug Dependence National Institute of Drug Abuse. Bethesda, MD: Research Monograph Series (Harris LS ed) pp 47–50, U.S. Department of Health and Human Services, National Institutes of Health. NIH Publication No (ADM)00-4737.

69. Mayer P, Hollt V. Allelic and somatic variations in the endogenous opioid system of humans. *Pharmacol Ther* 2001;91:167–177.

70. Dean RA, Christian CD, Sample RH, Bosron WF. Human liver cocaine esterases: ethanol-mediated formation of ethylcocaine. *FASEB J* 1991;5:2735–2739.

71. Kolbrich EA, Barnes AJ, Gorelick DA, Boyd SJ, Cone EJ, Huestis MA. Major and minor metabolites of cocaine in human plasma following controlled subcutaneous cocaine administration. *J Analytical Toxicol* 2006;30:501–510.

72. Lester BM, ElSohly M, Wright LL, Smeriglio VL, Verter J, Bauer CR, et al. The Maternal Lifestyle Study: drug use by meconium toxicology and maternal self-report. *Pediatrics* 2001;107:309–317.

73. Moore C, Negrusz A, Lewis D. Determination of drugs of abuse in meconium. *J Chromatography* 1998;713:137–146.

74. Cone EJ, Sampson-Cone AH, Darwin WD, Huestis MA, Oyler JM. Urine testing for cocaine abuse: metabolic and excretion patterns following different routes of administration and methods for detection of false-negative results. *J Analyt Toxicol* 2003;27:386–401.

75. Klette KL, Poch GK, Czarny R, Lau CO. Simultaneous GC-MS analysis of meta- and para-hydroxybenzoylecgonine and norbenzoylecgonine: a secondary method to corroborate cocaine ingestion using nonhydrolytic metabolites. *J Analyt Toxicol* 2000;24:482–488.

76. Hearn WL, Flynn DD, Hime GW, Rose S, Cofino JC, Mantero-Atienza E, et al. Cocaethylene: a unique cocaine metabolite displays high affinity for the dopamine transporter. *J Neurochemistry* 1991;56:698–701.

77. Jatlow P, Elsworth JD, Bradberry CW, Winger G, Taylor JR, Russell R, Roth RH. Cocaethylene: a neuropharmacologically active metabolite associated with concurrent cocaine-ethanol ingestion. *Life Sci* 1991;48:1787–1794.

78. Brzezinski MR, Abraham TL, Stone CL, Dean RA, Bosron WF. Purification and characterization of a human liver cocaine carboxylesterase that catalyzes the production of benzoylecgonine and the formation of cocaethylene from alcohol and cocaine. *Biochem Pharmacol* 1994;48:1747–1755.

79. Hart CL, Jatlow P, Sevarino KA, McCance-Katz EF. Comparison of intravenous cocaethylene and cocaine in humans. *Psychopharmacology* 2000;149:153–162.

80. Perez-Reyes M, Jeffcoat AR, Myers M, Sihler K, Cook CE. Comparison in humans of the potency and pharmacokinetics of intravenously injected cocaethylene and cocaine. *Psychopharmacology* 1994;116:428–432.

81. Farren CK, Hameedi FA, Rosen MA, Woods S, Jatlow P, Kosten TR. Significant interaction between clozapine and cocaine in cocaine addicts. *Drug Alcohol Dependence* 2000;59:153–163.

82. Jacob P, Mendelson JE, Jones RT, Benowitz N. Formation and elimination kinetics of cocaethylene in humans. *Clin Pharmacol Ther* 1993;53:174.

83. McCance-Katz EF, Price LH, McDougle CJ, Kosten TR, Black JE, Jatlow PI. Concurrent cocaine-ethanol ingestion in humans: pharmacology, physiology, behavior, and the role of cocaethylene. *Psychopharmacology* 1993;111:39–46.

84. Andrews P. Cocaethylene toxicity. *J Addictive Dis* 1997;16:75–84.

85. Vanek VW, Dickey-White HI, Signs SA, Schechter MD, Buss T, Kulics AT. Concurrent use of cocaine and alcohol by patients treated in the emergency department. *Ann Emerg Med* 1996;28:508–514.

86. Kloss MW, Rosen GM, Rauckman EJ. N-demethylation of cocaine to norcocaine. Evidence for participation by cytochrome P-450 and FAD-containing monooxygenase. *Mol Pharmacol* 1983;23:482–485.

87. Kloss MW, Rosen GM, Rauckman EJ. Cocaine-mediated hepatotoxicity. A critical review. *Biochem Pharmacol* 1984;33:169–173.

88. Boess F, Ndikum-Moffor FM, Boelsterli UA, Roberts SM. Effects of cocaine and its oxidative metabolites on mitochondrial respiration and generation of reactive oxygen species. *Biochem Pharmacol* 2000;60:615–623.

89. Inaba T, Stewart DJ, Kalow W. Metabolism of cocaine in man. *Clin Pharmacol Ther* 1978;23:547–552.

90. Farre M, de la Torre R, Llorente M, Lamas X, Ugena B, Segura J, Cami J. Alcohol and cocaine interactions in humans. *J Pharmacol Exp Ther* 1993;266:1364–1373.

91. Cone EJ, Hillsgrove M, Darwin WD. Simultaneous measurement of cocaine, cocaethylene, their metabolites, and "crack" pyrolysis products by gas chromatography-mass spectrometry. *Clin Chem* 1994;40:1299–1305.

92. Jenkins AJ, Goldberger BA. Identification of unique cocaine metabolites and smoking by-products in postmortem blood and urine specimens. *J Forensic Sci* 1997;42:824–827.

93. Satoh T. Role of carboxylesterase in xenobiotic metabolism. *Rev Biochem Toxicol* 1987;8:155–181.

94. Redinbo MR, Bencharit S, Potter PM. Human carboxylesterase 1: from drug metabolism to drug discovery. *Biochem Soc Trans* 2003;31:620–624.

95. Satoh T, Hosokawa M. The mammalian carboxylesterases: from molecules to functions. *Ann Rev Pharmacol Toxicol* 1998;38:257–288.

96. Bencharit S, Morton CL, Xue Y, Potter PM, Redinbo MR. Structural basis of heroin and cocaine metabolism by a promiscuous human drug-processing enzyme. *Nature Struct Biol* 2003;10:349–356.

97. Pennings EJ, Leccese AP, Wolff FA. Effects of concurrent use of alcohol and cocaine. *Addiction* (Abingdon, England) 2002;97:773–783.

98. Zhu HJ, Patrick KS, Yuan HJ, Wang JS, Donovan JL, DeVane CL, et al. Two CES1 gene mutations lead to dysfunctional carboxylesterase 1 activity in man: clinical significance and molecular basis. *Am J Hum Genet* 2008;82:1241–1248.

99. Weber WW. Human drug-metabolizing enzyme variants. In: Weber WW, ed. *Pharmacogenetics*, New York: Oxford University Press, 1997:181–186.

100. Kalow W, Genest K. A method for the detection of atypical forms of human serum cholinesterase; determination of dibucaine numbers. *Can J Biochem Physiol* 1957;35:339–346.

101. Harris H, Hopkinson DA, Robson EB, Whittaker M. Genetical studies on a new variant of serum cholinesterase detected by electrophoresis. *Ann Hum Genet* 1963;26:359–382.

102. Bartels CF, Jensen FS, Lockridge O, van der Spek AF, Rubinstein HM, Lubrano T, La Du BN. DNA mutation associated with the human butyrylcholinesterase K-variant and its linkage to the atypical variant mutation and other polymorphic sites. *Am J Hum Genet* 1992;50:1086–1103.

103. Harris H, Whittaker M. Differential inhibition of human serum cholinesterase with fluoride: recognition of two new phenotypes. *Nature* 1961;191:496–498.

104. Lando G, Mosca A, Bonora R, Azzario F, Penco S, Marocchi A, et al. Frequency of butyrylcholinesterase gene mutations in individuals with abnormal inhibition numbers: an Italian-population study. *Pharmacogenetics* 2003;13:265–270.

105. McGuire MC, Nogueira CP, Bartels CF, Lightstone H, Hajra A, Van der Spek AF, et al. Identification of the structural mutation responsible for the dibucaine-resistant (atypical) variant form of human serum cholinesterase. *Proc Natl Acad Sci USA* 1989;86:953–957.

106. Cone EJ. Pharmacokinetics and pharmacodynamics of cocaine. *J Analytical Toxicol* 1995;19:459–478.

107. Duysen EG, Lockridge O. Prolonged toxic effects after cocaine challenge in butyrylcholinesterase/plasma carboxylesterase double knockout mice: a model for butyrylcholinesterase-deficient humans. *Drug Metab Dispos* 2011;39:1321–1323.

108. Karch SB, Stephens B, Ho CH. Relating cocaine blood concentrations to toxicity—an autopsy study of 99 cases. *J Forensic Sci* 1998;43:41–45.

109. Om A, Ellahham S, Ornato JP, Picone C, Theogaraj J, Corretjer GP, Vetrovec GW. Medical complications of cocaine: possible relationship to low plasma cholinesterase enzyme. *Am Heart J* 1993;125:1114–1117.

110. Blaho K, Logan B, Winbery S, Park L, Schwilke E. Blood cocaine and metabolite concentrations, clinical findings, and outcome of patients presenting to an ED. *Am J Emerg Med* 2000;18:593–598.

111. Pellinen P, Kulmala L, Konttila J, Auriola S, Pasanen M, Juvonen R. Kinetic characteristics of norcocaine N-hydroxylation in mouse and human liver microsomes: involvement of CYP enzymes. *Arch Toxicol* 2000;74:511–520.

112. Li B, Duysen EG, Carlson M, Lockridge O. The butyrylcholinesterase knockout mouse as a model for human butyrylcholinesterase deficiency. *J Pharmacol Exp Therapeutics* 2008;324:1146–1154.
113. Duysen EG, Li B, Carlson M, Li YF, Wieseler S, Hinrichs SH, Lockridge O. Increased hepatotoxicity and cardiac fibrosis in cocaine-treated butyrylcholinesterase knockout mice. *Basic Clin Pharmacol Toxicol* 2008;103:514–521.
114. Hoffman RS, Henry GC, Howland MA, Weisman RS, Weil L, Goldfrank LR. Association between life-threatening cocaine toxicity and plasma cholinesterase activity. *Ann Emerg Med* 1992;21:247–253.
115. Asojo OA, Asojo OA, Ngamelue MN, Homma K, Lockridge O. Cocrystallization studies of full-length recombinant butyrylcholinesterase (BChE) with cocaine. *Acta Crystallographica* 2011;67:434–437.
116. Brimijoin S, Gao Y, Anker JJ, Gliddon LA, Lafleur D, Shah R, et al. A cocaine hydrolase engineered from human butyrylcholinesterase selectively blocks cocaine toxicity and reinstatement of drug seeking in rats. *Neuropsychopharmacology* 2008;33:2715–2725.
117. Duysen EG, Bartels CF, Lockridge O. Wild-type and A328W mutant human butyrylcholinesterase tetramers expressed in Chinese hamster ovary cells have a 16-hour half-life in the circulation and protect mice from cocaine toxicity. *J Pharmacol Exp Therapeutics* 2002;302:751–758.
118. Gao Y, LaFleur D, Shah R, Zhao Q, Singh M, Brimijoin S. An albumin-butyrylcholinesterase for cocaine toxicity and addiction: catalytic and pharmacokinetic properties. *Chemico-biol Interact* 2008;175:83–87.
119. Xie W, Altamirano CV, Bartels CF, Speirs RJ, Cashman JR, Lockridge O. An improved cocaine hydrolase: the A328Y mutant of human butyrylcholinesterase is 4-fold more efficient. *Mol Pharmacol* 1999;55:83–91.
120. Whittaker M. Cholinesterase. *Monogr Hum Genet* 1986;11.
121. Lockridge O. Genetic variants of human serum cholinesterase influence metabolism of the muscle relaxant succinylcholine. *Pharmacol Therapeutics* 1990;47:35–60.
122. Kalow W, Gunn DR. Some statistical data on atypical cholinesterase of human serum. *Ann Hum Genet* 1959;23:239–250.
123. Nogueira CP, Bartels CF, McGuire MC, Adkins S, Lubrano T, Rubinstein HM, et al. Identification of two different point mutations associated with the fluoride-resistant phenotype for human butyrylcholinesterase. *Am Journal Hum Genet* 1992;51:821–828.
124. Bartels CF, James K, La Du BN. DNA mutations associated with the human butyrylcholinesterase J-variant. *Am J Hum Genet* 1992;50:1104–1114.
125. Liddell J, Lehmann H, Silk E. A "silent" pseudo-cholinesterase gene. *Nature* 1962;193:561–562.
126. Nogueira CP, McGuire MC, Graeser C, Bartels CF, Arpagaus M, Van der Spek AF, et al. Identification of a frameshift mutation responsible for the silent phenotype of human serum cholinesterase, Gly 117 (GGT—GAG). *Am J Hum Genet* 1990;46:934–942.

7 Genetic Aspect of Marijuana Metabolism and Abuse

Pradip Datta

CONTENTS

7.1 Introduction 155
7.2 Marijuana Absorption... 156
7.3 Marijuana Metabolism... 157
    7.3.1 Pharmacogenomics of Marijuana Metabolism............................ 158
7.4 Marijuana: Mechanism of Action... 159
    7.4.1 Marijuana Receptors... 159
    7.4.2 Polymorphisms of Cannabinoid Binding Receptors 161
    7.4.3 Application of Marijuana Pharmacogenomics 162
7.5 Conclusions... 163
References... 163

7.1 INTRODUCTION

Marijuana and hashish contain many components, Δ9-tetrahydrocannabinol (Δ9-THC) being the principal psychoactive ingredient. The chemical structure of marijuana is given in Figure 7.1. Marijuana has recently come to the forefront of the news as more U.S. states legalize "medical marijuana," and as the U.S. Food and Drug Administration (FDA) approved the first Δ9-THC-containing drug, Marinol (or Dronabinol) [1]. Marinol is used to treat nausea and vomiting caused by chemotherapy and loss of appetite and weight loss in people who have acquired immunodeficiency syndrome (AIDS). The FDA recently approved a newer drug, Sativex, containing Δ9-THC and cannabidiol, to treat spasticity due to multiple sclerosis; various researchers have studied off-label applications of the drug to other diseases like Crohn's disease, ulcerative colitis, colorectal cancer, posttraumatic stress disorder, glaucoma, Tourette syndrome, ADHD, clinical depression, epilepsy, bipolar disorder, and schizophrenia [2]. Δ9-THC and its pharmacologically active analogs produce other characteristic behaviors in humans and animals as well, including memory impairment, antinociception, and locomotor and psychoactive effects. It also stimulates appetite [3]. However, tolerance to and dependence on cannabinoids develop after chronic use, as demonstrated both clinically and in animal models. The potential therapeutic

155

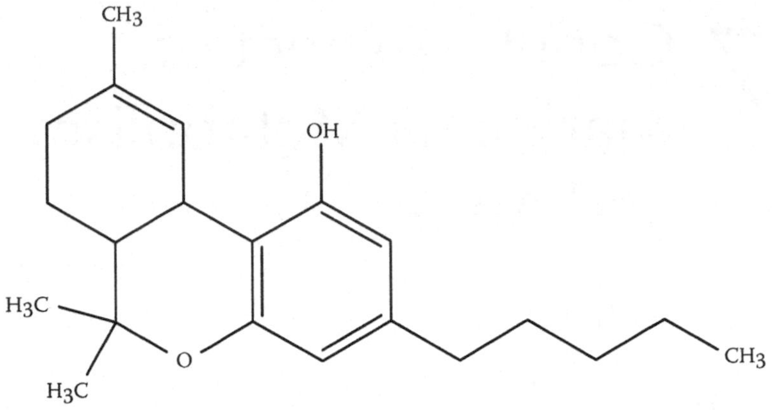

FIGURE 7.1 Chemical structure of marijuana.

benefits of certain cannabinoid-mediated effects, as well as the use of marijuana for its psychoactive properties and creating dependence, has raised interest in understanding the cellular adaptations and their pharmacogenetics, produced by chronic administration of this class of drugs. Pharmacogenomics of cannabinoids not only will assist in therapeutic applications of Δ9-THC drugs like Marinol, but also will allow newer personalized treatments of marijuana dependence using micro-RNA.

There are various genetic components that may introduce individual variation in the effect of the drug: absorption in the bloodstream, metabolism, crossing the blood-brain barrier, binding to target neurons, and the resulting effects. Δ9-THC is absorbed via inhalation or the gastrointestinal tract and is almost completely metabolized by liver enzymes. The various active and inactive components pass the blood-brain barrier and bind to specific receptors in the brain to produce the psychoactivity.

7.2   MARIJUANA ABSORPTION

Marijuana can be administered via various routes, smoking being most common. It is also orally taken. The drug Marinol comes as a pill for oral administration. Recent application of a marijuana patch also has been of interest [4].

If breathed in, Δ9-THC is rapidly transferred from lungs to blood during smoking. Δ9-THC bioavailability averages 30%. In one study, with a 3.55% Δ9-THC cigarette, a peak plasma level of 86.3 ng/mL of the drug occurred approximately 10 minutes after inhalation [5]. One study shows that the "style" of smoking marijuana cigarettes influences the THC concentration absorbed in blood [6]: the longer the "breath holding time," the more of THC is absorbed. Puff volume, on the other hand, had no effect on serum THC levels. Thus there is a genetic component in marijuana smoking—the longer one can hold the smoke from the drug, the more drug is absorbed. Furthermore, interindividual variability in obtaining the peak psychologic "high" was observed after inhalation. The duration of the "high" also varied considerably, ranging from 10 to 140 minutes (average peak "high" of 70 minutes). Lemburger et al. demonstrated that this variability correlated well with the peak plasma levels of metabolites of Δ9-THC [7].

Oral THC, on the other hand, is only 4% to 12% bioavailable, and absorption is highly variable [4]. The psychologic effects and plasma levels of metabolites of Δ9-THC peaked at 3 hours after oral administration [6].

Even though interindividual variability in marijuana absorption is acknowledged and documented, no definite pharmacogenetic study has been conducted in this regard. This topic needs to be explored further.

7.3   MARIJUANA METABOLISM

Δ9-THC is eliminated from plasma in a multiphasic manner, with low amounts detectable for over 1 week after dosing. A major active metabolite, 11-hydroxy-Δ-9-tetrahydrocannabinol (11-OH-THC), is formed after both inhalation and oral dosing (20% and 100% of parent, respectively). Δ9-THC is widely distributed, particularly to fatty tissues, but less than 1% of an administered dose reaches the brain, while the spleen and body fat are long-term storage sites. The elimination of Δ9-THC and its many metabolites (from all routes) occurs via the feces and urine. Metabolites persist in the urine and feces for several weeks [5].

Δ9-THC is almost completely metabolized by the liver microsomal enzymes yielding the active metabolite, 11-OH-THC, and the inactive metabolite, 11-nor-9-carboxy-delta 9-tetrahydrocannabinol (THCCOOH). In addition, 7-hydroxy-THC and 6-beta-hydroxy-THC are formed at almost equal rates. Furthermore, $1\alpha,2\alpha$-epoxyhexahydrocannabinol (EHHC) was formed at approximately one-third the rate of 7-hydroxy- and 6β-hydroxy-THC, and small amounts of 6α-hydroxy- and 6-keto-THC were also found [6]. Immunoinhibition studies with antibodies raised against human hepatic P-450 2C9, or a mouse hepatic P-450 isozyme belonging to the P-450 3A subfamily, revealed that P-450 2C9 catalyzed the formation of 7-hydroxy-THC, whereas P-450 3A catalyzed the formation of 6β-hydroxy-THC, EHHC, and the relatively minor metabolites. In contrast, antibodies raised against human P-450 2C8 had no effect on human microsomal THC hydroxylation. Excellent correlations were found between hepatic microsomal P-450 2C9 and 3A content and 7-hydroxy- and 6β-hydroxy-THC formation, respectively. In addition, purified P-450 2C9 cata-lyzed the formation of 7-hydroxy-THC at a sevenfold higher rate than that observed with microsomes. Microsomal 7-hydroxy-THC formation varied less than fivefold between the livers, suggesting that this activity is normally expressed and probably not subject to environmental influences [8].

Another study incorporated a highly automated gas chromatography mass spectrometry (GCMS) procedure for Δ9-THC, 11-OH THC, and THCCOOH to sample blood and to capture rapid drug-level changes during and following smoking. Human subjects smoked one marijuana cigarette (placebo, 1.75%, or 3.55% Δ9-THC) once a week according to a randomized, crossover, double-blind Latin square design. Δ9-THC levels increased rapidly, peaked prior to the end of smoking, and quickly dissipated. Mean peak 11-OH-THC levels were substantially lower than Δ9-THC levels and occurred immediately after the end of smoking. THCCOOH levels increased slowly and plateaued for an extended period. The mean peak time for THCCOOH was 113 minutes [9].

In another study, the human liver enzyme microsomal alcohol oxygenase was able to oxidize both 7α- and 7β-hydroxy-Δ8-tetrahydrocannabinol (7α- and 7β-hydroxy-Δ8-THC) to 7-oxo-Δ8-THC. The oxidative activity was determined by using a panel of 12 individual cDNA-expressed human cytochrome P450s (CYPs) (1A1, 1A2, 2A6, 2B6, 2C8, 2C9-Arg, 2C9-Cys, 2C19, 2D6-Met, 2D6-Val, 2E1, and 3A4). Among the CYP isoforms examined, *CYP3A4* showed the highest activity for both of substrates. The metabolism of 7α- and 7β-hydroxy-Δ8-THC to 7-oxo-Δ8-THC was also detected for CYPs 1A1 (4.8% of *CYP3A4*), 1A2 (4.7%), 2A6 (2.3%), 2C8 (16.6%), and 2C9-Cys (5.4%), and CYPs 1A1 (0.4%), 2C8 (1.3%), 2C9-Arg (4.3%), and 2C9-Cys (0.9%), respectively [10].

The importance of various isozyme expressions toward the rate and formation of various cannabinoid metabolites have been reviewed well in the literature [11]. However, no specific reference was found about their relevant pharmacogenomics. The major metabolites of Δ9-THC become detectable in urine within hours after exposure. Urine concentrations depend on the total amount of THC absorbed, frequency of abuse, rate of release from fatty tissue, and time lapse between exposure and specimen collection. Detection of Δ9-THC and its metabolites in blood and urine can be done by high-performance liquid chromatography (HPLC) or by immunoassay.

7.3.1 PHARMACOGENOMICS OF MARIJUANA METABOLISM

CYP3A4 and CYP2C9 enzymes are primarily involved in Δ9-THC metabolism. The *CYP3A4* gene is part of a cluster of cytochrome P450 genes on chromosome 7q21.1 [12]. Over 28 single nucleotide polymorphisms (SNPs) have been identified in the *CYP3A4* gene; for example, 2 polymorphisms, CYP3A4*1B and *3, were connected to pharmacokinetics of the calcineurin inhibitors, cyclosporine and tacrolimus [13]. Variability in *CYP3A4* function can be determined noninvasively by the erythromycin breath test (ERMBT) [14]. The ERMBT estimates *in vivo CYP3A4* activity by measuring the radiolabeled carbon dioxide exhaled after an intravenous dose of (14C-*N*-methyl)-erythromycin. However, none of these polymorphisms have been established to translate into significant interindividual variability of Δ9-THC metabolism *in vivo*. It can be supposed that this may be due to the induction of *CYP3A4* on exposure to substrates.

Similarly multiple polymorphisms of the *CYP2C9* gene have been studied in metabolism of drugs like the anticoagulant warfarin and the anti-epileptic phenytoin. In fact clinical pharmacogenomic analysis of this and other associated genes are now recommended for safe and individualized therapy of the warfarin. Such studies are needed for Δ9-THC metabolism.

Another source of genetic variability, *CYP3A4* is induced by a wide variety of ligands. These ligands bind to the pregnane X receptor (PXR). The activated PXR complex forms a heterodimer with the retinoid X receptor (RXR) that binds to the XREM region of the *CYP3A4* gene. XREM is a regulatory region of the *CYP3A4* gene, and binding causes a cooperative interaction with proximal promoter regions of the

TABLE 7.1
Genes of Possible Pharmacokinetic Relevance to D⁹-Tetrahydrocannabinol (THC)

Gene	Activity
CYP3A4	THC Metabolism
CYP2C9	THC Metabolism
CNR1	Cannabinoid receptor
CNR2	Cannabinoid receptor
FAAH	Endocannabinoid metabolism

gene, resulting in increased transcription and expression of *CYP3A4* [15]. But again no specific studies have been done to show this variability to affect Δ9-THC metabolism.

In summary, *CYP3A4* and *CYP2C9* pharmacogenomics are expected to affect marijuana metabolism and its pharmacological effects. However, more such studies are warranted in this area. Genes of possible pharmacokinetic relevance to THC are summarized in Table 7.1.

7.4 MARIJUANA: MECHANISM OF ACTION

Although marijuana is fat soluble, less than 1% of marijuana active components reach the brain passing the blood-brain barrier. Even though interindividual variability of the blood-brain barrier toward cannabinoid absorption will affect various aspects of the drug's pharmacological effects, there is no literature available for it. As with all psychoactive drugs, it is suspected that the cannabinoids will express their activity by binding to receptors, where endogenous, cannabinoid-like "endorcannabinoids" bind too, probably as part of the body's pain sensing and processing pathway.

7.4.1 MARIJUANA RECEPTORS

As expected, once in the brain, Δ9-THC binds to specific G-protein-coupled cell membrane receptors called cannabinoid receptors (CB) present at the neurons [16–18]. There are currently two known subtypes, termed *central cannabinoid receptor*, CB1, and the peripheral cannabinoid receptor, CB2 [19]. The protein sequences of CB1 and CB2 receptors are about 44% similar [19]. In both are proteins with seven transmembrane spanning domains [20]. These receptors were originally found in rat brain and spleen, respectively. The CB1 receptor is expressed mainly in the central nervous system (CNS); it is also found to a lesser extent in numerous other tissues, such as vas deferens, adrenal gland, heart, lungs, liver, prostate, uterus, ovary, testis, bone marrow, thymus, tonsils, and kidneys [16–21]. The CB2 receptor is mainly expressed in the immune system and hematopoietic cells, specially the B cells [16–21].

The CNS effects of Δ9-THC are mediated by CB1 receptors, which couple primarily to inhibitory G-proteins. The involvement of CB1 receptors in the pain pathway may explain the analgesic effects of the cannabinoids [22–23]. These receptors play an important and not yet fully understood role in hypothalamic and peripheral regulation of food intake, obesity, and metabolism. High levels of CB1 receptors are found in the basal ganglia, hippocampus, cortex, and cerebellum, consistent with the profile of behavioral effects. Studies over the past decade have determined that CB1 receptors undergo downregulation and desensitization following chronic administration of THC or synthetic cannabinoid agonists. In general, these adaptations are regionally widespread and of considerable magnitude, and are thought to contribute to tolerance to cannabinoid-mediated behavioral effects. Adaptation at the effector level has been more difficult to characterize, although it appears that alterations in cyclic adenosine monophosphate (cAMP) and protein kinase A (PKA) activity may be particularly important in cannabinoid dependence. Thus, receptor binding may activate the receptor-mediated adenyl cyclases, expressing the ligands' activity through the second messenger cAMP. A striking characteristic of CB1 receptor adaptation is the region dependence of the magnitude and rate of development of downregulation and desensitization. These regional differences may provide interesting insights into the mechanisms of CB1 receptor signaling in different brain regions. Moreover, region-specific adaptations in CB1 receptors following chronic cannabinoid administration may produce differential adaptations at the *in vivo* level [24].

Cannabinoid receptor type 1 (CB1) receptors are thought to be the most widely expressed G-protein-coupled receptors in the brain. This is due to endocannabinoid-mediated depolarization-induced suppression of inhibition, a very common form of short-term plasticity in which the depolarization of a single neuron induces a reduction in GABA-mediated neurotransmission. Endocannabinoid released from the depolarized neuron binds to CB1 receptors in the presynaptic neuron and causes a reduction in gamma-aminobutyric acid (GABA) release [18]. These receptors are also found in other parts of the body. For instance, in the liver, activation of the CB1 receptor is known to increase *de novo* lipogenesis associated with obesity [25]. Activation of presynaptic CB1 receptors is also known to inhibit sympathetic innervation of blood vessels and contributes to the suppression of the neurogenic vasopressor response in septic shock [26]. A study done on CB1 knockout mice (genetically altered mice that cannot produce CB1) showed an increase in mortality rate. They also displayed suppressed locomotor activity as well as hypoalgesia (decreased pain sensitivity). The CB1 knockout mice responded to Δ9-THC. The anti-inflammatory effects of the cannabinoids, on the other hand, have been suggested to arise from CB2 receptor binding [27]. CB2 receptors are expressed primarily on T cells of the immune system, on macrophages and B cells, and in hematopoietic cells. They also have a function in keratinocytes, and are expressed on mouse preimplantation embryos. It is also expressed on peripheral nerve terminals. Current research suggests that these receptors may also play a role in nociception. In the brain, they are mainly expressed by microglial cells, where their role remains unclear. Structure–activity relationships as regards binding to CB1 versus CB2 show that different cannabinoid analogs bind differently to the two receptors, which may explain the analogs' different pharmacological effects The physical and genetic localization of the CB1 and CB2 genes, *CNR1* and *CNR2,* have been mapped to chromosomes 6 and 1, respectively [28–30].

New evidence suggests that there are additional novel cannabinoid receptors—that is, non-CB1 and non-CB2—that are expressed in endothelial cells and in the CNS. In one study the authors showed the activation of capsaicin-sensitive sensory nerves by Δ9-THC, independent of the CB1 and CB2 receptors [28]. Recent molecular biology research suggested that the orphan receptor GPR55 should in fact be characterized as a cannabinoid receptor, on the basis of sequence homology at the binding site. Subsequent studies showed that GPR55 does respond to cannabinoid ligands. This profile, as a distinct non-CB1/CB2 receptor that responds to a variety of both endogenous and exogenous cannabinoid ligands, has led some groups to suggest GPR55 should be categorized as the CB3 receptor, and this reclassification may follow in time [31]. The existence of several additional CB receptors, including GPR119, has recently been proposed. In addition, minor variations in each receptor have been identified. Cannabinoids bind reversibly and stereoselectively to the cannabinoid receptors. The affinity of an individual cannabinoid to each receptor determines the effect of that cannabinoid. Obviously, cannabinoids that bind more selectively to certain receptors are more desirable for medical usage.

7.4.2 POLYMORPHISMS OF CANNABINOID BINDING RECEPTORS

Many genetic polymorphisms of CB1 receptor gene *(CNR)* and their physiological effects have been described. A number of variations in *CNR* genes have been associated with human disorders including osteoporosis, attention deficit-hyperactivity disorder (ADHD), posttraumatic stress disorder (PTSD), drug dependency, obesity, and depression. Another family of lipid receptors including vanilloid (VR1) and lysophosphatidic acid (LPA) receptors appears to be related to the *CNR*s at the phylogenetic level [32].

Because the abuse of cannabis is frequent among the young and is suspected to precipitate schizophrenia in vulnerable subjects, this association was examined in a French Caucasian population [33]. The CB1 receptor is particularly concentrated in dopamine-modulated areas of the nervous system, and an association between an AAT polymorphism of the CB1 gene *(CNR1)* and intravenous drug abuse was reported. The authors compared the distribution of a single-base polymorphism revealed by *Msp*I within the first exon of the *CNR1* gene in patients with schizophrenia and ethnic- and gender-matched controls. No significant difference was seen in the allele or genotype distribution between the whole sample of schizophrenic patients and controls. However, a borderline lack of allele g and a significant lack of gg genotype in the non-substance-abusing patients compared to substance-abusing patients were found. The authors concluded that further studies are needed to confirm and explore the precise role of the cannabinoid system in schizophrenia.

Another study examined the polymorphisms of CB1 receptor gene in people with Parkinson's disease (PD), because depression is a common symptom in this disease (present in up to 40% of patients). The cause of depression in PD is thought to be related to disturbance of monoamine neurotransmission. The endogenous cannabinoid system mediates different brain processes that play a role in the control of behavior and emotions. Cannabinoid function may be altered in neuropsychiatry diseases, directly or through interactions with monoamine, GABA, and glutamate systems. Depression was more frequent in patients with PD than in controls with osteoarthritis. The presence of depression did not correlate with the stage of the disease, but it was more frequent in patients with pure akinetic syndrome than in those with tremoric or mixed type PD. The CB1 receptor gene polymorphism (AAT) *n* is considered to modify the transcription of the gene; therefore, it may have functional relevance. The authors analyzed the length of the polymorphic triplet (AAT) *n* of the gene that encodes CB1 (CNR1) receptor in 89 subjects (48 PD patients and 41 controls). In patients with PD, the presence of two long alleles, with more than 16 repeated AAT trinucleotides in the CNR1 gene, was associated with a reduced prevalence of depression (Fisher's exact test: $P = 0.003$). This association did not reach significant differences in the control group, but the number of control individuals with depression was too small to allow for statistical analysis. Because the alleles with long expansions may have functional impact in cannabinoid neurotransmission, the results suggest that the pharmacological manipulation of cannabinoid neurotransmission could open a new therapeutic approach for the treatment of depression in PD and possibly in other conditions [34].

Because the CB1 receptor is involved in the involvement of the endogenous cannabinoid system in brain reward mechanisms, another study explored the association of a silent polymorphism (1359G/A; Thr453Thr) in the single coding exon of the CB1 gene (*CNR1*) in 121 severely affected Caucasian alcoholics and 136 most likely nonalcoholic controls [35]. The authors observed a frequency of the A allele was 31.2% for controls and 42.1% for alcoholics with severe withdrawal syndromes ($P = 0.010$). Post hoc exploration indicated that this allelic association resulted from an excess of the homozygous A/A genotype in patients with a history of alcohol delirium ($P = 0.031$, DF 2), suggesting an increased risk of delirium (OR = 2.45, 95% CI 1.14 to 5.25). Thus the homozygous genotype *CNR1* 1359A/A confers vulnerability to alcohol withdrawal delirium.

Another frequent single nucleotide polymorphism (SNP) was identified in the fatty acid amide hydrolase (*FAAH*), a key degradation enzyme of endocannabinoid: P129T polymorphism. This polymorphism was proved to be correlated to a higher body mass index (BMI) in a group of black and white Americans. Aberle et al. explored the association of both the 1359 G/A variant of the *CNR1* and P129T variant in *FAAH* in a group of 451 obese and dyslipidemic participants and observed the biometric and metabolic outcome of a 6-week low-fat diet [36]. While no significance was seen for the 1359 G/A variant, carriers of the P129T mutation in *FAAH* had a significantly greater decrease in triglycerides and total cholesterol as compared to wild type. The authors hypothesized that a hepatic downregulation of endocannabinoid tone may contribute to the observed outcome in studied subjects.

7.4.3 APPLICATION OF MARIJUANA PHARMACOGENOMICS

The pharmacogenomics variability of various aspects of marijuana absorption, disposition, and activity discussed so far opens up enormous prospects of safe and effective treatments not only for marijuana dependence, but also use of cannabinoids as drugs in many diseases. Thus, depending on the SNP-pattern, one could choose among the various cannabinoid antagonists to treat cannabinoid dependence. An antagonist of the CB1 receptor, the experimental drug, SR141716A (Rimonabant), could be used in treating marijuana dependence. SR141716A is used for weight reduction and smoking cessation. Other antagonists suggested are the natural endocannabinoids themselves, for example, anandamide.

7.5 CONCLUSIONS

Marijuana is one of the most commonly abused drugs, and Marinol, the synthetic marijuana, is clinically used in appetite stimulation in terminal cancer patients as well as patients with HIV infection. Marijuana is fat soluble but only a small fraction crosses the blood-brain barrier for exertion of pharmacological activity. Marijuana receptors in the brain play a major role in the mechanism of action of marijuana, and there are significant pharmacogenomics variations in metabolism as well as pharmacological activity of marijuana.

REFERENCES

1. http://www.ncbi.nlm.nih.gov/pubmedhealth/PMH0000403 (2008).
2. Sastre-Garriga J, Villa C, Clissold S, Montalban X. THC and CBD oromucosal spray (Sativex) in the management of spasticity associated with multiple sclerosis. *Expert Rev Neurother* 2011;11:627–637.
3. Williams CM, Kirkham TC. Observational analysis of feeding induced by Δ^9-THC and anandamide. *Physiol Behav* 2002;76:241–250.
4. Hu W. Marijuana patch draws interest of researchers. *New York Times.* January 30, 2000.
5. McGilveray IJ. Pharmacokinetics of cannabinoids. *Pain Res Manage* 2005;10 Suppl A:15A–22A.
6. Tashkin DP, Gliederer F, Rose J, et al. Effects of varying marijuana smoking profile on deposition of tar and absorption of CO and delta-9-THC *Pharmacol Biochem Behav* 1991;40:651–656.
7. Lemberger L, Weiss JL, Watanabe AM, et al. Delta-9-tetrahydrocannabinol—temporal correlation of the psychologic effects and blood levels after various routes of administration. *N Engl J Med* 1972;286:685–688.
8. Bornheim LM, Lasker JM, Raucy JL. Human hepatic microsomal metabolism of delta 1-tetrahydrocannabinol. *DMD* 1992;20:241–246.
9. Huestis MA, Henningfield JE, Cone EJ. Absorption of THC and formation of 11-OH-THC and THCCOOH during and after smoking marijuana. *J Anal Toxicol* 1992;16:276–282.
10. Matsunaga T, Kishi N, Higuchi S, et al. CYP3A4 is a major isoform responsible for oxidation of 7-hydroxy-Δ8-tetrahydrocannabinol to 7-oxo-Δ8-tetrahydrocannabinol in human liver microsomes. *DMD* 2000 November 1;28(11):1291–1296.
11. Maurer HH, Sauer C, Theobald DS. Toxicokinetics of drugs of abuse: current knowledge of the isozymes involved in the human metabolism of tetrahydrocannabinol, cocaine, heroin, morphine, and codeine. *Ther Drug Monit* 2006;28:447–453.
12. Inoue K, Inazawa J, Nakagawa H, et al. Assignment of the human cytochrome P-450 nifedipine oxidase gene (CYP3A4) to chromosome 7 at band q22.1 by fluorescence in situ hybridization. *Jpn J Hum Genet* 1992;37:133–138.
13. Hesselink DA, van Schaik RHN, van der Heiden IP, et al. Genetic polymorphisms of the CYP3A4, CYP3A5, and MDR-1 genes and pharmacokinetics of the calcineurin inhibitors cyclosporine and tacrolimus. *Clin Pharmacol Ther* 2003;74:245–254.
14. Watkins P. Noninvasive tests of CYP3A enzymes. *Pharmacogenetics* 1994;4:171–184.
15. Yang J, Liao M, Shou M, et al. Cytochrome P450 turnover: regulation of synthesis and degradation, methods for determining rates, and implications for the prediction of drug interactions. *Current Drug Metab* 2008;9:384–393.
16. Howlett AC. The cannabinoid receptors. *Prostaglandins Other Lipid Mediat* 2002;68–69:619–631.

17. Mackie K. Cannabinoid receptors: where they are and what they do. *J Neuroendocrinol* 2008;20 Suppl 1:10–14.
18. Graham ES, Ashton JC, Glass M. Cannabinoid receptors: a brief history and "what's hot." *Front Biosci* 2009;14:944–957.
19. Begg M, Pacher P, Bátkai S, et al. Evidence for novel cannabinoid receptors. *Pharmacol Ther* 2005;106:133–145.
20. Matsuda LA, Lolait SJ, Brownstein MJ, et al. Structure of cannabinoid receptor and functional expression in cloned cDNA. *Nature* 1990;346:561–564.
21. Munro S, Thomas KL, Abu-Shaar M. Molecular characterization of a peripheral receptor for cannabinoids. *Nature* 1993;365:61–65.
22. Pertwee RG. Cannabinoid receptors and pain. *Prog Neurobiol* 2001;63:569–611.
23. Morisset V, Ahluwalia J, Nagy I, Urban L. Possible mechanisms of cannabinoid-induced antinociception in the spinal cord. *Eur J Pharmacol* 2001;429:93–100.
24. Sim-Selley LJ. Regulation of cannabinoid CB1 receptors in the central nervous system by chronic cannabinoids. *Crit Rev Neurobiol* 2003;15:91–119.
25. Osei-Hyiaman D, DePetrillo M, Pacher P, et al. Endocannabinoid activation at hepatic CB_1 receptors stimulates fatty acid synthesis and contributes to diet-induced obesity. *J Clin Invest* 2005;115:1298–1305.
26. Godlewski G, Malinowska B, Schlicker E. Presynaptic cannabinoid CB_1 receptors are involved in the inhibition of the neurogenic vasopressor response during septic shock in pithed rats. *Br J Pharmacol* 2004;142:701–708.
27. Galiegue S, Mary S, Marchand J, et al. Expression of central and peripheral cannabinoid receptors in human immune tissues and leukocyte subpopulations. *Eur J Biochem* 1995;232:55–61.
28. Melvin LS, Milne GM, Ross JM, et al. Structure-activity relationships defining the ACD-tricyclic cannabinoids: cannabinoid receptor binding and analgesic activity. *Drug Des Discov* 1995;13:155–166.
29. Rhee M-H, Vogel Z, Barg J, et al. Cannabinol derivatives: binding to cannabinoid receptors and inhibition of adenylcyclase. *J Med Chem* 1997;40:3228–3233.
30. Zygmunt PM, Andersson DA, Hogestatt ED. Δ9-Tetrahydrocannabinol and cannabinol activate capsaicin-sensitive sensory nerves via a CB1 and CB2 cannabinoid receptor-independent mechanism. *J Neurosci* 2002;22:4720–4727.
31. Overton HA, Babbs AJ, Doel SM, et al. Deorphanization of a G protein-coupled receptor for oleoylethanolamide and its use in the discovery of small-molecule hypophagic agents. *Cell Metab* 2006;3:167–175.
32. Onaivi ES. Cannabinoid receptors in brain: pharmacogenetics, neuropharmacology, neurotoxicology, and potential therapeutic applications. *Int Rev of Neurobiol* 2009;88:335–369.
33. Brown AJ. Novel cannabinoid receptors. *Br J Pharmacol* 2007;152:567–575.
34. Barrero FJ, Ampuero I, Morales B, et al. Depression in Parkinson's disease is related to a genetic polymorphism of the cannabinoid receptor gene (CNR1). *Pharmacogenomics J* 2005;5:135–141.
35. Schmidt LG, Samochowiec J, Finckh U, et al. Association of a CB1 cannabinoid receptor gene (*CNR1*) polymorphism with severe alcohol dependence. *Drug Alcohol Dependence* 2002;65:221–224.
36. Aberle J, Fedderwitz I, Klages N, et al. Genetic variation in two proteins of the endocannabinoid system and their influence on body mass index and metabolism under low fat diet. *Horm Metab Res* 2007;39:395–397.

8 Genetic Aspect of Opiate Metabolism and Addiction

Jorge L. Sepulveda

CONTENTS

8.1   Introduction 165
8.2   Structure of Opioids... 166
8.3   Pharmacokinetics of Opioids... 170
        8.3.1   Heroin and Morphine 173
        8.3.2   Codeine.. 174
        8.3.3   Dihydrocodeine 174
        8.3.4   Ethylmorphine 175
        8.3.5   Hydrocodone and Hydromorphone 175
        8.3.6   Oxycodone and Oxymorphone.. 175
        8.3.7   Methadone 176
        8.3.8   Meperidine... 177
        8.3.9   Fentanyl and Analogs 178
        8.3.10 Tramadol.. 178
        8.3.11 Buprenorphine 179
8.4   Pharmacodynamics of Opioids... 179
8.5   Conclusions... 185
References.. 186

8.1   INTRODUCTION

Opiates are alkaloids derived from the poppy plant *Papaver somniferum,* named by Linnaeus in 1753 for its well-known sleep-inducing properties. The first evidence for the use of the opium plant comes from a fourth millennium BC Neolithic site in Switzerland, where *P. somniferum* seeds were found in crop sites. It is likely that these poppies were used for their oil and possibly for their analgesic effects. From these early sites, knowledge of opium spread to the Middle East and subsequently to the East. Although the evidence for use of opium by the Sumerians is highly controversial (1), there is ample mention of opium in the Egyptian "Therapeutic Papyrus of Thebes," dated from 1552 BC, and in an Assyrian Herbal tablet from the seventh century BC. Alexander the Great first and then Arab traders and conquerors spread the opium poppy to east Asia, reaching India in the fourth century BC and China by the fourth century AD. However, recreational use was not prominent in China until the sixteenth and seventeenth centuries, when Portuguese merchants used the

165

more efficient maritime route to significantly expand the volume of opium trade, especially transporting it from India, where it had become widely cultivated, to China.

By the beginning of the nineteenth century, opium addiction was pervasive among the Chinese population, and attempts by the government to restrict the opium trade caused the two Opium wars (1840 to 1842 and 1856 to 1860), won by the British and their Western allies. As a result of these wars, an epidemic of opium addiction quickly spread to affect over 25% of the Chinese population.

The addictive properties of opium were already feared in the fifth century BC by Erasistratus, who proscribed the use of opium. Hippocrates (460 to 357 BC) was well aware of its benefits and limitations and advocated its limited use. In the modern world, by 1878 the British had restricted the selling of opium, and in 1906 the U.S. Congress banned the sale of opium and required contents labeling on patent medicines, many of which contained opium derivatives. During U.S. involvement in Vietnam, the American support for anti-communist groups in Laos and Burma lead to the development of the "Golden Triangle" of opium production, comprising the mountainous regions of Burma, Laos, Vietnam, and Thailand, which became the largest world producer of opium until the Afghan War. Subsequently, illegal heroin flooded the U.S., resulting in an estimated prevalence of 750,000 heroin addicts. In another example of opium links to war, opium growth and trade have increased since the start of the war in Afghanistan and currently underlies the economy of Afghanistan, which currently produces 95% of the world opium.

This historical perspective highlights the main issues about opioid use: on the one hand, these drugs are potent analgesics and became a mainstay of pain treatment; on the other hand, their high potential for secondary effects and addiction mandates caution in their use. Key concepts in opioid pharmacology and pharmacogenetics are summarized in this chapter. While pharmacogenetic testing for opioid treatment is not widely used in clinical practice in the present time, better understanding of common genetic polymorphisms affecting opioid pharmacokinetics and pharmacodynamics will certainly contribute to progress in optimizing the use of these difficult drugs.

8.2   STRUCTURE OF OPIOIDS

The general definition of *opioids* consists of all substances able to interact with the opioid receptors, including the naturally occurring, semisynthetic, or fully synthetic alkaloids, as well as endogenous neuropeptides such as endorphins, enkephalins, dynorphins, and endomorphins, which will not be further discussed. The alkaloid opioids are characterized by at least two aromatic rings and their alkaline nature, typically resulting in a protonated molecule at physiologic pH. *Opiates* are those naturally occurring alkaloids found in the poppy resin, primarily morphine, codeine, and thebaine and their semisynthetic derivatives. By extension, naturally occurring opioids found in other plants, such as salvinorin A, may also be considered opiates, although they are rarely analyzed. Semisynthetic opiates are derived from natural opiates by chemical modification. The first semisynthetic opioid was synthesized in 1874 by heating morphine with acetyl anhydride to create diacetylmorphine, which was named *heroin*. Acetylation of morphine significantly increased the ability to penetrate the blood-brain barrier, thereby imparting higher analgesic and psychotropic effects to heroin. Other semisynthetic and synthetic opioids are described in Table 8.1. Relatively small changes in chemical structure significantly affect the pharmacokinetics (disposition and metabolism, see Table 8.2) and pharmacodynamics (receptor interaction, potency, see Table 8.5), of opioids.

8.3 PHARMACOKINETICS OF OPIOIDS

Pharmacokinetics describes the absorption, disposition, metabolism, and elimination of a drug. Some opioids have low oral bioavailability (<30%), including morphine, fentanyl and its derivatives, and buprenorphine, which require parenteric routes (e.g., sublingual, subcutaneous, transdermal) to achieve therapeutic levels. The absorption and distribution of opioids is affected by intestinal transporters, including P-glycoprotein, or multidrug resistance protein (mdr), coded by the *ABCB1* gene. This protein is a member of the ATP-binding cassette (ABC) class of transmembrane transporters, which play a role in drug absorption by promoting drug efflux into the intestinal lumen (2). In addition, and perhaps more importantly, these efflux transporters are also present in the brain capillaries and affect the distribution of the drug into the brain tissue, therefore contributing to the blood-brain barrier. This effect is clearly demonstrated in the negative correlation between the analgesic effects of morphine, fentanyl, and other opioids, and the abundance of *ABCB1* in model animals (3–6). These central effects explain the observation that polymorphisms in the *ABCB1* gene affect the response to some opioids, even those administered by parenteric route, such as fentanyl (7,8). *ABCB1* polymorphisms appear to make a small contribution to the wide interindividual variability in methadone efficacy (9–12). Opioids with decreased P-glycoprotein interaction, such as buprenorphine, may be preferable in patients with opioid dependence or tolerance (13,14), especially those with *ABCB1* polymorphisms associated with increased P-glycoprotein activity (15).

The three most common single nucleotide polymorphisms (SNPs) in the *ABCB1* gene (1236C>T, 2677G>T, and 3435C>T) have been associated with changes in the activity of P-glycoprotein, with the 1236T/2677G/3435T haplotype carriers demonstrating higher activity than noncarriers (15). In agreement with this finding, patients with homozygous 1236T/2677G/3435T haplotype, associated with higher P-glycoprotein efflux activity, showed up to fivefold increased probability of high methadone dose requirements (12). In contrast with several publications describing a contribution of *ABCB1* to methadone response (9–12), other studies did not show an effect of *ABCB1* polymorphisms on methadone dose requirements (9,16,17). To add to the difficulty in interpreting genotyping findings, the activity of *ABCB1* is subject to a multitude of inducers and repressors (18). For example, inhibition of ABCB1 by quetiapine resulted in increased levels of methadone, particularly in those patients with the most active alleles of *ABCB1* (19).

All opioids are metabolized to varying extents (see [20] for an excellent review on the metabolism of opioids), and the rates of metabolism and elimination through the renal and hepatobiliary routes determine the half-life of the drugs in circulation (Table 8.2). Practical clinical consequences of understanding opioid metabolism and pharmacokinetics include knowledge of the duration of therapeutic effect, detectability of the drug in urine screen tests, and proper interpretation of detected

TABLE 8.1

Opioid Classification

Opioid Class	Common Name	Selected Trade® and Street Names	Drug Enforcement Administration (DEA) Schedule	Chemical Class
Natural opiates	Morphine	Duramorph®; M, Junk, Morpho, Morphia, White Stuff, Methadose	II	Phenanthrene
	Codeine	®: Empirin, Robitussin, etc.; Rabo, School Boy	II[b]	
Semisynthetic opiates	Heroin	Diacetylmorphine, Horse, Stuff, Smack, Junk, Skag	I	
	Hydromorphone	Dihydromorphinone; ®: Dilaudid, Dimorphone, Hymorphan, Hydal, Laudicon, Novolaudon, Palladone	II	
	Hydrocodone	Dihydrocodeinone; ®: Vicodin, Dicodid, Hydromet, Hydrocet, Lortab, Orthoxycol, Norco, Zydone, etc.	II[a], III	
	Dihydrocodeine	6-α-Hydrocodol; ®: Codhydrine, Dehacodin, Dico, Fortuss, Hydrocodin, Nadeine, Novicodin, Panlor,etc.	III	
	Oxycodone	OxyContin®, Percodan®; Oxy, Oc, Oxycotton, Killer	II	
	Oxymorphone	Numorphan®, Opana®; Blues, Blue Morphine, Biscuits	II	
	Buprenorphine	®: Buprenex, Leptan, Suboxone, Subutex, Temgesic	III	
	Dihydromorphine	Paramorfan®	I	
	Desomorphine	Permonid®	I	
	Nicomorphine	Vilan®	I	
	Dipropionyl-morphine	Dipropanoylmorphine	I	
	Benzylmorphine	Peronine®	I	

TABLE 8.1 (continued)
Opioid Classification

Opioid Class	Common Name	Selected Trade® and Street Names	Drug Enforcement Administration (DEA) Schedule	Chemical Class
	Ethylmorphine	Codethyline, Dionine, Ethyl Morphine	II[a]	
	Ethorphine	Immobilon®, M99®	I[c]	
Synthetic opioids	Fentanyl	®: Duragesic, Fentanest, Fentora, Ionsys, Leptanal, Pentanyl, Sublimaze	II	Phenyl-piperidines
	Sufentanyl	Sufenta®	II	
	Carfentanyl	Wildnil®	I[c]	
	Meperidine	Fortis®, Demies®	II	
	Methadone	®: Symoron, Dolophine, Amidone, Methadose, Physeptone, Heptadon, Phy	II	Phenylheptylamines
	Propoxyphene	Darvon®, Dolocap®, Proxagesic®; Yellow, Footballs	IV	
	Levorphanol	®: Levorphan, Lemoran, Racemorphan, Methorphinan, Dromoran	II	Morphinans
	Butorphanol	Beforal®, Moradol®, Torbugesic®, various generics	IV	
	Nalbuphine	Nubain®	None[d]	
	Tramadol	®: Topalgic, Tramal, Tramol, Ultram, Zamudol, etc.	None[d]	Cyclohexanol
	Tapentadol	Nucynta®	II	Phenol
	Pentazocine	®: Fortral, Fortwin, Talwin, Talcen; T's	IV	Benzomorphan

[a] In bulk quantities or stand-alone product.
[b] In doses ≥90 mg.
[c] Allowed for veterinary use only.
[d] Scheduled in some states.

TABLE 8.2

Some Pharmacokinetic Parameters for Common Opioids

Common Name	$t_{1/2}$ (hours)	Detectability in Urine (Days)		Metabolites[a]
		Single Use	Chronic Use	
Heroin	1.7–6 min	0.3–1 (6AM)		**Unch** (0.1), **6AM** (1.3), **3AM, Morphine** (4, 1st 60′), Morphine© (38), Normorphine (minor)
Morphine	1–7	1–2	4	**Unch** (10), M3G (75); **M6G** (<1); Sulfate (<1); **Hydromorphone** (<3%, only if large doses); Nor (1); Nor© (4)
Codeine	2–4	1–2		Unch (5–20), © (32–46), Nor© (10–21); **Morphine** (5–13), **M6G, Hydrocodone** (minor)
Dihydrocodeine	3–5	1–2		**Unch** (31), © (28), **Nor** (19), **Dihydromorphine** (9), **Dihydronormorphine** (1.8), **Hydrocodone** (0.2)
Ethylmorphine	2–3			**Unch** (4), EM6G (41), M3G (12), Norethyl-M6G (5), Normorphine (5), Norethylmorphine (4), **M6G** (3), **Morphine** (0.8)
Hydrocodone	3–9	1–2	4	**Unch** (12); **Hydromorphone** (4), **Nor** (5), **Dihydrocodeine** (3)
Hydromorphone	1.5–4	1–2	4	**Unch** (6); Glucuronides (30), **Dihydrocodeine** (trace)
Oxycodone	2–3 IR 4–8 CR	1–1.5 IR 1.5–3 CR	4	**Unch** (13–19), © (7–29), Oxymorphone© (14), Nor (?)
Oxymorphone	3–12 IR 7–21 CR	1.5–2.5 IR 1–4 CR		**Unch** (2), © (44), 6α-Oxymorphol (3)
Methadone	12–55	3–11		**Unch** (5, Acid Urine: 22–50); EDDP (5–25, Acid Urine: 2); EMDP (<1)
Meperidine	2–5	0.5–1	3	**Unch** (7); **Nor** (4–23), Meperidinic acid, Normeperidinic acid
Fentanyl	3–29	1–3		**Unch** (0.4–6), Nor (26–55), Despropionyl (plasma only), Hydroxy (?), Hydroxynor (?)
Sufentanyl	1.5–6	0.3–1.5		**Unch** (2), N-Desalkyl + O-Desmethyl (2)
Tramadol	4–7	0.3–2		**Unch** (29), **O-Desmethyl** (20), O-Desmethyl-Nor (20), Nor (17), Dinor (0.4), O-Desmethyldinor (1)
Buprenorphine	1–7	0.5–1		**Unch** (1), **Nor** (3); © (10) ©-Nor (10)
Propoxyphene	3–24	0.3–2	5	**Unch Dextro** (1), **Nordextro** (13), Dinorpropoxyphene (0.7)
Levorphanol	11–16	2–3		**Unch** (7), **Nor** (2.5), 3-Glucuronide (31), Nor© (14)

TABLE 8.2 (continued)
Some Pharmacokinetic Parameters for Common Opioids

		Detectability in Urine (Days)		
		---	---	
Common Name	$t_{1/2}$ (hours)	Single Use	Chronic Use	Metabolites[a]
Butorphanol	3–8	1–2		**Unch** (5–10), Hydroxy (36), **Nor** (4), Nor© (4)
Nalbuphine	2–8	0.3–2		**Unch** + Nor + © (71)
Tapentadol	4	1–2		**Unch** (3), © (70), *N*-Desmethyl (13), Hydroxy (2)
Pentazocine	2–4	1–2		Free and ©: **Unch** (10), Hydroxy (12), Carboxy (40)

Note: Parameters can be altered with dosing amount, frequency, and route, tolerance, genetic variation, concurrent medications, and various pathologies, such as kidney or liver disease.

[a] **Bold,** active metabolite; numbers in parenthesis represent typical percentage (%) of a dose in urine; data adapted predominantly from Baselt, R.C., *Disposition of Toxic Drugs and Chemicals in Man*, 8 ed., Biomedical Publications, Foster City, CA, 2008; $t_{1/2}$ = half-life of parent drug in blood. *Detectability* is the time interval that the urine concentration of drug or metabolites remains above a typical cutoff level in an average individual. *Unch*, unchanged, free parent drug; *M3G*, morphine-3-glucuronide; *M6G*, morphine-6-glucuronide; *6AM*, 6-acetylmorphine; *3AM*, 3-acetylmorphine; *EM3G*, ethylmorphine-6-glucoronide; *EDDP*, 2-ethylidene-1,5-dimethyl-3,3-diphenylpyrrolidine; *EMDP*, 2-ethyl-5-methyl-3,3-diphenylpyrroline; *Nor*, Nor-metabolite of the parent drug; other prefixes likewise refer to metabolites of the parent drug; ©, glucuronide and sulfate conjugates. *IR*, immediate release tablets; *CR*, controlled release tablets.

urine metabolites (Table 8.2). In general, metabolism of drugs can be classified into two phases:

Phase I—The drug undergoes nonsynthetic chemical modification, typically oxidation, reduction, or hydrolysis. The most important enzymes catalyzing these reactions are the P450 cytochrome containing enzymes coded by the *CYP450* genes, which are most abundant in the liver microsomes. Table 8.3 summarizes CYP450 enzymes of importance in opioid metabolism.

Phase II—The drug is conjugated in a synthetic reaction with an endogenous moiety (glucuronate, glycine, and sulfate) that facilitates renal elimination by increasing solubility. Opioids such as morphine, codeine, hydromorphone, and oxymorphone are largely eliminated by glucuronide conjugation, predominantly dependent on the glucuronyl-transferase coded by *UGT2B7* (Table 8.3).

The major *CYP450* genes involved in opioid metabolism include *CYP2A4* and *CYP2D6*. The *CYP3A4* gene is highly inducible by pregnane X receptor (PXR) ligands, which include a large variety of substances, such as steroids, bile acids,

TABLE 8.3
Genes of Interest in Opioid Metabolism

Common Name	3A4	2D6	2B6	1A2	Glucuronidases	Others
Heroin						Esterases
Morphine	(↓)				↓	2C8 (↓)
Codeine	(↓)	↑			↓	
Dihydrocodeine	↓	↑			↓	
Ethylmorphine	↓	↑			↓	
Hydrocodone	(↓)	↑				
Hydromorphone	↓				↓	
Oxycodone	↓	↑			↓	
Oxymorphone	↓				↓	
Methadone	↓	(↓)	↓[a]	(↓)		2C19 (↓)[b]
Meperidine	(↓)		↓			2C19 (↓)
Fentanyl, Sufentanyl, Alfentanyl	↓					
Tramadol	↓	↑	(↓)			
Buprenorphine	↓				↓	2C8 (↓)
Propoxyphene	↓					
Levorphanol	(↓)				↓	
Nalbuphine					↓	
Tapentadol					↓	
Pentazocine					↓	1A2 (↓) ?

Notes: (↓) small role in inactivation; ↓ major inactivating enzyme; ↑ activation or conversion to a more active metabolite.

[a] S-methadone only.

[b] R-methadone only.

St. John's Wort herbs, rifampicin, phenytoin, carbamazepine, barbiturates, and reverse transcriptase inhibitors such as efavirenz and nevirapine, and modafinil. On the other hand, *CYP3A4* is potently inhibited by grapefruit juice and drugs such as nefazodone, protease inhibitors (ritonavir, indinavir, and nelfinavir), and macrolides such as erythromycin, chloramphenicol, azole antifungals, aprepitant, and verapamil. These effects account for the large number of potential drug interactions among *CYP3A4* substrates. In contrast, *CYP3A4* polymorphisms do not appear to be of major concern in predicting opioid response, although recent reports have associated the *CYP3A4*1G* (20230G>A), a hypofunctional intronic SNP present in about 18% to 26% of Asians, with lower fentanyl dose requirements (21,22).

With a spectrum of substrates similar to CYP3A4, the related CYP3A5 and CYP3A7 enzymes also contribute to CYP3A activity. For example, both of these enzymes play a role in N-demethylation of codeine, ethylmorphine, and alfentanyl (23,24). CYP3A7 is predominant in the fetal liver, but it can play a minor role in the adult as well. This redundancy of genes contributing to CYP3A4 activity is one reason why genetic variability in *CYP3A4* does not commonly result in significant variability in the metabolism of CYP3A substrates. On the other hand, the CYP3A enzymes tend to be coinhibited by the same spectrum of drugs; therefore drug interactions become major factors in predicting the pharmacokinetics of CYP3A4 substrates.

In contrast to *CYP3A4*, the *CYP2D6* gene is highly polymorphic, with 126 allelic variations acknowledged by the Human Cytochrome P450 Allele Nomenclature Committee (http://www.cypalleles.ki.se/cyp2a6.htm). *CYP2D6* variants can be classified into null alleles, with no enzymatic activity (*3–8, *11–16, *18–21, *38, *40, *42, *44, *56, and *62), partial functional (*9, *10, *17, *29, *36, and *41), and fully functional alleles (*1, *2, *35). The *17 variant can have decreased activity against some drugs, such as dextromethorphan, but normal activity against other drugs, such as codeine (25). Depending on the activity of the CYP2D6 enzyme, individuals are grouped into poor metabolizers (PMs), who have no active alleles; intermediate metabolizers (IMs), with one or two partially active alleles; extensive metabolizers (EMs), with one or two active alleles; and ultrarapid metabolizers (UMs), with more than two active alleles. CYP2D6 can be inhibited by drugs such as quinidine, cinacalcet, citalopram, fluoxetine, paroxetine, and sertraline (in high doses only), but in contrast to CYP3A4, CYP2D6 is not significantly inducible. The closely related *CYP2D7* gene is generally considered a pseudogene; however, a rare frame shift mutation can cause an active enzyme to be produced that is able to metabolize codeine to morphine (26).

The UGT2B7 glucuronyl transferase is the major enzyme responsible for opioid glucuronidation (27), with UGT1A1 and UGT1A3 making a small contribution, particularly in the case of buprenorphine glucuronidation (28). The *UGT2B7-840G* allele was associated with significantly reduced glucuronidation of morphine (29) and the 802C>T (*2) genotype was associated with lower morphine levels, higher morphine-glucuronide concentrations, and decreased adverse effects (30,31), although other studies showed lack of an effect of *UGT2B7* polymorphisms on morphine glucuronidation rates (32,33). One possible explanation for these contradictory results is that only certain haplotypes associated with *UGT2B7* mRNA expression result in changes in enzymatic efficiency (34). In addition to the potential role of polymorphisms, the UGT2B7 enzyme is inhibited by drugs such as ketoconazole and fluconazole and induced by a variety of xenobiotics, including polycyclic aromatic hydrocarbons and antioxidants.

8.3.1 HEROIN AND MORPHINE

Heroin is converted to active metabolites 6-acetylmorphine and morphine through the action of various plasma, liver, and brain esterases. It is not well understood to what extent genetic variability in esterase activity affects the response to heroin, although one report described the lack of heroin hydrolysis in the plasma of a cholinesterase-null individual (35). Morphine is predominantly eliminated by glucuronide conjugation into the inactive metabolite morphine-3-glucuronide and to a minor extent into the active metabolite morphine-6-glucuronide (Table 8.2) by the action of UGT2B7 with a minor contribution from UGT1A1. The role of genetic variability in these two glucuronyl-transferases in the response to heroin and morphine is controversial (see above). A small amount of morphine is *N*-demethylated to inactive normorphine by CYP3A4 and CYP2C8 (36) and in large doses, up to 3% of morphine can be metabolized to hydromorphone by an unknown mechanism (Table 8.2). Interestingly, small amounts of endogenous morphine are synthesized by the leukocytes, liver, and brain, from tyramine, norlaudanosoline, reticulin, and codeine through the action of CYP2D6 and CYP3A4 (37). Polymorphisms in CYP450 enzymes do not appear to significantly affect the response to morphine or heroin.

8.3.2 CODEINE

The majority of codeine is inactivated by glucuronidation to codeine-6-glucuronide by the action of UGT2B7. About 0.5% to 10% of codeine is demethylated by CYP2D6 to morphine, a step that appears required for the full analgesic effects of codeine. CYP2D6 variants significantly affect the morphine-to-codeine (M/C) ratio, with PMs showing urinary excretion ratios at 12 hours between 0.17% and 0.4% while EMs have M/C ratios of 2% to 9%, and UMs have ratios up to 17% (38). After 12 hours, UMs and EMs may have M/C ratios exceeding one (39), which can also be seen in individuals consuming heroin. Local conversion of codeine to morphine in the brain via CYP2D6 may explain the analgesic effect of codeine despite low plasma concentrations of morphine. CYP2D6 poor metabolizers are unable to convert codeine to morphine efficiently and therefore may not experience pain relief (40). Ultrametabolizers may convert codeine too efficiently into morphine leading to possible central nervous system (CNS) and respiratory depression, especially when combined with other predisposing factors such as renal failure and CYP3A4 inhibition (41), or with UGT2B7 deficiency in mothers of breast-feeding infants (42). The effect of CYP3A4 is explained by the fact that a minor amount of codeine is *N*-demethylated to inactive norcodeine via CYP3A4, and therefore inhibition of CYP3A4 activity (usually by coadministered drugs that act on CYP3A4), in conjunction with CYP2D6-catalyzed enhanced conversion to morphine, can increase the toxicity of codeine (41).

Genotyping for CYP2D6 can correctly predict the rate of morphine formation, especially when supplemented by knowledge of coadministered drugs and CYP2D6 phenotyping by determining the dextromethorphan metabolic ratio (43), but explains only about 50% of patients with codeine ineffectiveness. This lack of predictive value of *CYP2D6* genotyping could be due to the possibility that the major codeine metabolite, codeine-6-glucuronide, is an active metabolite, similarly to the morphine-6-glucuronide (44). However, due to the risk of toxicity, it is reasonable to avoid codeine in CYP2D6 ultrarapid metabolizers, especially in patients with renal insufficiency, with coadministered CYP3A4 inhibitors, and in breast-feeding mothers (41,42).

8.3.3 DIHYDROCODEINE

The metabolism of dihydrocodeine is similar to codeine: a small amount of dihydrocodeine is *O*-demethylated by CYP2D6 to dihydromorphine, and excretion of this metabolite is impaired in CYP2D6 PMs (45). However, the main urinary metabolites are composed of the parent drug and its conjugates, as well as of norhydrocodeine produced by *N*-demethylation catalyzed by CYP3A4 (46). A small amount of hydrocodone is also produced from dihydrocodeine (Table 8.2). Given the various elimination pathways, genetic variation in CYP450 enzymes does not appear to significantly affect the pharmacokinetics of dihydrocodeine.

8.3.4 ETHYLMORPHINE

Ethylmorphine is converted to morphine by *O*-de-ethylation through the action of CYP2D6, and consequently PMs have higher ethylmorphine-to-morphine ratios than EMs (47). Regardless of the CYP2D6 genotype, more morphine is generated from ethylmorphine than from an identical dose of codeine (47). Ethylmorphine is also converted by *N*-demethylation to inactive norethylmorphine by the action of CYP3A4 (48).

8.3.5 HYDROCODONE AND HYDROMORPHONE

Hydrocodone is converted to hydromorphone by CYP2D6 (49), and while both drugs show analgesic effects, hydromorphone is about five times more potent at binding the μ opioid receptor (MOP). In addition, hydrocodone is metabolized by CYP3A4 to the inactive metabolite norhydrocodone, and a small amount is converted to dihydrocodeine. Very little hydrocodone is eliminated by glucuronide conjugation. The importance of CYP2D6 activity in the elimination of hydrocodone, by conversion to hydromorphone and subsequent glucuronidation, is highlighted by the case of fatal hydrocodone toxicity in a 5-year-old child with a CYP2D6 PM genotype (*2A/*41) who also received a potent CYP3A4 inhibitor, clarithromycin (50). The lack of CYP2D6 activity, combined with reduced CYP3A4 activity, presumably resulted in accumulation of hydrocodone and consequent toxicity. On the other hand, CYP2D6 ultrametabolizers may also experience increased toxicity due to enhanced conversion to hydromorphone and increased opioid effects (51).

8.3.6 OXYCODONE AND OXYMORPHONE

Oxycodone is similar to hydrocodone and is likewise metabolized by CYP2D6 to oxymorphone, a stronger analgesic, and by CYP3A4 to noroxycodone, an inactive or very weak analgesic. Oxymorphone is also converted to inactive noroxymorphone by CYP3A4 and eliminated by glucuronide conjugation. Conversion to oxymorphone may reduce oxycodone toxicity, as individuals with poor CYP2D6 metabolism, either with a PM genotype or cotreated with inhibitors such as paroxetine or fluoxetine, are overrepresented in cases of fatal oxycodone overdose (52). One reported case described a patient with CYP2D6 PM genotype who had poor tolerance and limited response to oxycodone therapy, associated with higher than expected levels of oxycodone and noroxycodone and with lower levels of oxymorphone (53). In contrast, a randomized study failed to demonstrate a critical role of CYP2D6 inhibition alone in oxycodone pharmacokinetics, while it showed that combined inhibition of CYP2D6 and CYP3A4 did result in substantial increases in oxycodone levels (54,55). Another study showed that inhibition of CYP3A4 resulted in increased pharmacodynamic response to oxycodone, an effect probably mediated by increased conversion to oxymorphone (56). Finally, a double-blind placebo-controlled study in 10 healthy volunteers genotyped for *CYP2D6* showed that the level of oxymorphone, decreased by CYP2D6 inhibition and increased in CYP2D6 UMs or CYP3A4 inhibited individuals, was the major determinant of the analgesic response to oxycodone (57).

8.3.7 METHADONE

Methadone maintenance therapy is extensively used for the treatment of opioid dependence, and given its safety, it is widely regarded to outweigh the harms of illicit drug use. However, methadone's use is complicated and potentially dangerous because of the variability of its metabolism in populations, its interaction with other medications (e.g., antidepressants and anticonvulsants) commonly used to treat chronic pain (58), its association with potential cardiac effects (59), and the lack of clearly developed guidelines to help convert from other opioids to methadone therapy. Therapeutic plasma methadone concentrations greater than 200 to 400 ng/mL are usually necessary for therapeutic effects and prevention of opioid withdrawal symptoms. However, there is up to a 17-fold variation in plasma concentrations of methadone after a given dose, some of it depending on variation in metabolism by CYP450 enzymes in the liver (60). Clearly, it is of major interest to fully understand the genetic and environmental factors contributing to variability in response to methadone therapy.

Methadone can be metabolized *in vitro* to the inactive metabolite 2-ethylidene-1,5-dimethyl-3,3-diphenylpyrrolidine (EDDP) by *N*-demethylation by liver CYP2A1, 2C9, 2C19, 2B6, 2D6, 3A4, and 3A5 (61). Because methadone is typically administered as a racemic mixture of (R)- and (S)-enantiomers, it is of interest that CYP2C19 metabolizes only the (R)-enantiomer, while CYP2B6 acts predominantly on the inactive (S)-enantiomer (62). A pharmacogenetics study in Switzerland (63) suggested that CYP3A4 and CYP2B6 are the major contributors to plasma levels of methadone, while CYP2D6 is a minor contributor to *in vivo* metabolism. Polymorphisms in other CYP enzymes (1A2, 2C9, 2C19) and other genes implicated in methadone metabolism (ABCB1, UGT2B7) did not affect serum methadone levels in this study. A role for hepatic CYP3A4/5 activity on methadone metabolism was suggested by the correlation between higher methadone levels and metabolism of midazolam, a CYP3A4 and CYP3A5 substrate (63). Due to the rarity of hypofunctional alleles of *CYP3A4* in the studied populations, there is no reported correlation between *CYP3A4* genotypes and (R)-methadone metabolism, although Crettol et al. showed a 1.5-fold increase in S-methadone levels in individuals with the *CYP3A4*1B* genotype, a rare allele that is associated with decreased expression of the enzyme (63).

The contribution of CYP2D6 variation to methadone metabolism is controversial, with Coller et al. (64) finding no effect, while Crettol et al. (63) found lower levels in CYP2D6 UMs. In a recent study, CYP2D6 UMs responded poorly to methadone and required higher doses, while PMs required slightly lower doses despite doubled methadone steady-state levels (9). Importantly, this enzyme has been associated with higher levels of methadone when inhibited by other coadministered drugs, such as paroxetine (58), commonly used to treat depression in patients with chronic pain. This effect may be augmented in patients with hypofunctional alleles of *CYP2D6*, especially in combination with hypofunctional alleles of other CYP enzymes involved in methadone metabolism. *CYP2D6* PM alleles are underrepresented in oral opioid-dependent patients (65), suggesting that these patients need high levels of opioids to overcome their rapid metabolism and achieve effective pain relief. Alternatively, addictive behavior may be associated with high pain sensitivity related to low levels of endogenous opiates resulting from their fast metabolism, or to decreased synthesis of endogenous morphine mediated by CYP2D6 (66). On the other hand, the frequency of PM alleles is increased in methadone and other opioid-related fatalities (67), suggesting that CYP2D6 metabolism is protective against accumulation of toxic levels of opioids.

CYP2B6 is more selective toward the inactive S-methadone enantiomer and does not play a major role in dose requirements, although Fonseca et al. showed that the slow metabolizer *CYP2B6*6/*6* genotype was associated with higher levels of (S)-methadone and lower dose requirements (9). Another study showed that *CYP2D6 *6/*6* individuals were overrepresented in methadone-associated fatalities and had higher methadone levels (68). It is possible that patients with high S-methadone have increased adverse effects associated with this enantiomer (tension, fatigue, confusion) and do not tolerate higher doses of the racemic methadone.

An important issue in methadone pharmacogenetics is the prediction of cardiac arrhythmias associated with QT prolongation. The potassium voltage-gated channel subunit alpha (KCNH2) channel, also known as hERG, mediates the "rapid" delayed rectifier K+ current (IKr), which plays an important role in ventricular repolarization. Mutations in this gene or inhibition by various drugs, including antipsychotic agents, quinolones, and macrolides, can result in cardiac arrhythmias, such as prolonged QT and *torsades de pointes*. Methadone directly inhibits KCNH2 (MIRK) channels (69), a postulated mechanism for its association with the long QT syndrome (70). Overall, methadone causes moderate prolongation of the QT interval, but in a minority of patients (around 2%) it may result in increased risk for severe arrhythmias and death, resulting in the recommendation that EKG screening should be performed premethadone treatment and subsequently (71). Interestingly, the S-enantiomer appears to have a 3.5-fold higher inhibitory effect than the R-enantiomer, and CYP2B6 poor metabolizers were associated with increased heart-rate-corrected QT interval (QTc) (72), which can be explained by the fact that CYP2B6 metabolizes only the S-enantiomer. These results suggest that *CYP2B6* genotyping may predict QT prolongation by racemic methadone therapy and that pure (R)-methadone may be a safer alternative. On the other hand, genotyping for *KCNH2* and *CYP2B6* polymorphisms may provide another predictor of risk for prolonged QT syndrome in patients with normal or borderline QT.

8.3.8 MEPERIDINE

Meperidine is metabolized in the liver by hydrolysis to meperidinic acid, and by *N*-demethylation to normeperidine through the action of CYP2B6 with a smaller contribution by CYP2C19 and CYP3A4 (73). Normeperidine as an analgesic is half as potent as meperidine, but it has significantly higher CNS stimulation properties potentially leading to toxic effects such as convulsive seizures. The impact of *CYP2B6* polymorphisms in meperidine efficacy and toxicity has not been extensively studied, although in one published dissertation describing pharmacogenetic studies in 49 children, *CYP2B6* variants appeared to affect the effectiveness of meperidine for oral sedation (74).

8.3.9 FENTANYL AND ANALOGS

Fentanyl and its analogs alfentanyl and sufentanyl are predominantly metabolized by oxidative *N*-dealkylation by CYP3A enzymes to the inactive and nontoxic metabolites norfentanyl, noralfentanyl, and norsufentanyl, respectively (75). A polymorphism in the *CYP3A4* gene (*1G), particularly when linked to a *CYP3A5*3* hypofunctional allele, was associated with lower fentanyl dose requirements in Asian populations (21,22,76). In agreement with these data, a pharmacogenetic study of fentanyl-related fatalities in Milwaukee County showed higher mean fentanyl levels in homozygous *CYP3A5*3* cases (77). The importance of drug interactions inhibiting CYP3A activity is highlighted by a report of a case of delirium induced by fentanyl in a patient with concomitantly administered diltiazem (78). In contrast, the short-acting drug remifentanyl is not metabolized by CYP3A enzymes, but rather by liver carboxylases, and less interindividual variation is expected (79).

8.3.10 TRAMADOL

Tramadol is available as a racemic mixture of the (1R,2R)-(+) and (1S,2S)-(–) enantiomers, similarly to methadone, but in contrast to methadone, analgesic activity requires *O*-demethylation of tramadol by CYP2D6 to form the active compound M1 (*O*-desmethyl-tramadol), which binds the MOP with 200- to 300-fold higher affinity than the parent drug. Although both enantiomers are metabolized, CYP2D6 has some preference for the (+)-enantiomer (80). The (+) M1 metabolite is about 10-fold more potent as an analgesic than the (–)-enantiomer. Interestingly, the parent drug (+)-tramadol is more potent as a serotonin reuptake inhibitor, and (–)-tramadol is more active as a norepinephrine reuptake inhibitor than the M1 metabolites, and these effects contribute to analgesia by inhibiting pain transmission in the spinal cord. These properties of tramadol lead to an increased risk of serotonin toxicity with concomitant administration of serotoninergic drugs or in individuals carrying polymorphisms in the serotonin transporter *SLC6A4* (81), and may be a consideration in individuals with genetic changes in *CYP2D6* affecting the ratios of tramadol/M1.

In competition with the *O*-demethylation reaction to form the active compound, the (+) enantiomer of tramadol also undergoes *N*-demethylation to M2 (*N*-desmethyl-tramadol), mediated by CYP3A4 and CYP2B6 (82). The M1 and M2 metabolites are further inactivated by CYP2D6 and phase II conjugation to sulfates and glucuronides. CYP2D6 PMs have increased levels of (+) and (–) tramadol, M1 and M2 metabolites, and tramadol/M1 ratios (83). In a study in the Faroe Islands, where the prevalence of CYP2D6 PMs is double that of other Europeans, individuals with the *CYP2D6* PM genotype had levels of (+)-M1 and (+)-M1/(+)-tramadol ratios 14 times lower than EMs (84). This decreased metabolism to the active drug by impaired CYP2D6 activity resulted in poor analgesic response (85), higher dose requirements, and increased adverse effects, as observed in carriers of the *CYP2D6*10* allele (86,87). In a contrasting report, CYP2D6 UMs were more sensitive to the analgesic effects of tramadol and had higher propensity for nausea, an adverse effect related to opioid receptor stimulation (88). In another study, the risk of nausea and vomiting was increased in CYP2D6 EMs versus intermediate metabolizers, and the effect was amplified by OPRM1 polymorphisms associated with higher levels of the MOP (89). A small role for ABCB1 polymorphism in determining the bioavailability of tramadol has also been demonstrated (90). In summary, it appears that the CYP2D6 genotype, in conjunction with modifying genes, is a major determinant of tramadol effectiveness and toxicity.

8.3.11 BUPRENORPHINE

Buprenorphine is a partial MOP agonist and a κ opioid receptor (KOP) antagonist with a good safety profile. It is a very potent analgesic at lower doses, but its analgesic potential decreases at higher doses due to nociceptin receptor (NOP) stimulation (91). Buprenorphine is metabolized by *N*-dealkylation via CYP3A4 with a minor contribution of CYP2C8 into norbuprenorphine, which is also active at the MOP, δ opioid receptor (DOP), and NOP, and a KOP antagonist. Importantly, norbuprenorphine is 10 times more active as a respiratory depressant. Although CYP3A4 is the major enzyme responsible for buprenorphine metabolism, inhibition of CYP3A4 with flunitrazepam (92) or ketoconazole (93) does not appear to significantly affect buprenorphine concentrations. In contrast, induction of CYP3A4 with antiretrovirals such as efavirenz and delavirdine resulted in a decrease in buprenorphine levels, but the effect may not be clinically significant (94). Despite these reassuring findings, caution is recommended when coadministering buprenorphine with drugs that affect CYP3A4 activity (95). Both buprenorphine and norbuprenorphine are glucuronide conjugated by a variety of enzymes including UGT1A1, UGT1A3, UGT1B7, and UGT1A8, and polymorphisms in *UGT1A1* and UGT2B7 affect the rate of glucuronidation of buprenorphine *in vitro* (28).

8.4 PHARMACODYNAMICS OF OPIOIDS

Opioids act though inhibitory G-protein-coupled, seven transmembrane receptors, which are classified in four types (see Table 8.4). An additional receptor, the sigma receptor, was formerly classified as an opioid receptor, because it mediates some effects of opioids such as pentazocine and nalbuphine, but is encoded by a completely unrelated gene (*SIGMAR1*), and appears to interact with the endogenous hallucinogen dimethyltryptamine. Sigma receptors may mediate some of the effects of heroin, cocaine, methamphetamines, and phencyclidine, such as CNS excitation, hallucinations, respiratory and vasomotor excitation, and mydriasis (96). These receptors may also mediate the antitussic effects of opioids, including those that do not interact with μ, κ, or δ receptors, such as the (+) stereoisomer of pentazocine. The interactions of selected opioids with the different opioid receptors are summarized in Table 8.5.

TABLE 8.4
Opioid Receptor Types

Receptor	Gene	Protein	Subtypes[a]	Presumed Endogenous Ligands	Effects
Mu	OPRM1	MOP	μ1, μ2, μ3	β-Endorphin Enkephalins Endomorphin-1 Endomorphin-2	μ1: Supraspinal analgesia, some euphoria; μ2: spinal analgesia, respiratory depression, pruritus, miosis, prolactin release, reduced gastrointestinal (GI) motility, and anorexia; both μ1 and μ2: physical tolerance and dependence
Kappa	OPRK1	KOP	κ1, κ2, κ3	Enkephalins β-endorphin	Spinal analgesia, sedation, miosis, inhibition of antidiuretic hormone release, respiratory depression, intense dysphoria, and psychomimetic effects; minor tolerance and dependence liability
Delta	OPRD1	DOP	δ1, δ2	Dynorphin A, B α-neoendorphin	Spinal analgesia, euphoria, antidepressant, psychomimetic, and cardiovascular effects
Nociceptin receptor	OPRL1	NOP		Nociceptin	Anxiety, depression, appetite, inhibition of mu (μ) agonists, hyperalgesia

[a] Receptor subtypes identified by pharmacological methods, (e.g., μ1, μ2, μ3), probably derive from the same gene product by posttranscriptional modification. For example, different splice variants of the OPRM1 gene code for receptors with significantly different affinities for morphine versus fentanyl. In addition, posttranslation modifications, heterodimerization, and interaction with different regulatory proteins at different locations contribute to receptor heterogeneity.

The μ-type receptor (MOP), coded by *OPRM1*, is central to the analgesic activity of morphine and most other opioids. The gene is highly polymorphic, with the A118G SNP being the most extensively studied polymorphism. This nucleotide change causes a substitution of the asparagine residue at position 40 with an aspartate residue, resulting in removal of a highly conserved *N*-glycosylation site in the extracellular domain of the protein, and appears to be associated with 10-fold lower levels of the receptor (97). Despite this postulated effect, the association results are contradictory, with four studies showing higher prevalence of the G allele in opioid-addicted patients while three other studies showed the converse (98). The association with addictive behavior seems stronger for alcohol than for opioid drugs (99).

TABLE 8.5

Pharmacodynamic Effects of Opioids

Common Name	μ	κ	δ	σ	Potency[a]	Additional Targets/Comments
Heroin	+++			++	1–5 (IV)	After conversion to 6-acetylmorphine and morphine
Morphine	+++	+	+	+	1 (IV = 4)	Morphine-6-glucuronide metabolite is 2× potent
Dihydromorphine	+++				0.5–1.2	
Desomorphine	+++				10	
Nicomorphine	+++				2–3	
Dipropionylmorphine	+++				1.1	
Benzylmorphine	+				0.1	
Ethylmorphine	+				0.1	
Ethorphine	+++	+	+		1000–3000	NOP agonist
Codeine	+		+		0.1	After conversion to morphine
Dihydrocodeine	++				0.3	
Hydrocodone	++				0.6–2	Hydromorphone is a metabolite with higher potency
Hydromorphone	+++				3–10	
Oxycodone	++	+	+		1–2	Oxymorphone is a metabolite with higher potency
Oxymorphone	+++				7–15	
Methadone	+++				0.5–1	(R)-methadone: analgesia; (S)-methadone: NMDA, and nicotinic α3β4 antagonist, SNRI
Meperidine	++	++			0.1–0.4	NMDA-antagonist
Fentanyl	+++				50–200	
Sufentanyl	+++				500–1000	
Carfentanyl	+++				10^4–10^5	
Tramadol	+				0.1–0.2	SNRI, NMDA-antagonist; O-Desmethyl-metabolite 2-10× more potent analgesic than tramadol
Buprenorphine	+/–	–	+		25–50	Partial μ agonist, opioid κ antagonist, NOP agonist; norbuprenorphine is full δ, μ, NOP agonist
Propoxyphene	++				<0.1	Only D-isomer is analgesic; SNRI, nicotinic α3β4 antagonist
Levorphanol	+++	+++	+		4–8	SNRI, NMDA-antagonist
Butorphanol	–	+++			4–8	Opioid μ antagonist
Nalbuphine	+/–	+++		+++	0.5–1	Opioid μ partial agonist

continued

TABLE 8.5 (continued)
Pharmacodynamic Effects of Opioids

Common Name	μ	κ	δ	σ	Potency[a]	Additional Targets/Comments
Tapentadol	+				0.3–0.5	SNRI
Pentazocine	+/–	++		++	0.2–0.3	(–)-pentazocine: κ agonist; (+)-pentazocine: σ agonist

Notes: μ, κ, δ, σ, extent of agonist activity at the μ, κ, δ, and σ opioid receptors (+++ strong, ++ intermediate, + weak, +/– partial agonist, – antagonist). NMDA, *N*-methyl-D-aspartate receptor; SNRI, serotonin and norepinephrine reuptake inhibitor; NOP, nociceptin receptor, also known as ORL1, orphanin FQ, and kappa-3 opioid receptor.

[a] Relative analgesic potency (morphine = 1).

Randomized trials showed a significant effect of the 118G allele in the response to OPRM1 antagonist naltrexone in alcoholics (100–102). On the other hand, a meta-analysis of 22 studies concluded that this particular polymorphism had no significant association with substance dependence (98). These discrepancies may be due to the different populations studied and to linkage disequilibrium with other polymorphisms in the *OPRM1* gene. A recent study disclosed an association between certain *OPRM1* haplotypes and addictive behavior (103). *OPRM1* polymorphisms did correlate with methadone response as measured by pupil diameter (104) and pain perception (105), but the preponderance of evidence is insufficient to recommend personalized opioid treatment based on *OPRM1* genotyping at this time.

The other opioid receptors have not been as extensively studied for association of polymorphisms with opioid response or drug addiction. Various polymorphisms have been reported in the *OPRK1* gene, and preliminary studies showed the G36T SNP may be associated with opiate addiction (106), while an insertion/deletion in the promoter region has been associated with alcohol dependence (107). The *OPRD1* gene is well conserved, but a polymorphism in the promoter region associated with increased transcription (rs569356) (108) is in close linkage disequilibrium with the C80T SNP, which has been linked with opioid dependence (109).

In addition to the opioid receptors, polymorphisms in the endogenous opioid agonists have been associated with addiction disorders. Striatal enkephalin opioid pathways mediate reward and a 3′-untranslated region dinucleotide repeat polymorphism in the proenkephalin (*PENK*) gene was associated with higher PENK levels in the nucleus accumbens and with opioid dependence (110,111). The pituitary secretes pro-opiomelanocortin (POMC), which can be cleaved into a variety of hormonally active substances including beta-endorphin, which interacts with mu- and kappa-opioid receptors, adrenocorticotropin (ACTH) in the anterior pituitary, and melanocortin (MSH) in the intermediate lobe of the pituitary. MSH interacts with melanocortin receptors to stimulate melanin production in melanocytes, and also exerts pronociceptive effects in spinal neurons. Interestingly, a polymorphism in the melanocortin-1-receptor (*MC1R*) gene was associated with red hair, fair skin, and increased analgesia by the kappa-opioid agonist pentazocine, especially in women (112). The

POMC gene is a highly polymorphic gene, particularly in the beta-endorphin region, and the minor allele of the rs1866146 SNP has been associated with alcohol, cocaine, and opioid dependence (111,113).

Several other genes are modifiers of the opioid response. Therefore it is illustrative to summarize some of the pathways affected by opioids, as polymorphisms in some of the genes involved may affect the opioid response and become targets for pharmacogenetic testing. Typical opioid agonists cause inhibition of nociceptive pathways. On sensory neurons, opioid receptors reduce release of pain neurotransmitters such as neuropeptide substance P, glutamate, calcitonin gene-related peptide (CGRP), and the noxious heat transducer TRPV1. On spinal presynaptic axons of nociceptive type C and A-delta fibers, opioid receptor stimulation causes inhibition of voltage-gated calcium channels (VDCC) to reduce sensory input from nociceptive receptors.

On postsynaptic locations the MOP activates inward-rectifying potassium channels (GIRK) to reduce excitability. Interestingly, the rs2070995 AA polymorphism in one of the GIRK genes (*KCNJ6 = GIRK2 = Kir3.2*) has been associated with increased opioid and methadone requirements (114). Presynaptic opioid receptors also inhibit central GABA neurons, resulting in increased dopaminergic activity in the nucleus accumbens and stimulation of the pleasure and reward pathways. Some opioids also antagonize the *N*-methyl-aspartate (NMDA) receptors, which are involved in some nociceptive pathways and development of opioid tolerance, and act as inhibitors of serotonin and norepinephrine reuptake (SNRI), with resulting anxiolytic and antidepressive effects.

Opioid receptor signal transduction also involves inhibitory G-protein (G_i) inactivation of adenylate cyclase followed by reduction in cAMP-dependent phosphorylation of various transcription factors and consequent gene expression changes underlying long-term effects of opioids. Opioid receptor deactivation occurs acutely by the effect of β-arrestin 1 and 2 and RGS proteins (115). Interestingly, a common allele in the beta-arrestin-2 (*ARRB2*) gene was associated with decreased morphine response in cancer patients (116), and a haplotype block in the *ARRB2* gene was associated with methadone nonresponsiveness (117). Long-term desensitization to opioids occurs by reduced expression of the MOP at the neuronal surface. Other mechanisms of desensitization involve activation of glutamate and other anti-opioid pathways.

The opioid antagonist naloxone, better referred to as a reverse agonist, blocks the activity of μ, κ, or δ agonists but not of σ, or NOP agonists. The NOP receptor has very different effects than the classic opioid receptors, mediating a variety of physiologic functions depending on its anatomic location. In general, NOP receptor agonists tend to create anxiety and increased pain perception and to antagonize the analgesic properties of opioids.

Monoaminergic neurotransmitters are relevant for the opioid response in that they affect pain perception and response, as well as the reward mechanisms involved in potential for addiction, and there is significant cross-talk between monoaminergic pathways and the opioid pathways. The main monoamine neurotransmitters are dopamine, norepinephrine, and serotonin. Dopaminergic pathways are involved in cognition, mood, motivation, reward, sleep, fine motor control, and inhibition of prolactin release. The dopamine receptor D2 (DRD2) is a G-protein coupled receptor located in postsynaptic dopaminergic neurons, and binding of dopamine results in stimulation of motivation and reward-mediating mesocorticolimbic pathways (118). Polymorphisms in this gene (particularly the A1 allele) have been associated with myoclonal dystonia, schizophrenia, and predisposition to addictive behavior, including cocaine, alcohol, and opioid abuse. In Chinese and Australian populations, the A1 allele was a strong predictor of heroin abuse (119,120), while the converse was observed in a German population (121). In Australian patients, the A1 allele was associated with poor outcomes of methadone therapy (119) and with fewer withdrawal symptoms (122), while in Spain the A1 allele correlated with heroin dependence only in males (123). A recent study in 238 methadone maintenance therapy patients concluded that while the A1 allele was not significantly associated with response to methadone, another polymorphism (957C>T) was more frequent in nonresponders (124). These discrepancies may be in part explained by the fact that the A1 polymorphism is actually located 3′ of the *DRD2* gene, in the closely linked *ANKK1* gene. In a study of 85 methadone-substituted Caucasian patients, certain *DRD2/ANKK1* haplotypes were associated with the risk of opiate addiction and the methadone dose requirements (125). The *ANKK1* gene codes for a serine-threonine kinase that presumably regulates signal transduction pathways. Interestingly, the *DRD2 A1* polymorphism is closely linked to the SNP rs7118900, which changes the alanine at position 239 to threonine in the ANKK1 protein. This change results in altered nuclear-cytoplasmic transport and affects the response to the dopaminergic agonist apomorphine (126), suggesting that functional interactions between DRD2 and ANNK1 may be conserved at the haplotype level.

Other dopaminergic receptor genes have been related to addictive behavior. The -521 C/T SNP in the *DRD4* gene has been associated with heroin dependence and may predispose individuals to chronic opioid use through modulation of cold-pain responses (127). The *DRD4* exon III 7-repeat allele has also been associated with novelty seeking behavior and opioid addiction (128).

The catechol-*O*-methyltransferase (COMT) enzyme degrades catecholamines, including dopamine and norepinephrine, and is involved in pain pathways. A polymorphism (val158met) in the *COMT* gene correlated with decreased mu-opioid response to pain, with higher pain perception and more negative emotional effects (129). Analysis of haplotypes improves the predictive value of *COMT* genotyping for pain sensitivity and response to opioids (130).

Serotoninergic pathways play an important role in various psychiatric and addiction disorders. Polymorphisms in the tryptophan hydroxylase genes (*TPH1* and *TPH2*) and coding for a critical enzyme for serotonin synthesis correlated with heroin addiction (131). The serotonin receptor 5-HT$_{1B}$, coded by the *HTR1B* gene, has also been associated with drug and alcohol dependence, and the 1180G allele correlated with a protective effect from heroin addiction (132). Similarly, polymorphisms in *HTR3A* and *HTR3B* were associated with cocaine and heroin dependence, possibly through increased synaptic serotonin responsiveness leading to enhanced dopamine transmission in the reward pathway, which increases the addictive risk (133). Finally, the serotonin transporter SLC6A4 has been associated with addictive behavior toward alcohol, nicotine, and opioids, and modulates the analgesic effect of opioids (134).

The endocannabinoid pathways are involved in analgesia, memory extinction, and appetite, which are well-known effects of cannabinoid use. Interestingly, polymorphisms in endocannabinoid pathway genes have been associated with opioid dependence. The major endocannabinoids are arachidonic acid derivatives, anandamide and 2-arachidonoyl glycerol, which bind to the central (CB1) and peripheral (CB2) cannabinoid receptors, respectively encoded by the *CNR1* and *CNR2* genes. Anandamide is rapidly degraded by the fatty acid amide hydrolase (FAAH) enzyme. Polymorphisms in *CNR1* have been associated with opioid dependence (135), while certain *FAAH* haplotypes have been associated with vulnerability to addiction to multiple drugs (136).

Glutamate is the most abundant neurotransmitter in the brain, and together with the opioid receptor system, is involved in synaptic plasticity in limbic circuits induced by chronic substance abuse. Polymorphisms in the metabotropic glutamate receptors GRM6 and GRM8 were associated with heroin addiction (137), with GRM6 involved in modulating interindividual variations in the response to methadone maintenance therapy (138).

8.5 CONCLUSIONS

In this chapter several examples of genetic polymorphisms affecting metabolism, response, and tolerability of various opioids are presented. Many studies show significant differences in the pharmacokinetics of opioids among individuals with different polymorphisms in drug-metabolizing enzymes, with the most significant examples being the effects of CYP2D6 in the metabolism of codeine, dihydrocodeine, ethylmorphine, hydrocodone, oxycodone, and tramadol. Genetic variation in other drug disposition enzymes, such as CYP2B6, CYP3A4, CYP3A5, and UGT2B7, as well as a variety of genes involved in opioid pharmacodynamics, such as the OPRM1 and DDR2 receptors, has also been demonstrated to play a significant role in the response to opioid treatment and to influence the potential for addiction to these drugs. On the other hand, several studies simultaneously looking at multiple genetic polymorphisms failed to show significant effects. For example, a multicenter study of the effects of polymorphisms in *OPRM1, COMT, MC1R, ABCB1*, and *CYP2D6* in outpatients treated with opioids showed only minor effects for *ABCB1* and *OPRM1* variants (139). The lack of a positive result has to be regarded with some caution, as the probability of type II errors (failing to demonstrate an effect when it is present) is largely affected by the sample size and the power of the study. Another issue potentially contributing to negative results may be the lack of complete haplotype and genetic interaction analyses, and there are several examples of studies including haplotype analysis that show positive results missed by isolated SNP analysis. Further pharmacogenomic studies, ideally using deep sequencing of the entire genome, should resolve these discrepancies and ultimately lead to a full understanding of all of the genetic contributors to the response to opioids.

While the predictive value of genotyping to forecast the outcome of opioid therapy is generally low at present, some practical applications of current knowledge can be recommended. For example, patients with known CYP2D6 deficiency should avoid the opioids metabolized by this enzyme. Conversely, patients showing difficulties in the management of their opioid treatment, either because of lack of analgesic response or unexpected secondary effects, may benefit from genotyping for *CYP2D6* and potentially other genes of interest, such as *UGT2B7, CYP3A5, OPRM1, DRD2,* and so forth. Knowledge of the genetic variants of the various enzymes involved in opioid metabolism and their inhibitors and inducers is critical to avoid potentially harmful drug interactions. As illustrated in the various examples provided, in several cases genotyping provides a plausible explanation for the therapeutic difficulties and will help exclude alternative hypothesis, such as noncompliance or overdose.

REFERENCES

1. Krikorian, AD. 1975: Were the opium poppy and opium known in the ancient Near East? *J Hist Biol* 8 (1): 96–114.
2. Christians, U, Schmitz, V, Haschke, M. 2005: Functional interactions between P-glycoprotein and CYP3A in drug metabolism. *Expert Opin Drug Metab Toxicol* 1 (4): 641–654.
3. Hamabe, W, Maeda, T, Kiguchi, N, Yamamoto, C, et al. 2007: Negative relationship between morphine analgesia and P-glycoprotein expression levels in the brain. *J Pharmacol Sci* 105 (4): 353–360.
4. Kalvass, JC, Olson, ER, Pollack, GM. 2007: Pharmacokinetics and pharmacodynamics of alfentanil in P-glycoprotein-competent and P-glycoprotein-deficient mice: P-glycoprotein efflux alters alfentanil brain disposition and antinociception. *Drug Metab Dispos* 35 (3): 455–459.
5. Wandel, C, Kim, R, Wood, M, Wood, A. 2002: Interaction of morphine, fentanyl, sufentanil, alfentanil, and loperamide with the efflux drug transporter P-glycoprotein. *Anesthesiology* 96 (4): 913–920.
6. Thompson, SJ, Koszdin, K, Bernards, CM. 2000: Opiate-induced analgesia is increased and prolonged in mice lacking P-glycoprotein. *Anesthesiology* 92 (5): 1392–1399.
7. Park, HJ, Shinn, HK, Ryu, SH, Lee, HS, et al. 2007: Genetic polymorphisms in the ABCB1 gene and the effects of fentanyl in Koreans. *Clin Pharmacol Ther* 81 (4): 539–546.
8. Campa, D, Gioia, A, Tomei, A, Poli, P, Barale, R. 2008: Association of ABCB1/MDR1 and OPRM1 gene polymorphisms with morphine pain relief. *Clin Pharmacol Ther* 83 (4): 559–566.
9. Fonseca, F, de la Torre, R, Diaz, L, Pastor, A, et al. 2011: Contribution of cytochrome P450 and ABCB1 genetic variability on methadone pharmacokinetics, dose requirements, and response. *PLoS One* 6 (5): e19527.
10. Coles, LD, Lee, IJ, Hassan, HE, Eddington, ND. 2009: Distribution of saquinavir, methadone, and buprenorphine in maternal brain, placenta, and fetus during two different gestational stages of pregnancy in mice. *J Pharm Sci* 98 (8): 2832–2846.
11. Kharasch, ED, Hoffer, C, Whittington, D, Walker, A, Bedynek, PS. 2009: Methadone pharmacokinetics are independent of cytochrome P4503A (CYP3A) activity and gastrointestinal drug transport: insights from methadone interactions with ritonavir/indinavir. *Anesthesiology* 110 (3): 660–672.
12. Levran, O, O'Hara, K, Peles, E, Li, D, et al. 2008: ABCB1 (MDR1) genetic variants are associated with methadone doses required for effective treatment of heroin dependence. *Hum Mol Genet* 17 (14): 2219–2227.
13. Mercer, SL, Coop, A. 2011: Opioid analgesics and P-glycoprotein efflux transporters: a potential systems-level contribution to analgesic tolerance. *Curr Top Med Chem* 11 (9): 1157–1164.

14. Hassan, HE, Myers, AL, Coop, A, Eddington, ND. 2009: Differential involvement of P-glycoprotein (ABCB1) in permeability, tissue distribution, and antinociceptive activity of methadone, buprenorphine, and diprenorphine: in vitro and in vivo evaluation. *J Pharm Sci* 98 (12): 4928–4940.

15. Kim, RB, Leake, BF, Choo, EF, Dresser, GK, et al. 2001: Identification of functionally variant MDR1 alleles among European Americans and African Americans. *Clin Pharmacol Ther* 70 (2): 189–199.

16. Klepstad, P, Fladvad, T, Skorpen, F, Bjordal, K, et al. 2011: Influence from genetic variability on opioid use for cancer pain: a European genetic association study of 2294 cancer pain patients. *Pain* 152 (5): 1139–1145.

17. Crettol, S, Deglon, JJ, Besson, J, Croquette-Krokar, M, et al. 2008: No influence of ABCB1 haplotypes on methadone dosage requirement. *Clin Pharmacol Ther* 83 (5): 668–669; author reply 669–670.

18. Hodges, LM, Markova, SM, Chinn, LW, Gow, JM, et al. 2011: Very important pharmacogene summary: ABCB1 (MDR1, P-glycoprotein). *Pharmacogenet Genomics* 21 (3): 152–161.

19. Uehlinger, C, Crettol, S, Chassot, P, Brocard, M, et al. 2007: Increased (R)-methadone plasma concentrations by quetiapine in cytochrome P450s and ABCB1 genotyped patients. *J Clin Psychopharmacol* 27 (3): 273–278.

20. Smith, HS. 2011: The metabolism of opioid agents and the clinical impact of their active metabolites. *Clin J Pain* 27 (9): 824–838.

21. Yuan, R, Zhang, X, Deng, Q, Wu, Y, Xiang, G. 2011: Impact of CYP3A4*1G polymorphism on metabolism of fentanyl in Chinese patients undergoing lower abdominal surgery. *Clin Chim Acta* 412 (9–10): 755–760.

22. Zhang, W, Chang, YZ, Kan, QC, Zhang, LR, et al. 2010: CYP3A4*1G genetic polymorphism influences CYP3A activity and response to fentanyl in Chinese gynecologic patients. *Eur J Clin Pharmacol* 66 (1): 61–66.

23. Klees, TM, Sheffels, P, Dale, O, Kharasch, ED. 2005: Metabolism of alfentanil by cytochrome p4503a (cyp3a) enzymes. *Drug Metab Dispos* 33 (3): 303–311.

24. Gillam, EM, Guo, Z, Ueng, YF, Yamazaki, H, et al. 1995: Expression of cytochrome P450 3A5 in *Escherichia coli*: effects of 5′ modification, purification, spectral characterization, reconstitution conditions, and catalytic activities. *Arch Biochem Biophys* 317 (2): 374–384.

25. Wennerholm, A, Dandara, C, Sayi, J, Svensson, JO, et al. 2002: The African-specific CYP2D617 allele encodes an enzyme with changed substrate specificity. *Clin Pharmacol Ther* 71 (1): 77–88.

26. Pai, HV, Kommaddi, RP, Chinta, SJ, Mori, T, et al. 2004: A frameshift mutation and alternate splicing in human brain generate a functional form of the pseudogene cytochrome P4502D7 that demethylates codeine to morphine. *J Biol Chem* 279 (26): 27383–27389.

27. Coffman, BL, Rios, GR, King, CD, Tephly, TR. 1997: Human UGT2B7 catalyzes morphine glucuronidation. *Drug Metab Dispos* 25 (1): 1–4.

28. Rouguieg, K, Picard, N, Sauvage, FL, Gaulier, JM, Marquet, P. 2010: Contribution of the different UDP-glucuronosyltransferase (UGT) isoforms to buprenorphine and norbuprenorphine metabolism and relationship with the main UGT polymorphisms in a bank of human liver microsomes. *Drug Metab Dispos* 38 (1): 40–45.

29. Darbari, DS, van Schaik, RH, Capparelli, EV, Rana, S, et al. 2008: UGT2B7 promoter variant −840G>A contributes to the variability in hepatic clearance of morphine in patients with sickle cell disease. *Am J Hematol* 83 (3): 200–202.

30. Fujita, K, Ando, Y, Yamamoto, W, Miya, T, et al. 2010: Association of UGT2B7 and ABCB1 genotypes with morphine-induced adverse drug reactions in Japanese patients with cancer. *Cancer Chemother Pharmacol* 65 (2): 251–258.

31. Sawyer, MB, Innocenti, F, Das, S, Cheng, C, et al. 2003: A pharmacogenetic study of uridine diphosphate-glucuronosyltransferase 2B7 in patients receiving morphine. *Clin Pharmacol Ther* 73 (6): 566–574.

32. Holthe, M, Klepstad, P, Zahlsen, K, Borchgrevink, PC, et al. 2002: Morphine glucuronide-to-morphine plasma ratios are unaffected by the UGT2B7 H268Y and UGT1A1*28 polymorphisms in cancer patients on chronic morphine therapy. *Eur J Clin Pharmacol* 58 (5): 353–356.

33. Bhasker, CR, McKinnon, W, Stone, A, Lo, AC, et al. 2000: Genetic polymorphism of UDP-glucuronosyltransferase 2B7 (UGT2B7) at amino acid 268: ethnic diversity of alleles and potential clinical significance. *Pharmacogenetics* 10 (8): 679–685.

34. Innocenti, F, Liu, W, Fackenthal, D, Ramirez, J, et al. 2008: Single nucleotide polymorphism discovery and functional assessment of variation in the UDP-glucuronosyltransferase 2B7 gene. *Pharmacogenet Genomics* 18 (8): 683–697.

35. Lockridge, O, Mottershaw-Jackson, N, Eckerson, HW, La Du, BN. 1980: Hydrolysis of diacetylmorphine (heroin) by human serum cholinesterase. *J Pharmacol Exp Ther* 215 (1): 1–8.

36. Projean, D, Morin, PE, Tu, TM, Ducharme, J. 2003: Identification of CYP3A4 and CYP2C8 as the major cytochrome P450 s responsible for morphine *N*-demethylation in human liver microsomes. *Xenobiotica* 33 (8): 841–854.

37. Grobe, N, Zhang, B, Fisinger, U, Kutchan, TM, et al. 2009: Mammalian cytochrome P450 enzymes catalyze the phenol-coupling step in endogenous morphine biosynthesis. *J Biol Chem* 284 (36): 24425–24431.

38. Kirchheiner, J, Schmidt, H, Tzvetkov, M, Keulen, JT, et al. 2007: Pharmacokinetics of codeine and its metabolite morphine in ultra-rapid metabolizers due to CYP2D6 duplication. *Pharmacogenomics J* 7 (4): 257–265.

39. He, YJ, Brockmoller, J, Schmidt, H, Roots, I, Kirchheiner, J. 2008: CYP2D6 ultrarapid metabolism and morphine/codeine ratios in blood: was it codeine or heroin? *J Anal Toxicol* 32 (2): 178–182.

40. Desmeules, J, Gascon, MP, Dayer, P, Magistris, M. 1991: Impact of environmental and genetic factors on codeine analgesia. *Eur J Clin Pharmacol* 41 (1): 23–26.

41. Gasche, Y, Daali, Y, Fathi, M, Chiappe, A, et al. 2004: Codeine intoxication associated with ultrarapid CYP2D6 metabolism. *N Engl J Med* 351 (27): 2827–2831.

42. Madadi, P, Ross, CJ, Hayden, MR, Carleton, BC, et al. 2009: Pharmacogenetics of neonatal opioid toxicity following maternal use of codeine during breastfeeding: a case-control study. *Clin Pharmacol Ther* 85 (1): 31–35.

43. Lotsch, J, Rohrbacher, M, Schmidt, H, Doehring, A, et al. 2009: Can extremely low or high morphine formation from codeine be predicted prior to therapy initiation? *Pain* 144 (1–2): 119–124.

44. Vree, TB, van Dongen, RT, Koopman-Kimenai, PM. 2000: Codeine analgesia is due to codeine-6-glucuronide, not morphine. *Int J Clin Pract* 54 (6): 395–398.

45. Fromm, MF, Hofmann, U, Griese, EU, Mikus, G. 1995: Dihydrocodeine: a new opioid substrate for the polymorphic CYP2D6 in humans. *Clin Pharmacol Ther* 58 (4): 374–382.

46. Kirkwood, LC, Nation, RL, Somogyi, AA. 1997: Characterization of the human cytochrome P450 enzymes involved in the metabolism of dihydrocodeine. *Br J Clin Pharmacol* 44 (6): 549–555.

47. Hedenmalm, K, Sundgren, M, Granberg, K, Spigset, O, Dahlqvist, R. 1997: Urinary excretion of codeine, ethylmorphine, and their metabolites: relation to the CYP2D6 activity. *Ther Drug Monit* 19 (6): 643–649.

48. Liu, Z, Mortimer, O, Smith, CA, Wolf, CR, Rane, A. 1995: Evidence for a role of cytochrome P450 2D6 and 3A4 in ethylmorphine metabolism. *Br J Clin Pharmacol* 39 (1): 77–80.

49. Otton, SV, Schadel, M, Cheung, SW, Kaplan, HL, et al. 1993: CYP2D6 phenotype determines the metabolic conversion of hydrocodone to hydromorphone. *Clin Pharmacol Ther* 54 (5): 463–472.
50. Madadi, P, Hildebrandt, D, Gong, IY, Schwarz, UI, et al. 2010: Fatal hydrocodone overdose in a child: pharmacogenetics and drug interactions. *Pediatrics* 126 (4): e986–e989.
51. de Leon, J, Dinsmore, L, Wedlund, P. 2003: Adverse drug reactions to oxycodone and hydrocodone in CYP2D6 ultrarapid metabolizers. *J Clin Psychopharmacol* 23 (4): 420–421.
52. Jannetto, PJ, Wong, SH, Gock, SB, Laleli-Sahin, E, et al. 2002: Pharmacogenomics as molecular autopsy for postmortem forensic toxicology: genotyping cytochrome P450 2D6 for oxycodone cases. *J Anal Toxicol* 26 (7): 438–447.
53. Foster, A, Mobley, E, Wang, Z. 2007: Complicated pain management in a CYP450 2D6 poor metabolizer. *Pain Pract* 7 (4): 352–356.
54. Gronlund, J, Saari, TI, Hagelberg, NM, Neuvonen, PJ, et al. 2011: Effect of inhibition of cytochrome P450 enzymes 2D6 and 3A4 on the pharmacokinetics of intravenous oxycodone: a randomized, three-phase, crossover, placebo-controlled study. *Clin Drug Investig* 31 (3): 143–153.
55. Gronlund, J, Saari, TI, Hagelberg, NM, Neuvonen, PJ, et al. 2010: Exposure to oral oxycodone is increased by concomitant inhibition of CYP2D6 and 3A4 pathways, but not by inhibition of CYP2D6 alone. *Br J Clin Pharmacol* 70 (1): 78–87.
56. Kummer, O, Hammann, F, Moser, C, Schaller, O, et al. 2011: Effect of the inhibition of CYP3A4 or CYP2D6 on the pharmacokinetics and pharmacodynamics of oxycodone. *Eur J Clin Pharmacol* 67 (1): 63–71.
57. Samer, CF, Daali, Y, Wagner, M, Hopfgartner, G, et al. 2010: Genetic polymorphisms and drug interactions modulating CYP2D6 and CYP3A activities have a major effect on oxycodone analgesic efficacy and safety. *Br J Pharmacol* 160 (4): 919–930.
58. Begre, S, von Bardeleben, U, Ladewig, D, Jaquet-Rochat, S, et al. 2002: Paroxetine increases steady-state concentrations of (R)-methadone in CYP2D6 extensive but not poor metabolizers. *J Clin Psychopharmacol* 22 (2): 211–215.
59. Peles, E, Bodner, G, Kreek, MJ, Rados, V, Adelson, M. 2007: Corrected-QT intervals as related to methadone dose and serum level in methadone maintenance treatment (MMT) patients: a cross-sectional study. *Addiction* 102 (2): 289–300.
60. Eap, CB, Buclin, T, Baumann, P. 2002: Interindividual variability of the clinical pharmacokinetics of methadone: implications for the treatment of opioid dependence. *Clin Pharmacokinet* 41 (14): 1153–1193.
61. Bomsien, S, Skopp, G. 2007: An *in vitro* approach to potential methadone metabolic-inhibition interactions. *Eur J Clin Pharmacol* 63 (9): 821–827.
62. Gerber, JG, Rhodes, RJ, Gal, J. 2004: Stereoselective metabolism of methadone *N*-demethylation by cytochrome P4502B6 and 2C19. *Chirality* 16 (1): 36–44.
63. Crettol, S, Deglon, JJ, Besson, J, Croquette-Krokar, M, et al. 2006: ABCB1 and cytochrome P450 genotypes and phenotypes: influence on methadone plasma levels and response to treatment. *Clin Pharmacol Ther* 80 (6): 668–681.
64. Coller, JK, Joergensen, C, Foster, DJ, James, H, et al. 2007: Lack of influence of CYP2D6 genotype on the clearance of (R)-, (S)- and racemic-methadone. *Int J Clin Pharmacol Ther* 45 (7): 410–417.
65. Tyndale, RF, Droll, KP, Sellers, EM. 1997: Genetically deficient CYP2D6 metabolism provides protection against oral opiate dependence. *Pharmacogenetics* 7 (5): 375–379.
66. Zhu, W, Cadet, P, Baggerman, G, Mantione, KJ, Stefano, GB. 2005: Human white blood cells synthesize morphine: CYP2D6 modulation. *J Immunol* 175 (11): 7357–7362.
67. Wong, S, Gock, S, Zhang-Shi, R, Jin, M, et al., Pharmacogenetics as an aspect of molecular autopsy for forensic pathology/toxicology, in *Pharmacogenomics and Proteomics*, S.H.Y. Wong, M.W. Linder, and R.J. Valdes, Editors. 2006: AACC Press: Washington, DC. pp. 311–318.

68. Bunten, H, Liang, WJ, Pounder, D, Seneviratne, C, Osselton, MD. 2011: CYP2B6 and OPRM1 gene variations predict methadone-related deaths. *Addict Biol* 16 (1): 142–144.
69. Jensen, KB, Lonsdorf, TB, Schalling, M, Kosek, E, Ingvar, M. 2009: Increased sensitivity to thermal pain following a single opiate dose is influenced by the COMT val(158) met polymorphism. *PLoS One* 4 (6): e6016.
70. Rodriguez-Martin, I, Braksator, E, Bailey, CP, Goodchild, S, et al. 2008: Methadone: does it really have low efficacy at micro-opioid receptors? *Neuroreport* 19 (5): 589–593.
71. Krantz, MJ, Martin, J, Stimmel, B, Mehta, D, Haigney, MC. 2009: QTc interval screening in methadone treatment. *Ann Intern Med* 150 (6): 387–395.
72. Eap, CB, Crettol, S, Rougier, JS, Schlapfer, J, et al. 2007: Stereoselective block of hERG channel by (S)-methadone and QT interval prolongation in CYP2B6 slow metabolizers. *Clin Pharmacol Ther* 81 (5): 719–728.
73. Ramirez, J, Innocenti, F, Schuetz, EG, Flockhart, DA, et al. 2004: CYP2B6, CYP3A4, and CYP2C19 are responsible for the *in vitro* N-demethylation of meperidine in human liver microsomes. *Drug Metab Dispos* 32 (9): 930–936.
74. Whitfield, HA, Genetic variations of CYP2B6 enzyme and the response to meperidine in oral sedation, Dissertations. 2010, Virginia Commonwealth University: Richmond, VA, p. 50.
75. Tateishi, T, Krivoruk, Y, Ueng, YF, Wood, AJ, et al. 1996: Identification of human liver cytochrome P-450 3A4 as the enzyme responsible for fentanyl and sufentanil N-dealkylation. *Anesth Analg* 82 (1): 167–172.
76. Zhang, W, Yuan, JJ, Kan, QC, Zhang, LR, et al. 2011: Influence of CYP3A5*3 polymorphism and interaction between CYP3A5*3 and CYP3A4*1G polymorphisms on postoperative fentanyl analgesia in Chinese patients undergoing gynaecological surgery. *Eur J Anaesthesiol* 28 (4): 245–250.
77. Jin, M, Gock, SB, Jannetto, PJ, Jentzen, JM, Wong, SH. 2005: Pharmacogenomics as molecular autopsy for forensic toxicology: genotyping cytochrome P450 3A4*1B and 3A5*3 for 25 fentanyl cases. *J Anal Toxicol* 29 (7): 590–598.
78. Levin, TT, Bakr, MH, Nikolova, T. 2010: Case report: delirium due to a diltiazem-fentanyl CYP3A4 drug interaction. *Gen Hosp Psychiatry* 32 (6): 648 e649–648 e610.
79. Wilhelm, W, Kreuer, S. 2008: The place for short-acting opioids: special emphasis on remifentanil. *Crit Care* 12 Suppl 3: S5.
80. Pedersen, RS, Damkier, P, Brosen, K. 2006: Enantioselective pharmacokinetics of tramadol in CYP2D6 extensive and poor metabolizers. *Eur J Clin Pharmacol* 62 (7): 513–521.
81. Fox, MA, Jensen, CL, Murphy, DL. 2009: Tramadol and another atypical opioid meperidine have exaggerated serotonin syndrome behavioural effects, but decreased analgesic effects, in genetically deficient serotonin transporter (SERT) mice. *Int J Neuropsychopharmacol* 12 (8): 1055–1065.
82. Subrahmanyam, V, Renwick, AB, Walters, DG, Young, PJ, et al. 2001: Identification of cytochrome P-450 isoforms responsible for cis-tramadol metabolism in human liver microsomes. *Drug Metab Dispos* 29 (8): 1146–1155.
83. Garcia-Quetglas, E, Azanza, JR, Sadaba, B, Munoz, MJ, et al. 2007: Pharmacokinetics of tramadol enantiomers and their respective phase I metabolites in relation to CYP2D6 phenotype. *Pharmacol Res* 55 (2): 122–130.
84. Halling, J, Weihe, P, Brosen, K. 2008: CYP2D6 polymorphism in relation to tramadol metabolism: a study of faroese patients. *Ther Drug Monit* 30 (3): 271–275.
85. Stamer, UM, Musshoff, F, Kobilay, M, Madea, B, et al. 2007: Concentrations of tramadol and O-desmethyltramadol enantiomers in different CYP2D6 genotypes. *Clin Pharmacol Ther* 82 (1): 41–47.

86. Wang, G, Zhang, H, He, F, Fang, X. 2006: Effect of the CYP2D6*10 C188T polymorphism on postoperative tramadol analgesia in a Chinese population. *Eur J Clin Pharmacol* 62 (11): 927–931.

87. Gan, SH, Ismail, R, Wan Adnan, WA, Zulmi, W. 2007: Impact of CYP2D6 genetic polymorphism on tramadol pharmacokinetics and pharmacodynamics. *Mol Diagn Ther* 11 (3): 171–181.

88. Kirchheiner, J, Keulen, JT, Bauer, S, Roots, I, Brockmoller, J. 2008: Effects of the CYP2D6 gene duplication on the pharmacokinetics and pharmacodynamics of tramadol. *J Clin Psychopharmacol* 28 (1): 78–83.

89. Kim, E, Choi, CB, Kang, C, Bae, SC. 2010: Adverse events in analgesic treatment with tramadol associated with CYP2D6 extensive-metaboliser and OPRM1 high-expression variants. *Ann Rheum Dis* 69 (10): 1889–1890.

90. Slanar, O, Nobilis, M, Kvetina, J, Matouskova, O, et al. 2007: Pharmacokinetics of tramadol is affected by MDR1 polymorphism C3435T. *Eur J Clin Pharmacol* 63 (4): 419–421.

91. Lutfy, K, Eitan, S, Bryant, CD, Yang, YC, et al. 2003: Buprenorphine-induced antinociception is mediated by mu-opioid receptors and compromised by concomitant activation of opioid receptor-like receptors. *J Neurosci* 23 (32): 10331–10337.

92. Kilicarslan, T, Sellers, EM. 2000: Lack of interaction of buprenorphine with flunitrazepam metabolism. *Am J Psychiatry* 157 (7): 1164–1166.

93. Noveck, R, Harris, S, El-Tahtawy, A, Kim, R, et al. 2005: Lack of effect of CYP3A4 inhibitor ketoconazole on transdermally administered buprenorphine. *Clin Pharmacol Ther* 77 (2): P78.

94. McCance-Katz, EF, Moody, DE, Morse, GD, Friedland, G, et al. 2006: Interactions between buprenorphine and antiretrovirals. I. The nonnucleoside reverse-transcriptase inhibitors efavirenz and delavirdine. *Clin Infect Dis* 43 (Suppl 4): S224–S234.

95. Bruce, RD, McCance-Katz, E, Kharasch, ED, Moody, DE, Morse, GD. 2006: Pharmacokinetic interactions between buprenorphine and antiretroviral medications. *Clin Infect Dis* 43 (Suppl 4): S216–S223.

96. Leonard, BE. 2004: Sigma receptors and sigma ligands: background to a pharmacological enigma. *Pharmacopsychiatry* 37 (Suppl 3): S166–S170.

97. Zhang, Y, Wang, D, Johnson, AD, Papp, AC, Sadee, W. 2005: Allelic expression imbalance of human mu opioid receptor (OPRM1) caused by variant A118G. *J Biol Chem* 280 (38): 32618–32624.

98. Arias, A, Feinn, R, Kranzler, HR. 2006: Association of an Asn40Asp (A118G) polymorphism in the mu-opioid receptor gene with substance dependence: a meta-analysis. *Drug Alcohol Depend* 83 (3): 262–268.

99. van den Wildenberg, E, Wiers, RW, Dessers, J, Janssen, RG, et al. 2007: A functional polymorphism of the mu-opioid receptor gene (OPRM1) influences cue-induced craving for alcohol in male heavy drinkers. *Alcohol Clin Exp Res* 31 (1): 1–10.

100. Oslin, DW, Berrettini, WH, O'Brien, CP. 2006: Targeting treatments for alcohol dependence: the pharmacogenetics of naltrexone. *Addict Biol* 11 (3–4): 397–403.

101. Anton, RF, Oroszi, G, O'Malley, S, Couper, D, et al. 2008: An evaluation of mu-opioid receptor (OPRM1) as a predictor of naltrexone response in the treatment of alcohol dependence: results from the Combined Pharmacotherapies and Behavioral Interventions for Alcohol Dependence (COMBINE) study. *Arch Gen Psychiatry* 65 (2): 135–144.

102. Oslin, DW, Berrettini, W, Kranzler, HR, Pettinati, H, et al. 2003: A functional polymorphism of the mu-opioid receptor gene is associated with naltrexone response in alcohol-dependent patients. *Neuropsychopharmacology* 28 (8): 1546–1552.

103. Zhang, H, Luo, X, Kranzler, HR, Lappalainen, J, et al. 2006: Association between two mu-opioid receptor gene (OPRM1) haplotype blocks and drug or alcohol dependence. *Hum Mol Genet* 15 (6): 807–819.

104. Lotsch, J, Skarke, C, Wieting, J, Oertel, BG, et al. 2006: Modulation of the central nervous effects of levomethadone by genetic polymorphisms potentially affecting its metabolism, distribution, and drug action. *Clin Pharmacol Ther* 79 (1): 72–89.

105. Shabalina, SA, Zaykin, DV, Gris, P, Ogurtsov, AY, et al. 2009: Expansion of the human mu-opioid receptor gene architecture: novel functional variants. *Hum Mol Genet* 18 (6): 1037–1051.

106. Yuferov, V, Fussell, D, LaForge, KS, Nielsen, DA, et al. 2004: Redefinition of the human kappa opioid receptor gene (OPRK1) structure and association of haplotypes with opiate addiction. *Pharmacogenetics* 14 (12): 793–804.

107. Edenberg, HJ, Wang, J, Tian, H, Pochareddy, S, et al. 2008: A regulatory variation in OPRK1, the gene encoding the kappa-opioid receptor, is associated with alcohol dependence. *Hum Mol Genet* 17 (12): 1783–1789.

108. Zhang, H, Gelernter, J, Gruen, JR, Kranzler, HR, et al. 2010: Functional impact of a single-nucleotide polymorphism in the OPRD1 promoter region. *J Hum Genet* 55 (5): 278–284.

109. Zhang, H, Kranzler, HR, Yang, BZ, Luo, X, Gelernter, J. 2008: The OPRD1 and OPRK1 loci in alcohol or drug dependence: OPRD1 variation modulates substance dependence risk. *Mol Psychiatry* 13 (5): 531–543.

110. Nikoshkov, A, Drakenberg, K, Wang, X, Horvath, MC, et al. 2008: Opioid neuropeptide genotypes in relation to heroin abuse: dopamine tone contributes to reversed mesolimbic proenkephalin expression. *Proc Natl Acad Sci USA* 105 (2): 786–791.

111. Xuei, X, Flury-Wetherill, L, Bierut, L, Dick, D, et al. 2007: The opioid system in alcohol and drug dependence: family-based association study. *Am J Med Genet B Neuropsychiatr Genet* 144B (7): 877–884.

112. Mogil, JS, Wilson, SG, Chesler, EJ, Rankin, AL, et al. 2003: The melanocortin-1 receptor gene mediates female-specific mechanisms of analgesia in mice and humans. *Proc Natl Acad Sci USA* 100 (8): 4867–4872.

113. Zhang, H, Kranzler, HR, Weiss, RD, Luo, X, et al. 2009: Pro-opiomelanocortin gene variation related to alcohol or drug dependence: evidence and replications across family- and population-based studies. *Biol Psychiatry* 66 (2): 128–136.

114. Lotsch, J, Pruss, H, Veh, RW, Doehring, A. 2010: A KCNJ6 (Kir3.2, GIRK2) gene polymorphism modulates opioid effects on analgesia and addiction but not on pupil size. *Pharmacogenet Genomics* 20 (5): 291–297.

115. Xie, GX, Palmer, PP. 2005: RGS proteins: new players in the field of opioid signaling and tolerance mechanisms. *Anesth Analg* 100 (4): 1034–1042.

116. Ross, JR, Rutter, D, Welsh, K, Joel, SP, et al. 2005: Clinical response to morphine in cancer patients and genetic variation in candidate genes. *Pharmacogenomics J* 5 (5): 324–336.

117. Oneda, B, Crettol, S, Bochud, M, Besson, J, et al. 2011: beta-Arrestin2 influences the response to methadone in opioid-dependent patients. *Pharmacogenomics J* 11 (4): 258–266.

118. Noble, EP. 2000: Addiction and its reward process through polymorphisms of the D2 dopamine receptor gene: a review. *Eur Psychiatry* 15 (2): 79–89.

119. Lawford, BR, Young, RM, Noble, EP, Sargent, J, et al. 2000: The D(2) dopamine receptor A(1) allele and opioid dependence: association with heroin use and response to methadone treatment. *Am J Med Genet* 96 (5): 592–598.

120. Li, T, Liu, X, Zhao, J, Hu, X, et al. 2002: Allelic association analysis of the dopamine D2, D3, 5-HT2A, and GABA(A)gamma2 receptors and serotonin transporter genes with heroin abuse in Chinese subjects. *Am J Med Genet* 114 (3): 329–335.

121. Xu, K, Lichtermann, D, Lipsky, RH, Franke, P, et al. 2004: Association of specific haplotypes of D2 dopamine receptor gene with vulnerability to heroin dependence in 2 distinct populations. *Arch Gen Psychiatry* 61 (6): 597–606.

122. Barratt, DT, Coller, JK, Somogyi, AA. 2006: Association between the DRD2 A1 allele and response to methadone and buprenorphine maintenance treatments. *Am J Med Genet B Neuropsychiatr Genet* 141B (4): 323–331.

123. Perez de los Cobos, J, Baiget, M, Trujols, J, Sinol, N, et al. 2007: Allelic and genotypic associations of DRD2 TaqI A polymorphism with heroin dependence in Spanish subjects: a case control study. *Behav Brain Funct* 3: 25.

124. Crettol, S, Besson, J, Croquette-Krokar, M, Hammig, R, et al. 2008: Association of dopamine and opioid receptor genetic polymorphisms with response to methadone maintenance treatment. *Prog Neuropsychopharmacol Biol Psychiatry* 32 (7): 1722–1727.

125. Doehring, A, Hentig, N, Graff, J, Salamat, S, et al. 2009: Genetic variants altering dopamine D2 receptor expression or function modulate the risk of opiate addiction and the dosage requirements of methadone substitution. *Pharmacogenet Genomics* 19 (6): 407–414.

126. Garrido, E, Palomo, T, Ponce, G, Garcia-Consuegra, I, et al. 2011: The ANKK1 protein associated with addictions has nuclear and cytoplasmic localization and shows a differential response of Ala239Thr to apomorphine. *Neurotox Res* 20 (1): 32–39.

127. Ho, AM, Tang, NL, Cheung, BK, Stadlin, A. 2008: Dopamine receptor D4 gene –521C/T polymorphism is associated with opioid dependence through cold-pain responses. *Ann NY Acad Sci* 1139: 20–26.

128. Kotler, M, Cohen, H, Segman, R, Gritsenko, I, et al. 1997: Excess dopamine D4 receptor (D4DR) exon III seven repeat allele in opioid-dependent subjects. *Mol Psychiatry* 2 (3): 251–254.

129. Zubieta, JK, Heitzeg, MM, Smith, YR, Bueller, JA, et al. 2003: COMT val158met genotype affects mu-opioid neurotransmitter responses to a pain stressor. *Science* 299 (5610): 1240–1243.

130. Rakvag, TT, Ross, JR, Sato, H, Skorpen, F, et al. 2008: Genetic variation in the catechol-O-methyltransferase (COMT) gene and morphine requirements in cancer patients with pain. *Mol Pain* 4: 64.

131. Nielsen, DA, Barral, S, Proudnikov, D, Kellogg, S, et al. 2008: TPH2 and TPH1: association of variants and interactions with heroin addiction. *Behav Genet* 38 (2): 133–150.

132. Proudnikov, D, LaForge, KS, Hofflich, H, Levenstien, M, et al. 2006: Association analysis of polymorphisms in serotonin 1B receptor (HTR1B) gene with heroin addiction: a comparison of molecular and statistically estimated haplotypes. *Pharmacogenet Genomics* 16 (1): 25–36.

133. Enoch, MA, Gorodetsky, E, Hodgkinson, C, Roy, A, Goldman, D. 2011: Functional genetic variants that increase synaptic serotonin and 5-HT3 receptor sensitivity predict alcohol and drug dependence. *Mol Psychiatry* 16: 1139–1146.

134. Kosek, E, Jensen, KB, Lonsdorf, TB, Schalling, M, Ingvar, M. 2009: Genetic variation in the serotonin transporter gene (5-HTTLPR, rs25531) influences the analgesic response to the short acting opioid Remifentanil in humans. *Mol Pain* 5: 37.

135. Proudnikov, D, Kroslak, T, Sipe, JC, Randesi, M, et al. 2010: Association of polymorphisms of the cannabinoid receptor (CNR1) and fatty acid amide hydrolase (FAAH) genes with heroin addiction: impact of long repeats of CNR1. *Pharmacogenomics J* 10 (3): 232–242.

136. Flanagan, JM, Gerber, AL, Cadet, JL, Beutler, E, Sipe, JC. 2006: The fatty acid amide hydrolase 385 A/A (P129T) variant: haplotype analysis of an ancient missense mutation and validation of risk for drug addiction. *Hum Genet* 120 (4): 581–588.

137. Nielsen, DA, Ji, F, Yuferov, V, Ho, A, et al. 2008: Genotype patterns that contribute to increased risk for or protection from developing heroin addiction. *Mol Psychiatry* 13 (4): 417–428.
138. Fonseca, F, Gratacos, M, Escaramis, G, De Cid, R, et al. 2010: Response to methadone maintenance treatment is associated with the MYOCD and GRM6 genes. *Mol Diagn Ther* 14 (3): 171–178.
139. Lotsch, J, von Hentig, N, Freynhagen, R, Griessinger, N, et al. 2009: Cross-sectional analysis of the influence of currently known pharmacogenetic modulators on opioid therapy in outpatient pain centers. *Pharmacogenet Genomics* 19 (6): 429–436.

9 Pharmacogenomics Aspects of Addiction Treatment

F. Gerard Moeller

CONTENTS

9.1 Introduction 195
9.2 Pharmcogenomics of Metabolism of Medications for the Treatment of
Addiction 196
9.3 Pharmacogenomics of Site of Action of Medications for Addiction 197
9.4 Pharmacogenomics of Biochemical Effects of Drugs of Abuse 203
9.5 Conclusions 205
References 206

9.1 INTRODUCTION

As with other pharmacogenomics studies, two potential targets for research on pharmacogenomics of medication treatment for drug addiction include genes encoding enzymes that metabolize the medications used in the treatment of addictions and genes encoding the putative neurochemical target of the pharmacotherapy. Examples of these two areas of research include genes encoding cytochrome P450 enzymes and genes encoding opiate receptors in studies of opiate agonists and antagonists for addiction. In addition to these two potential targets, drug addiction has a third potential area of research; drugs of abuse have known biochemical effects on the brain which could be altered by genetic variation in neurotransmitter receptors. Although these three areas could provide important lines of research in the development of medication for addiction, to date research in this area has not lived up to its potential, with the few studies that have been done on pharmacogenomics of medication treatment of addictions finding some inconsistent results. Recently genome-wide association studies (GWAS) have been done on the treatment response for medications for nicotine addiction with more consistent findings across studies. This chapter will review the research done to date on pharmacogenomics of addiction medication and discuss potential future areas of research based on promising findings in addiction treatment.

9.2 PHARMCOGENOMICS OF METABOLISM OF MEDICATIONS FOR THE TREATMENT OF ADDICTION

A large body of research supports the use of opiate agonists as treatments for opiate dependence. (Reviewed in Kosten.[1]) Two opiate agonists are currently approved by the U.S. Food and Drug Administration (FDA) for the treatment of opiate dependence. The first of the opiates to be approved by the FDA as an agonist or replacement therapy for opiate dependence is methadone. Methadone in the racemic mixture of R and S enantiomers undergoes extensive hepatic metabolism, primarily through CYP3A4 and to a lesser extent CYP2D6, and CYP2B6, with primarily inactive metabolites.[2] The analgesic and therapeutic effect for opiate maintenance treatment is thought to primarily occur through the R enantiomer of methadone. There is a large interindividual variability in the metabolism of methadone, which is thought to be based on variability in CYP enzyme activity.[3]

As discussed more extensively in Chapter 9, due to deletions or mutations in the gene coding for the CYP2D6 enzyme leading to a loss of CYP2D6 enzyme activity, a minority of Caucasian individuals are classified as poor metabolizers. Reviewed by Zhou,[2] studies that examined blood levels of methadone in extensive and poor metabolizers based on CYP2D6 activity found inconsistent results, with some studies finding higher plasma concentrations in poor metabolizers than in extensive metabolizers, and other studies finding no difference in plasma levels between groups.

In studies that examined the effect of CYP genotype on treatment outcome in methadone patients, Eap et al.[4] genotyped 256 patients who were undergoing treatment for opiate dependence using methadone maintenance to determine whether *CYP2D6* genotype was related to differences in treatment outcome and to blood levels of methadone. Results of that study showed that of the 256 patients studied, only 18 were found to be CYP2D6 poor metabolizers, with the rest of the subjects being either heterozygous extensive metabolizers (106 subjects) or homozygous extensive metabolizers (132 subjects). Within the extensive metabolizers, 10 subjects were found to be ultrarapid metabolizers due to extra functional copies of the *CYP2D6* gene. There was a significant difference between genotypes for methadone concentrations to dose-to-weight ratios (methadone concentrations divided by patient doses and patient weight), with the poor metabolizers having significantly higher levels than ultrarapid metabolizers. With respect to treatment outcome, 72% of the poor metabolizers were considered to be successful in treatment, whereas only 40% of the ultrarapid metabolizers were considered to be successful based on a chart review. Although the differences between groups were not statistically significant, these results suggest that *CYP2D6* genotype could affect treatment outcome in methadone-maintained patients. A second study examined the relationship between *CYP2D6* genotype and treatment satisfaction in methadone-maintained patients.[5] In that study 205 patients in methadone maintenance therapy were genotyped for *CYP2D6* alleles and completed a treatment satisfaction scale. Results of this study showed that 4.4% of patients were poor metabolizers, 90.2% of patients were extensive metabolizers, and 5.4% of patients were ultrarapid metabolizers. No significant differences were found between groups for either retention in treatment or methadone doses or plasma levels. The only significant difference between groups was a lower overall patient satisfaction score in ultrarapid metabolizers compared to the other groups.[5]

One possible explanation for the relatively inconsistent results regarding the association between *CYP2D6* genotype and either methadone plasma levels or treatment outcome is that methadone inhibits CYP2D6 enzyme activity, which could make patients who are genotypically extensive metabolizers phenotypically poor metabolizers, minimizing any difference between groups. This could be important clinically for the use of methadone with other drugs extensively metabolized by CYP2D6, as methadone could significantly increase blood levels of other drugs metabolized by this enzyme.[2]

In a study of other genes associated with methadone metabolism, 209 patients in methadone maintenance were genotyped for *CYP2B6, CYP2C9*, and *CYPC19*. Results of that study were that there was no effect of *CYP2C9* or *CYPC19* genotype on plasma methadone levels. *CYP2B6* genotype had a significant impact on plasma levels of the S-enantiomer of methadone, but because the majority of the clinical effect of methadone is mediated by the R-enantiomer, the authors concluded that *CYP2B6* genotype was unlikely to be of significance clinically.[3]

Buprenorphine, the other opiate agonist approved for the treatment of opiate addiction, is not significantly metabolized by the CYP2D6 enzyme;[6] thus any effect of *CYP2D6* genotype on this drug would be expected to be minimal.

9.3 PHARMACOGENOMICS OF SITE OF ACTION OF MEDICATIONS FOR ADDICTION

The μ opiate receptor is the primary binding site and primary site of action of both endogenous opiates, such as β-endorphin, and opiate drugs of abuse as well as opiate agonists used in the treatment of opiate abuse. A single nucelotide polymorphism (SNP) in the coding region of the μ opiate receptor gene (*OPRM1*) with a nucleotide substitution at position 118 (*A118G*) has been shown to alter β-endorphin binding, which could have a significant impact on treatment response in addictions.[7]

Although a number of studies have examined the frequency of this and other *OPRM1* SNPs in addictive disorders, few studies have examined the role of *OPRM1* gene polymorphisms in treatment response with opiate agonists. In a study of 169 methadone-maintained subjects, SNP variants in multiple candidate genes were examined in relation to treatment response, including *OPRM1* as well as genes for glutamate receptors *GRM6* and *GRM8*, the nuclear receptor *NR4A2*, the transcription factor *MYOCD*, and the enzyme *CRY1* chosen based on some studies showing an association between these genes and opiate dependence. Results of that study found no relationship between *OPRM1* genotype and treatment response. However, there was an association between *CRY1* and *MYOCD* genotype, and those subjects who carried the *AA* genotype of *CRY1* were more likely to be nonresponders to methadone, although this finding did not survive correction for multiple comparisons. The study also found that subjects who carried the *A* allele of the *MYOCD* gene were more likely to be nonresponders, but only if they also carried the *AG* genotype of *GRM6*.[8]

The other class of drugs approved for the treatment of opiate dependence is the opiate antagonists. Naltrexone pharmacotherapy for opiate dependence has a clear mechanism of action and a compelling rationale.[9] Naltrexone selectively competes for μ opiate receptors and prevents opiate-dependent individuals from achieving reinforcement from continued opiate use. Naltrexone has a high affinity for μ opiate receptors and is not psychoactive, thus minimizing the risk of diversion. It can be administered on an outpatient basis, and compliance is associated with favorable outcomes.[10,11] Naltrexone has been extensively evaluated in several trials and results are uniformly positive for patients who comply and remain in treatment. In addition to reducing relapse, patients typically show reductions in nonopiate drug use, improvement in employment status, and reduction in the social and legal problems associated with addiction.[12–14] It has been demonstrated that long-acting sustained-release injectable formulation of naltrexone has a higher efficacy for relapse prevention in heroin-dependent subjects,[15] and the FDA approved long-acting injectable naltrexone (Vivitrol) for opiate dependence. The major problem with naltrexone has been compliance. A number of controlled studies demonstrating safety and efficacy noted this problem.[16,17] Possibly because of the problem of compliance, no studies have been published on the role of *OPRM1* gene polymorphisms on treatment response to naltrexone in opiate dependence.

However, naltrexone has also been shown to reduce alcohol consumption, and naltrexone in oral and depot injectable formulations is approved by the FDA for the treatment of alcohol dependence. *OPRM1* gene polymorphisms have been studied extensively in the treatment response of naltrexone for alcoholism. In the initial study reported by Oslin et al.,[18] the relationship between two *OPRM1* gene polymorphisms (*A118G, Asn40Asp*), (*C17T, Ala6Val*) and treatment response to naltrexone for alcohol dependence were examined in 82 patients treated with naltrexone and 59 patients treated with placebo. As the frequency of the *A118G* SNP is very low in African Americans, and the frequency of the *C17T* SNP is very low in subjects with European descent, these SNPs were examined separately in the ethnic group in which they are most common. Results of that study showed that subjects of European descent who were homozygous or heterozygous for the *Asp40* allele treated with naltrexone had a significantly lower rate of relapse and a significantly longer time to return to heavy drinking than subjects with the *Asn40* allele.[18] A second study on the potential role of *OPRM1* genotype on treatment response to naltrexone for alcoholism was published by Gelernter et al.[19] In that study, 240 subjects were randomized to one of three treatment groups: naltrexone 50 mg/day for 12 months, naltrexone 50 mg/day for 3 months followed by 9 months of placebo, or 12 months of placebo. Seven different polymorphisms were examined in relation to treatment outcome: three *OPRM1* polymorphisms (*A188G, 2044C/A*, and *rs648893*), three *OPRD1* genotypes (*T921, F27C*, and *rs678849*), and one *OPRK1* genotype (*rs963549*). Results of that study found no significant interaction between genotype and naltrexone treatment response, with only the *OPRK1* polymorphism approaching statistical significance ($p = 0.06$). A second paper published the same year[20] examined the effect of naltrexone on alcohol-induced behavioral responses in the laboratory. In that study, 40 non-treatment-seeking heavy drinkers underwent intravenous alcohol challenge sessions after placebo and 50 mg of naltrexone. Results of that study showed a significant medication by genotype by breath alcohol concentration interaction, with a lower self-reported alcohol-induced high in subjects with at least one copy of the G allele of the *OPRM1 A118G* polymorphism.[20]

A third clinical trial (project COMBINE) with a larger sample size than the two previous studies examined the effect of *OPRM1* genotype on naltrexone treatment outcome in 604 non-Hispanic Caucasian alcohol-dependent subjects. Subjects were randomized to treatment with 100 mg per day of naltrexone or placebo and medication management (MM) alone or MM plus a combined behavioral intervention (CBI). Results of that study showed that in subjects who received MM alone there was a significant gene by treatment interaction on the trend for days abstinent from alcohol, with subjects who had at least one copy of the *Asp40* allele treated with naltrexone showing an increasing trend for abstinence days and heavy drinking days over time. Likewise, the best clinical outcome as rated by being abstinent or moderately drinking without problems was related to genotype, with naltrexone-treated subjects who carried the *Asp40* allele having the best clinical outcome.[21] The authors of this study suggest that the difference in findings between their study and the study of Gelertner et al. could be due at least in part to the dose of naltrexone used, because the Gelertner et al. study used a lower dose of naltrexone. A second study using a haplotype approach to examine the association between *OPRM1* genotype and naltrexone treatment outcome using the same pool of subjects as Anton et al. found a significant haplotype by medication interaction with one of two *OPRM1* haplotype blocks examined. This block contained the *Asp40* polymorphism reported previously to be associated with treatment outcome.[22]

In summary, the majority of clinical trials examining the role of *OPRM1 Asp40* alleles in naltrexone treatment response for alcoholism find that patients lacking the Asp40 allele do more poorly in treatment with naltrexone. These findings are consistent with a recent publication in rhesus monkeys showing that the ability of naltrexone to reduce alcohol consumption was related to a pharmacologically similar *OPRM1* genotype.[23]

Two other medications are approved by the FDA for the treatment of alcohol dependence: disulfiram (Antabuse) and acamprosate (Campral). Acamprosate was also included as a treatment in the project COMBINE study; however, in the study results reported by Anton et al.,[21] acamprosate did not show added benefits to naltrexone or show efficacy alone, leading the authors not to examine the effects of genotype on response to acamprosate separately. In a study of the pharmacogenetics of acamprosate and naltrexone related to the behavioral response to alcohol cue exposure, 52 subjects treated with naltrexone and 56 subjects treated with acamprosate were studied. In this study both treatment-seeking and non-treatment-seeking subjects were recruited. A number of genes were examined to determine if polymorphisms were related to behavioral effects of alcohol after treatment with naltrexone or acamprosate, including the *OPRM1, DRD1, DRD2, GRIN2B, GABRA6, GABRB2,* and *GABRG2* genes. Results of that study showed a significant interaction between *GABRB2* genotype and treatment for change in peak heart rate after alcohol cue exposure, with acamprosate outperforming naltrexone when the *GABRB2* C1412T polymorphism was homozygous for the T allele,[24] suggesting a potential role for the GABA A receptor beta 2 subunit gene and treatment response.

Although disulfiram is approved by the FDA for the treatment of alcohol dependence, research on the pharmacogenomics of treatment response to disulfiram for alcohol dependence is lacking. However, disulfiram has a number of biochemical effects that have led to its use as a treatment for other addictive disorders. Disulfiram inhibits the enzyme aldehyde dehydrogenase, leading to increased levels of acetaldehyde, and the adverse effects of nausea and vomiting after alcohol consumption. Based on disulfiram's efficacy at reducing alcohol consumption, a clinical trial of disulfiram was carried out for cocaine dependence in subjects with comorbid alcohol dependence.[25] Results of that trial showed that disulfiram improved treatment retention and produced a longer duration of abstinence from alcohol and cocaine use. A 1-year follow-up of subjects in this study found that cocaine use at follow-up was significantly less in subjects assigned to disulfiram versus subjects who were not assigned to disulfiram.[26] Based on these findings a second trial was executed in 121 cocaine-dependent subjects, with roughly half of the subjects also meeting criteria for current alcohol dependence. Results of this trial showed that disulfiram significantly reduced cocaine-positive urines compared to placebo, and that disulfiram was most effective in patients who were not alcohol dependent at baseline or who remained abstinent from alcohol during the trial.[27] This finding suggested that the effect of disulfiram on cocaine use was mediated by another mechanism than the effect on alcohol, and led investigators to seek an alternative mechanism through which disulfiram could be reducing cocaine consumption. In addition to disulfiram's effects on aldehyde dehydrogenase, it affects other enzymes presumably through its effects as a copper chelator. One of these enzymes is dopamine beta hydroxylase, which is a key enzyme in the conversion of dopamine to norepinephrine.[28] A functional polymorphism in close proximity to the dopamine beta hydroxylase gene has been identified, and this polymorphism has been suggested to be potentially meaningful in the treatment response of subjects treated with disulfiram for cocaine dependence;[28] however, to date no studies have been published on the pharmacogenetics of this polymorphism in the treatment response of disulfiram for cocaine dependence.

In the pharmacotherapy for stimulant abuse, there is as yet no FDA-approved medication for any of the psychostimulants. However, there are several studies that examined different classes of medications and found some reduction in stimulant use. The majority of placebo-controlled trials across a range of medications have not produced significant reductions in cocaine use (reviewed in Vocci and Ling.[29]) Citalopram is one of a handful of medications that have been shown in placebo-controlled trials to significantly reduce cocaine-positive urines in cocaine-dependent subjects. In a randomized, 12-week placebo controlled trial of citalopram, 40 cocaine-dependent subjects were treated with placebo plus contingent reinforcement for cocaine-negative urines, compared to 36 cocaine-dependent subjects treated with 20 mg of citalopram plus contingent reinforcement for cocaine-negative urines.[30] Results of that study showed that citalopram significantly reduced cocaine-positive urines compared to placebo, without causing clinically significant side effects or leading to greater subject dropout than placebo.[30] At the same time, studies using other selective serotonin reuptake inhibitors (SSRIs) have produced conflicting results. In a previous study, the SSRI fluoxetine 20 mg, 40 mg, or placebo were administered to 155 cocaine-dependent outpatients for 12 weeks. Results of that study showed a significant worsening in treatment retention in cocaine-dependent subjects treated with higher doses of fluoxetine, and no effect on urine toxicology results or craving compared to placebo.[31] In a study of fluoxetine carried out by another research group, cocaine-dependent subjects treated with fluoxetine 40 mg/day for 12 weeks (16 subjects) had improved treatment retention but no differences in cocaine-positive urines or craving compared to placebo (16 subjects).[32] A third study examining fluoxetine for cocaine dependence combined abstinence-based contingent reinforcement (CM) for cocaine-negative urines with placebo (29 subjects), fluoxetine 20 mg (23 subjects), or fluoxetine 40 mg (29 subjects), and found that cocaine-dependent subjects treated with 40 mg of fluoxetine had fewer cocaine-positive urines during treatment with CM than placebo or fluoxetine 20 mg treated subjects, with no difference in retention.[33] Other small sample studies have examined other SSRIs for treatment of cocaine dependence. As part of the CREST trial, 16 cocaine-dependent subjects were treated with the SSRI sertraline and compared with 17 subjects treated with placebo, 17 subjects with tiagabine, and 17 subjects with donepezil for 10 weeks.[34] Results of that study showed that the sertraline-treated group had a 68% decrease in urine benzoylecgonine levels, compared to a 32% increase in urine benzoylecgonine levels for the placebo-treated group; however, this difference was not statistically significant, possibly due to small sample size. A second CREST trial study compared 16 cocaine-dependent subjects treated with the SSRI paroxetine to 16 subjects treated with placebo, 16 subjects treated with pentozifylline, and 17 subjects treated with riluzole for 8 weeks. Results of that study were that no drug was significantly different from placebo at reducing cocaine-positive urines; however, the small sample size limits the findings.[35] Another study with the primary aim of using citalopram for opiate use in methadone-maintained subjects examined cocaine use as a secondary outcome. That study found no difference between citalopram (20 subjects), citalopram plus bupropion (20 subjects), and placebo (20 subjects) in the rate of cocaine-positive urines.[36] However, the study by Poling[36] is difficult to compare with other studies as it is unclear if subjects met criteria for cocaine dependence or whether there was sufficient variability in cocaine use to detect change over time. In summary, although there is some evidence to support the SSRI citalopram for cocaine dependence, further research is needed to confirm this finding.

Across different trials of SSRIs for cocaine dependence, there has been considerable heterogeneity in treatment response. It has been suggested that some of the heterogeneity in treatment response to psychotropic medication could be accounted for by pharmacogenetics.[37] Based on the known mechanism of action of citalopram as an SSRI, an important target of citalopram in the treatment of addiction is the serotonin transporter gene (*SLC6A4*). The 5-HT transporter, encoded by the *SLC6A4* gene, located at 17q11.2, is a central regulator of 5-HT turnover, transporting 5-HT from the synaptic cleft into the presynaptic neuron and terminating serotonin's action. A 20-23-bp repeat polymorphism in the *SLC6A4* promoter (termed *5-HTTLPR*) has been shown to affect transcriptional activity of this gene. The short or "S" allele with 14 repeats was shown to have lower transcriptional activity than the long or "L" allele with 16 repeats.[38-40] This repeat polymorphism was found to be associated with anxiety-related traits.[39] Recently, a single nucleotide polymorphism (*A/G*) was identified (rs25531) near the repeat sequence of the L-allele that subdivided the polymorphism further into the L_A and L_G alleles.[41] The L_G and S alleles were found to have similar low expression levels. This allows the genotypes to be classified as having high (L_A/L_A), intermediate (L_A/L_G, S_A/L_A), or low (S_A/S_A, L_G/S_A, L_G/L_G) transcriptional efficacy. In a recent study examining triallelic serotonin transporter genotype and treatment response to citalopram in dementia, results showed that low-expression alleles (S and L_G) had more side effects and early treatment discontinuation.[42]

More than 20 studies examined the relationship of *SLC6A4* variants and response to antidepressant treatment (reviewed in Horstmann and Binder[43]).

A meta-analysis of 1,400 subjects of Caucasian and Asian ethnicity from 15 studies supported worse response to antidepressant therapy in carriers of the S-allele.[44] However, to date no pharmacogenomic studies have been published on SSRI treatment outcome for addictions. In the most recent paper published on this topic to date in non-treatment-seeking alcohol-dependent subjects, 21 subjects with different *SLC6A4* variants were treated with sertraline or ondasetron followed by placebo followed by the other medication, so that all subjects received both medications and placebo in a crossover design. Subjects underwent an alcohol self-administration experiment while on active medication. Results of that study showed that ondansetron reduced alcohol self-administration compared to sertraline in subjects with the *L/L* genotype.[45]

Besides citalopram, a few other medications have shown some promise for cocaine dependence with at least one placebo-controlled trial showing a reduction in cocaine use. Among medications tested, "agonist"-like interventions that facilitate dopaminergic function in the central nervous system have shown reasonable success. Of these, some inhibit dopamine reuptake or metabolism (such as bupropion and disulfiram[27,46]) or replenish dopamine stores (such as levodopa[47]), while other medications indirectly enhance dopaminergic function via effects on dopamine release (such as dextroamphetamine[48-50] and methamphetamine[51]). The use of these medications is based on the effects of cocaine on the dopamine system. Cocaine is a dopamine transporter inhibitor, capable of increasing extracellular concentrations of dopamine substantially.[52,53] Chronic cocaine use results in dopamine depletion or a hypodopaminergic state.[54] It follows that medications acting on the dopamine system might be appropriate candidates for treatment of cocaine dependence.

Overall, medication development strategies involving dopamine-enhancing agents have been productive, promising, and worthy of further investigation.[55-57] None of the agents studied to date is without limitations. Stimulants such as d-amphetamine have shown some efficacy signal in terms of reducing cocaine-positive urines.[48,49] however the possibility of abuse or diversion of d-amphetamine has limited its widespread clinical use.[57]

Based on the dopamine-enhancing effects of these medications, one polymorphism that could be examined as a potential moderator of treatment response is the *COMT* gene. Catechol-O-methyltransferase (COMT) is an S-adenosylmethionine-dependent enzyme that methylates catecholamine neurotransmitters (dopamine and norepinephrine) as a step in the degradation of these neurotransmitters. COMT exists in membrane-bound and soluble forms, with the membrane-bound form being the predominant form in the brain. The *COMT* gene has a functional polymorphism

(*Val158Met*) in which substitution of valine 158 to methionine significantly alters the activity of the enzyme. As reviewed in Harrison and Tunbridge,[58] dopamine signaling is likely to be higher in *Met 158* carriers compared to *Val 158* carriers, because studies have shown that *Met 158* homozygotes had substantially lower COMT activity in human prefrontal cortex and erythrocytes. The *Val158Met COMT* polymorphism has been studied in relation to a number of psychiatric and other disorders and to cognitive function.[59,60]

To date, no published studies have examined the role of *COMT* Val158Met polymorphisms in treatment response to dopamine-enhancing medications for stimulant abuse. However, studies have examined the relationship between this polymorphism and treatment with stimulants for attention deficit-hyperactivity disorder (ADHD). In a study of 173 ADHD subjects and 284 sex-matched controls, the *Val* allele and the *Val* homozygous genotype were significantly more frequent in the ADHD group compared to controls. Further, in a subset of 122 ADHD subjects treated with methylphenidate, the *Val* allele showed a significant association with good treatment response, with the *Val* homozygous genotype being twice as frequent in the responder group than in the nonresponder group.[61] This finding was supported by a second study that examined several candidate genes in relation to methylphenidate treatment response. In that study of 82 subjects with ADHD, there was a trend for *COMT* genotype to be associated with ADHD symptom reduction with methylphenidate.[62] Based on these studies, *COMT* genotype could be a potential target for future pharmacogenomic studies of stimulant addiction.

9.4 PHARMACOGENOMICS OF BIOCHEMICAL EFFECTS OF DRUGS OF ABUSE

Prior studies using positron emission tomography (PET) have also shown evidence of reduced dopamine D2 receptor binding in cocaine-dependent subjects compared to non-drug-using controls which persists for days to weeks after cessation of cocaine use.[63–65] Based on this finding another potential target for pharmacogenomic studies of medication treatment for cocaine dependence would be dopamine D2 receptor polymorphisms. The dopamine D2 receptor is a G protein coupled receptor that inhibits adenylyl cyclase activity. In humans the dopamine D2 receptor is encoded by the *DRD2* gene. The *DRD2 Taq1* A1 variant was initially thought to be directly related to the *DRD2* gene; however, it was later found to be in the *ANKK1* gene 10kB downstream of the *DRD2* gene.[66] An insertion/deletion variant *DRD2-141C* has also been identified which alters transcriptional efficiency of the receptor,[67] and a single nucleotide polymorphism *C957T* in the *DRD2* gene that affects protein synthesis was described.[68] Although to date no studies have been published on the pharmacogenetics of cocaine treatment related to the *DRD2* gene, three *DRD2*-related polymorphisms have been studied in relation to treatment response for nicotine dependence. The rationale for these studies is that while the direct effects of nicotine are mediated though binding of nicotinic acetylcholine receptors, nicotine also increases dopamine levels in the brain, similar to other drugs of abuse.

David et al.[69] examined the effect of the *DRD2 TaqI* A1 variant on treatment response to bupropion for cigarette smoking in 29 smokers. Results of that study showed that within the bupropion treatment group, subjects with the *A2/A2* alleles demonstrated significant reductions in craving and irritability, which were not seen in subjects with *A1/A2* or *A1/A1* alleles. In a second larger-scale study, 755 subjects provided DNA for analysis of *DRD2* and dopamine beta hydroxylase (*DBH*) genotype to determine whether polymorphisms for these genes were related to treatment outcome using the nicotine patch for smoking cessation. Results of that study were that at week 1 there was a significant association between *DRD2* Taq1 A1 variant and treatment response, particularly when combined with *DBH* A1368.[70] However, in a follow-up study by the same research group in a larger sample of 804 subjects, there was no association between *DRD2 TaqI A* polymorphism and treatment response to nicotine replacement therapy, contrary to the previous finding.[71] Lerman et al.[72] examined the relationship between *DRD2* genotype and treatment response for nicotine dependence in two treatment studies. In a placebo-controlled trial of bupropion in 414 subjects, they found a significant relationship between *DRD2-141C Ins/Del* genotype, with smokers homozygous for the *Ins C* allele having a more favorable treatment response to bupropion than subjects carrying the *Del C* allele. In the second treatment study, an open-label trial of transdermal nicotine versus nicotine nasal spray in 368 subjects, subjects carrying the *Del C* allele had higher smoking cessation rates, regardless of route of administration, along with the *DRD2* C957T variant.[72] These authors interpreted their findings as suggesting that *DRD2-141C Ins/Del* genotype may be helpful in choice of bupropion or nicotine replacement therapy for smoking cessation; however these findings need to be replicated.

More recently, treatment response for nicotine or bupropion for smoking cessation has been studied using genome-wide association (GWA) methods. In GWA studies a large number (often over a thousand) of polymorphisms across DNA samples are examined for potential differences between groups, without a priori hypotheses about which genes may differ. In a study of 550 European American smokers treated with nicotine replacement therapy, bupropion, or placebo from three different samples of subjects, a number of single nucleotide polymorphisms from the three groups were found whose allele frequencies differed between successful abstainers and nonabstainers. Genes identified as being associated with treatment response included cell adhesion genes, genes involved in enzymes, receptors, channels, transporters, DNA/RNA processing, intracellular signaling pathways, and structural proteins.[73] In a second GWA study, 480 subjects provided DNA for analysis of polymorphisms related to smoking cessation. These subjects were recruited from community settings in an attempt to replicate the previous GWA results. Findings of this study were that the haplotypes identified as being associated with not being able to quit smoking partially overlapped with findings from the previous study, with some of the genes identified in both studies being *A2BP1, CSMD1,* and *DSCAM, PCDH15,* and *RARB.*[74] In another GWA study by the same group, nicotine abstinence genes were examined in 369 subjects who were randomized to nicotine replacement therapy initiated 2 weeks before target quit dates. For treatment outcome of 10-week continuous smoking abstinence, overlap with previous studies was found for genes associated with treatment outcome.[75] The authors of this study argue that the results of recent

GWA studies on smoking cessation treatment are on track to produce personalized approaches for smoking cessation treatment but that "We need to continue to work to apply an integrated sum of SNPs in the context of appropriate clinical information to match individuals with the best type and/or intensity of therapy to maximize benefits and minimize side effects in smoking cessation."[75] (p. 523)

9.5 CONCLUSIONS

Although there has been much discussion in the literature about the potential of pharmacogenomic research for addiction treatment across several different addictive disorders, to date the majority of the pharmacogenomic research on addictions has been done in alcohol and nicotine addiction treatment studies. Results of the candidate gene studies in nicotine addiction treatment have been conflicting, with some studies finding an association with *DRD2* gene polymorphisms and others not. More recently GWA studies have provided more consistent results, with several studies finding overlap in genes associated with smoking cessation.

Results of pharmacogenomics of treatment response in alcohol dependence have also shown conflicting results, with some but not all studies showing a differential treatment response to naltrexone based on *OPRM1* genotype. It has been argued that although not all studies found this association, the majority of the studies found an association between *OPRM1* genotype and naltrexone treatment response, and it is possible that the negative study used an insufficient naltrexone dose. One potential limitation of this research is that the most studied allele of the *OPRM1* gene, the *Asp40* allele, is uncommon in some ethnic groups such as African Americans, which could limit its usefulness outside of Caucasian patients. With this limitation in mind, there is some evidence to support this polymorphism as a potential moderator of treatment response with naltrexone for alcohol dependence.

With respect to treatment outcome for other addictions, few pharmacogenomic studies have been reported. The focus on alcohol and nicotine in pharmacogenomic research has occurred largely because these are two of the most common drugs of abuse, and there are FDA-approved medications for both disorders. Without a medication that has been shown to be effective in large-scale trials for other drugs of abuse, such as cocaine and marijuana, it is unlikely that research on these drugs will approach the level of pharmacogenomic research that has been done for nicotine or alcohol in the near future. Another approach that has been suggested is to use pharmacogenomics to help explain the substantial heterogeneity in treatment response in addictions such as cocaine. Based on the putative mechanism of action of medications that have shown some promise, several candidate genes could be examined as potential moderators of treatment response. Candidate genes discussed include dopamine (*DRD2* and *COMT*) and serotonin (*SLC6A4*) genes, but as yet there have not been studies published examining these genes as potential moderators of treatment response in stimulant addiction.

In short, pharmacogenomics research in addiction treatment to date has shown some promise of being important in treatment of addictions, but larger-scale trials with effective medications are needed before pharmacogenomics can be applied outside the research setting.

REFERENCES

1. Kosten TR. Current pharmacotherapies for opioid dependence. *Psychopharmacol Bull* 1990; 26(1): 69–74.
2. Zhou SF. Polymorphism of human cytochrome P450 2D6 and its clinical significance: part II. *Clin Pharmacokinet* 2009; 48(12): 761–804.
3. Crettol S, Deglon JJ, Besson J, Croquette-Krokkar M, Gothuey I, Hammig R, et al. Methadone enantiomer plasma levels, CYP2B6, CYP2C19, and CYP2C9 genotypes, and response to treatment. *Clin Pharmacol Ther* 2005; 78(6): 593–604.
4. Eap CB, Broly F, Mino A, Hammig R, Deglon JJ, Uehlinger C, et al. Cytochrome P450 2D6 genotype and methadone steady-state concentrations. *J Clin Psychopharmacol* 2001; 21(2): 229–234.
5. Perez de los Cobos J, Sinol N, Trujols J, del Rio E, Banuls E, Luquero E, et al. Association of CYP2D6 ultrarapid metabolizer genotype with deficient patient satisfaction regarding methadone maintenance treatment. *Drug Alcohol Depend* 2007; 89(2–3): 190–194.
6. McCance-Katz EF, Sullivan LE, Nallani S. Drug interactions of clinical importance among the opioids, methadone and buprenorphine, and other frequently prescribed medications: a review. *Am J Addict* 2010; 19(1): 4–16.
7. Bond C, LaForge KS, Tian M, Melia D, Zhang S, Borg L, et al. Single-nucleotide polymorphism in the human mu opioid receptor gene alters beta-endorphin binding and activity: possible implications for opiate addiction. *Proc Natl Acad Sci USA* 1998; 95(16): 9608–9613.
8. Fonseca F, Gratacos M, Escaramis G, De Cid R, Martin-Santos R, Fernandez-Espejo E, et al. Response to methadone maintenance treatment is associated with the MYOCD and GRM6 genes. *Mol Diagn Ther* 2010; 14(3): 171–178.
9. Rounsaville BJ. Can psychotherapy rescue naltrexone treatment of opioid addiction? *NIDA Res Monogr* 1995; 150: 37–52.
10. Rawson RA, Glazer M, Callahan EJ, Liberman RP. Naltrexone and behavior therapy for heroin addiction. *NIDA Res Monogr* 1979; (25): 26–43.
11. Kleber HD. Clinical aspects of the use of narcotic antagonists: the state of the art. *Int J Addict* 1977; 12(7): 857–861.
12. Greenstein RA, Resnick RB, Resnick E. Methadone and naltrexone in the treatment of heroin dependence. *Psychiatr Clin North Am* 1984; 7(4): 671–679.
13. O'Brien CP, Greenstein RA, Mintz J, Woody GE. Clinical experience with naltrexone. *Am J Drug Alcohol Abuse* 1975; 2(3–4): 365–377.
14. Shufman EN, Porat S, Witztum E, Gandacu D, Bar-Hamburger R, Ginath Y. The efficacy of naltrexone in preventing reabuse of heroin after detoxification. *Biol Psychiatry* 1994; 35(12): 935–945.
15. Comer SD, Sullivan MA, Yu E, Rothenberg JL, Kleber HD, Kampman K, et al. Injectable, sustained-release naltrexone for the treatment of opioid dependence: a randomized, placebo-controlled trial. *Arch Gen Psychiatry* 2006; 63(2): 210–218.
16. Kleber HD, Kosten TR. Naltrexone induction: psychologic and pharmacologic strategies. *J Clin Psychiatry* 1984; 45(9 Pt 2): 29–38.
17. Preston KL, Silverman K, Umbricht A, DeJesus A, Montoya ID, Schuster CR. Improvement in naltrexone treatment compliance with contingency management. *Drug Alcohol Depend* 1999; 54(2): 127–135.
18. Oslin DW, Berrettini W, Kranzler HR, Pettinati H, Gelernter J, Volpicelli JR et al. A functional polymorphism of the mu-opioid receptor gene is associated with naltrexone response in alcohol-dependent patients. *Neuropsychopharmacology* 2003; 28(8): 1546–1552.

19. Gelernter J, Gueorguieva R, Kranzler HR, Zhang H, Cramer J, Rosenheck R, et al. Opioid receptor gene (OPRM1, OPRK1, and OPRD1) variants and response to naltrexone treatment for alcohol dependence: results from the VA Cooperative Study. *Alcohol Clin Exp Res* 2007; 31(4): 555–563.

20. Ray LA, Hutchison KE. Effects of naltrexone on alcohol sensitivity and genetic moderators of medication response: a double-blind placebo-controlled study. *Arch Gen Psychiatry* 2007; 64(9): 1069–1077.

21. Anton RF, Oroszi G, O'Malley S, Couper D, Swift R, Pettinati H, et al. An evaluation of mu-opioid receptor (OPRM1) as a predictor of naltrexone response in the treatment of alcohol dependence: results from the Combined Pharmacotherapies and Behavioral Interventions for Alcohol Dependence (COMBINE) study. *Arch Gen Psychiatry* 2008; 65(2): 135–144.

22. Oroszi G, Anton RF, O'Malley S, Swift R, Pettinati H, Couper D, et al. OPRM1 Asn40Asp predicts response to naltrexone treatment: a haplotype-based approach. *Alcohol Clin Exp Res* 2009; 33(3): 383–393.

23. Vallender EJ, Ruedi-Bettschen D, Miller GM, Platt DM. A pharmacogenetic model of naltrexone-induced attenuation of alcohol consumption in rhesus monkeys. *Drug Alcohol Depend* 2010; 109(1–3): 252–256.

24. Ooteman W, Naassila M, Koeter MW, Verheul R, Schippers GM, Houchi H, et al. Predicting the effect of naltrexone and acamprosate in alcohol-dependent patients using genetic indicators. *Addict Biol* 2009; 14(3): 328–337.

25. Carroll KM, Nich C, Ball SA, McCance E, Rounsavile BJ. Treatment of cocaine and alcohol dependence with psychotherapy and disulfiram. *Addiction* 1998; 93(5): 713–727.

26. Carroll KM, Nich C, Ball SA, McCance E, Frankforter TL, Rounsaville BJ. One-year follow-up of disulfiram and psychotherapy for cocaine-alcohol users: sustained effects of treatment. *Addiction* 2000; 95(9): 1335–1349.

27. Carroll KM, Fenton LR, Ball SA, Nich C, Frankforter TL, Shi J, et al. Efficacy of disulfiram and cognitive behavior therapy in cocaine-dependent outpatients: a randomized placebo-controlled trial. *Arch Gen Psychiatry* 2004; 61(3): 264–272.

28. Haile CN, Kosten TR, Kosten TA. Pharmacogenetic treatments for drug addiction: cocaine, amphetamine and methamphetamine. *Am J Drug Alcohol Abuse* 2009; 35(3): 161–177.

29. Vocci F, Ling W. Medications development: successes and challenges. *Pharmacol Ther* 2005; 108(1): 94–108.

30. Moeller FG, Schmitz JM, Steinberg JL, Green CM, Reist C, Lai LY, et al. Citalopram combined with behavioral therapy reduces cocaine use: a double-blind, placebo-controlled trial. *Am J Drug Alcohol Abuse* 2007; 33(3): 367–378.

31. Grabowski J, Rhoades H, Elk R, Schmitz J, Davis C, Creson D, et al. Fluoxetine is ineffective for treatment of cocaine dependence or concurrent opiate and cocaine dependence: two placebo-controlled double-blind trials. *J Clin Psychopharmacol* 1995; 15(3): 163–174.

32. Batki SL, Washburn AM, Delucchi K, Jones RT. A controlled trial of fluoxetine in crack cocaine dependence. *Drug Alcohol Depend* 1996; 41(2): 137–142.

33. Schmitz JM, Rhoades HM, Elk R, Creson D, Hussein I, Grabowski J. Medication take-home doses and contingency management. *Exp Clin Psychopharmacol* 1998; 6(2): 162–168.

34. Winhusen TM, Somoza EC, Harrer JM, Mezinskis JP, Montgomery MA, Goldsmith RJ, et al. A placebo-controlled screening trial of tiagabine, sertraline and donepezil as cocaine dependence treatments. *Addiction* 2005; 100 Suppl 1: 68–77.

35. Ciraulo DA, Sarid-Segal O, Knapp CM, Ciraulo AM, LoCastro J, Bloch DA, et al. Efficacy screening trials of paroxetine, pentoxifylline, riluzole, pramipexole and venlafaxine in cocaine dependence. *Addiction* 2005; 100 Suppl 1: 12–22.

36. Poling J, Pruzinsky R, Kosten TR, Gonsai K, Sofuoglu M, Gonzalez G, et al. Clinical efficacy of citalopram alone or augmented with bupropion in methadone-stabilized patients. *Am J Addict* 2007; 16(3): 187–194.

37. Malhotra AK, Murphy GM, Jr., Kennedy JL. Pharmacogenetics of psychotropic drug response. *Am J Psychiatry* 2004; 161(5): 780–796.

38. Heils A, Teufel A, Petri S, Stober G, Riederer P, Bengel D, et al. Allelic variation of human serotonin transporter gene expression. *J Neurochem* 1996; 66(6): 2621–2624.

39. Lesch KP, Bengel D, Heils A, Sabol SZ, Greenberg BD, Petri S, et al. Association of anxiety-related traits with a polymorphism in the serotonin transporter gene regulatory region. *Science* 1996; 274(5292): 1527–1531.

40. Heils A, Mossner R, Lesch KP. The human serotonin transporter gene polymorphism—basic research and clinical implications. *J Neural Transm* 1997; 104(10): 1005–1014.

41. Hu XZ, Lipsky RH, Zhu G, Akhtar LA, Taubman J, Greenberg BD, et al. Serotonin transporter promoter gain-of-function genotypes are linked to obsessive-compulsive disorder. *Am J Hum Genet* 2006; 78(5): 815–826.

42. Dombrovski AY, Mulsant BH, Ferrell RE, Lotrich FE, Rosen JI, Wallace M, et al. Serotonin transporter triallelic genotype and response to citalopram and risperidone in dementia with behavioral symptoms. *Int Clin Psychopharmacol* 2010; 25(1): 37–45.

43. Horstmann S, Binder EB. Pharmacogenomics of antidepressant drugs. *Pharmacol Ther* 2009; 124(1): 57–73.

44. Serretti A, Kato M, De Ronchi D, Kinoshita T. Meta-analysis of serotonin transporter gene promoter polymorphism (5-HTTLPR) association with selective serotonin reuptake inhibitor efficacy in depressed patients. *Mol Psychiatry* 2007; 12(3): 247–257.

45. Kenna GA, Zywiak WH, McGeary JE, Leggio L, McGeary C, Wang S, et al. A within-group design of nontreatment seeking 5-HTTLPR genotyped alcohol-dependent subjects receiving ondansetron and sertraline. *Alcohol Clin Exp Res* 2009; 33(2): 315–323.

46. Poling J, Oliveto A, Petry N, Sofuoglu M, Gonsai K, Gonzalez G, et al. Six-month trial of bupropion with contingency management for cocaine dependence in a methadone-maintained population. *Arch Gen Psychiatry* 2006; 63(2): 219–228.

47. Schmitz JM, Mooney ME, Moeller FG, Stotts AL, Green C, Grabowski J. Levodopa pharmacotherapy for cocaine dependence: choosing the optimal behavioral therapy platform. *Drug Alcohol Depend* 2008; 94(1–3): 142–150.

48. Grabowski J, Rhoades H, Schmitz J, Stotts A, Daruzska LA, Creson D, et al. Dextroamphetamine for cocaine-dependence treatment: a double-blind randomized clinical trial. *J Clin Psychopharmacol* 2001; 21(5): 522–526.

49. Grabowski J, Rhoades H, Stotts A, Cowan K, Kopecky C, Dougherty A, et al. Agonist-like or antagonist-like treatment for cocaine dependence with methadone for heroin dependence: two double-blind randomized clinical trials. *Neuropsychopharmacology* 2004; 29(5): 969–981.

50. Shearer J, Wodak A, van Beek I, Mattick RP, Lewis J. Pilot randomized double blind placebo-controlled study of dexamphetamine for cocaine dependence. *Addiction* 2003; 98(8): 1137–1141.

51. Mooney ME, Herin DV, Schmitz JM, Moukaddam N, Green CE, Grabowski J. Effects of oral methamphetamine on cocaine use: a randomized, double-blind, placebo-controlled trial. *Drug Alcohol Depend* 2009; 101(1–2): 34–41.

52. Bradberry CW, Barrett-Larimore RL, Jatlow P, Rubino SR. Impact of self-administered cocaine and cocaine cues on extracellular dopamine in mesolimbic and sensorimotor striatum in rhesus monkeys. *J Neurosci* 2000; 20(10): 3874–3883.

53. Tsukada H, Harada N, Nishiyama S, Ohba H, Kakiuchi T. Dose-response and duration effects of acute administrations of cocaine and GBR12909 on dopamine synthesis and transporter in the conscious monkey brain: PET studies combined with microdialysis. *Brain Res* 2000; 860(1–2): 141–148.

54. Parsons LH, Smith AD, Justice JB, Jr. Basal extracellular dopamine is decreased in the rat nucleus accumbens during abstinence from chronic cocaine. *Synapse* 1991; 9(1): 60–65.

55. Castells X, Casas M, Vidal X, Bosch R, Roncero C, Ramos-Quiroga JA, et al. Efficacy of central nervous system stimulant treatment for cocaine dependence: a systematic review and meta-analysis of randomized controlled clinical trials. *Addiction* 2007; 102(12): 1871–1887.

56. Vocci FJ, Elkashef A. Pharmacotherapy and other treatments for cocaine abuse and dependence. *Curr Opin Psychiatry* 2005; 18(3): 265–270.

57. Moeller FG, Schmitz JM, Herin D, Kjome KL. Use of stimulants to treat cocaine and methamphetamine abuse. *Curr Psychiatry Rep* 2008; 10(5): 385–391.

58. Harrison PJ, Tunbridge EM. Catechol-*O*-methyltransferase (COMT): a gene contributing to sex differences in brain function, and to sexual dimorphism in the predisposition to psychiatric disorders. *Neuropsychopharmacology* 2008; 33(13): 3037–3045.

59. Oosterhuis BE, LaForge KS, Proudnikov D, Ho A, Nielsen DA, Gianotti R, et al. Catechol-*O*-methyltransferase (COMT) gene variants: possible association of the Val158Met variant with opiate addiction in Hispanic women. *Am J Med Genet B Neuropsychiatr Genet* 2008; 147B(6): 793–798.

60. Cusin C, Serretti A, Lattuada E, Lilli R, Lorenzi C, Smeraldi E. Association study of MAO-A, COMT, 5-HT2A, DRD2, and DRD4 polymorphisms with illness time course in mood disorders. *Am J Med Genet* 2002; 114(4): 380–390.

61. Kereszturi E, Tarnok Z, Bognar E, Lakatos K, Farkas L, Gadoros J, et al. Catechol-*O*-methyltransferase Val158Met polymorphism is associated with methylphenidate response in ADHD children. *Am J Med Genet B Neuropsychiatr Genet* 2008; 147B(8): 1431–1435.

62. McGough JJ, McCracken JT, Loo SK, Manganiello M, Leung MC, Tietjens JR, et al. A candidate gene analysis of methylphenidate response in attention-deficit/hyperactivity disorder. *J Am Acad Child Adolesc Psychiatry* 2009; 48(12): 1155–1164.

63. Volkow ND, Fowler JS, Wolf AP, Schlyer D, Shiue CY, Alpert R, et al. Effects of chronic cocaine abuse on postsynaptic dopamine receptors. *Am J Psychiatry* 1990; 147(6): 719–724.

64. Martinez D, Broft A, Foltin RW, Slifstein M, Hwang DR, Huang Y, et al. Cocaine dependence and d2 receptor availability in the functional subdivisions of the striatum: relationship with cocaine-seeking behavior. *Neuropsychopharmacology* 2004; 29(6): 1190–1202.

65. Volkow ND, Fowler JS, Wang GJ, Hitzemann R, Logan J, Schlyer DJ, et al. Decreased dopamine D2 receptor availability is associated with reduced frontal metabolism in cocaine abusers. *Synapse* 1993; 14(2): 169–177.

66. Neville MJ, Johnstone EC, Walton RT. Identification and characterization of ANKK1: a novel kinase gene closely linked to DRD2 on chromosome band 11q23.1. *Hum Mutat* 2004; 23(6): 540–545.

67. Arinami T, Gao M, Hamaguchi H, Toru M. A functional polymorphism in the promoter region of the dopamine D2 receptor gene is associated with schizophrenia. *Hum Mol Genet* 1997; 6(4): 577–582.

68. Duan J, Wainwright MS, Comeron JM, Saitou N, Sanders AR, Gelernter J, et al. Synonymous mutations in the human dopamine receptor D2 (DRD2) affect mRNA stability and synthesis of the receptor. *Hum Mol Genet* 2003; 12(3): 205–216.

69. David SP, Niaura R, Papandonatos GD, Shadel WG, Burkholder GJ, Britt DM, et al. Does the DRD2-Taq1 A polymorphism influence treatment response to bupropion hydrochloride for reduction of the nicotine withdrawal syndrome? *Nicotine Tob Res* 2003; 5(6): 935–942.

70. Johnstone EC, Yudkin PL, Hey K, Roberts SJ, Welch SJ, Murphy MF, et al. Genetic variation in dopaminergic pathways and short-term effectiveness of the nicotine patch. *Pharmacogenetics* 2004; 14(2): 83–90.

71. Munafo MR, Johnstone EC, Murphy MF, Aveyard P. Lack of association of DRD2 rs1800497 (Taq1A) polymorphism with smoking cessation in a nicotine replacement therapy randomized trial. *Nicotine Tob Res* 2009; 11(4): 404–407.

72. Lerman C, Jepson C, Wileyto EP, Epstein LH, Rukstalis M, Patterson F, et al. Role of functional genetic variation in the dopamine D2 receptor (DRD2) in response to bupropion and nicotine replacement therapy for tobacco dependence: results of two randomized clinical trials. *Neuropsychopharmacology* 2006; 31(1): 231–242.

73. Uhl GR, Liu QR, Drgon T, Johnson C, Walther D, Rose JE, et al. Molecular genetics of successful smoking cessation: convergent genome-wide association study results. *Arch Gen Psychiatry* 2008; 65(6): 683–693.

74. Drgon T, Montoya I, Johnson C, Liu QR, Walther D, Hamer D, et al. Genome-wide association for nicotine dependence and smoking cessation success in NIH research volunteers. *Mol Med* 2009; 15(1–2): 21–27.

75. Uhl GR, Drgon T, Johnson C, Ramoni MF, Behm FM, Rose JE. Genome-wide association for smoking cessation success in a trial of precessation nicotine replacement. *Mol Med* 2010; 16(11–12): 513–526.

10 Methodologies in Pharmacogenetics Testing

Jorge L. Sepulveda

CONTENTS

10.1 Introduction 211
10.2 Low Multiplexing Assays 212
10.3 Mid-Multiplexing Assays 219
10.4 High Multiplexing and Next-Generation Sequencing Systems 222
    10.4.1 High-Plex Single Nucleotide Polymorphism (SNP) Assays 222
    10.4.2 High-Throughput Library Preparation and Polymerase Chain
        Reaction (PCR) Systems... 222
    10.4.3 Next-Generation Sequencing 224
10.5 Conclusions.. 227
References... 227

10.1 INTRODUCTION

Twenty-first century medical practice can be predicted to benefit from a confluence of scientific and technologic advances leading toward truly rational and personalized health care. These advances include thorough understanding of the human genome including sequence–function relationships, the role of regulatory regions, and the effects of genetic variation on phenotypic expression; comprehensive structural modeling of DNA, ribonucleic acids, proteins, small molecules, and their specific interactions so that rational drug design can be perfected; systems biology approaches to understanding the complex networks of metabolic, signal transduction, and regulatory pathways at the organelle, cellular, organ, and organism levels so that targeted therapies can be optimized with minimal side effects; and exponential growth in computer hardware and software to process the increasing complexity of the above developments. A particular aspect of personalized medicine is the study of the effects of genetic variation in the response to drug therapy, which can be labeled as *pharmacogenomics* when the focus is on large-scale analyses of the genome, or *pharmacogenetics*, when the focus is on a relatively restricted number of genes. This distinction is becoming fuzzier as newer technologies for genetic analysis exhibit higher multiplexing abilities and lower cost per assay and if whole genomic testing becomes routine and replaces focused genetic testing. This chapter briefly describes a variety of methods used to identify DNA variation affecting genes involved in drug metabolism (pharmacokinetics) or drug target genes (pharmacodynamics), starting from currently available methods focused on a restricted set of genes and ending on high-throughput next-generation sequencing. Despite the rapid development of technologic advances and the multitude of commercially available research-use-only reagents for pharmacogenetic testing, in June 2011 only nine pharmacogenetics assays were U.S. Food and Drug Administration (FDA)-approved, of which five are used for Warfarin PGx (Table 10.1). It is important that the legal and regulatory environment keeps up with the rapid pace of development in genetics to fully realize its benefits to human health.

10.2   LOW MULTIPLEXING ASSAYS

In this category we briefly describe methods designed to simultaneously identify a restricted number of variations, usually assaying for one or two alleles at each position. The first widely used method to identify genetic variation was the Southern blot [1]. This labor-intensive method involves digestion of the DNA with site-specific restriction endonucleases, separation of the resulting DNA fragments by agarose gel electrophoresis, transfer of the separated DNA to a membrane, probing the membrane with radiolabeled DNA (or RNA) probes complementary to the sequences of interest, and visualization of the bands by autoradiography. Sequence variants that affect the recognition site for a restriction enzyme result in the presence or absence of cleavage and therefore on changes in restriction fragment size. Additionally, the Southern blot can identify deletions and insertions in the DNA sequence that change the size of each restriction fragment. Collectively, these genetic variations that result in band size changes are known as restriction fragment length polymorphisms (RFLP).

Another major advance in genetic testing was the invention of the *polymerase chain reaction* (PCR) amplification technique. The basic principle of this method is that DNA sequences can be amplified from very small amounts of DNA using a set of two oligonucleotide primers complementary to the 5′ and 3′ ends of the region of interest. The method involves denaturation of the DNA, annealing of the primers to each of the strands, and extension of the primer sequences using a thermostable DNA polymerase and the complementary sequence as a template. The process is repeated for 25 to 40 cycles, each cycle allowing the primers to bind to the recently synthesized strands. With maximal amplification efficiency, the amount of target sequence (amplicon) is effectively doubled with each cycle, for a total amount of DNA $= 2^{\text{number of cycles}}$, which for 30 cycles amounts to 10^9-fold. Other amplification techniques have been developed, but PCR remains a main stem of genetic analysis methods, because it can accomplish massive amplification and enrichment of specific target sequences.

The PCR method can be easily adapted to measure RFLPs: after PCR amplification and subsequent purification of the sequence of interest, the DNA is digested with restriction enzymes and the resulting fragments are separated by gel electrophoresis and typically visualized with fluorescent DNA-intercalating dyes such as ethidium bromide.

Given the high complexity and labor involved in the Southern blot assay or in sequencing large DNA fragments, screening methods to detect mutations in genes of

TABLE 10.1

U.S. Food and Drug Administration (FDA)-Approved Pharmacogenetics Assays

Target Gene(s)	Target Drug(s)	Product Name	Method	Platform	Company	Approval Date
UGT1A1	Irinotecan	INVADER UGT1A1	Invader assay	Any plate fluorimeter	Third wave/hologic	8/18/2005
CYP2D6 + CYP2C19	Various	AMPLICHIP CYP450	Microarray hybridization	GeneChip 3000Dx	Roche	1/10/2005, 12/17/2004
CYP2C19	Various	CYP2C19 assay	Microarray hybridization	Infiniti analyzer	Autogenomics	10/25/2010
CYP2D6	Various	XTAG CYP2D6	Hybridization + bead flow-cytometry detection	xMAP flow-cytometer	Luminex	8/26/2010
CYP2C9 + VKORC1	Warfarin	EQ-PRC LC	Real-time polymerase chain reaction (PCR) with high-resolution melting analysis	Light Cycler (Roche)	Trimgen	2/6/2009
		eSensor warfarin assay	Hybridization + electrochemical detection	eSensor XT-8	Osmetech/GenMark	7/17/2008
		CYP2C9 and VKORC1 genotyping	Hybridization to molecular-beacon probes	SmartCycler (Cepheid)	ParagonDx	4/28/2008
		2C9 and VKORC1 warfarin	Microarray hybridization	Infiniti analyzer	Autogenomics	1/23/2008
		VERIGENE warfarin metabolism	Hybridization + chemical detection	Verigene ID Image analyzer	Nanosphere	9/17/2007

interest have been developed. These methods typically rely on the physical properties determined by the structure of the different DNA alleles. For example, analysis
of *single stranded conformation polymorphisms* (SSCP) relies on small changes in
the electrophoretic mobility of single-stranded DNA after PCR amplification with a
biotinylated primer, denaturation, and purification of one strand with streptavidin.
Another screening method uses denaturation followed by reannealing of the two
DNA strands. If there are differences in one or more nucleotides, the resulting rean-
nealed molecules will have perfectly matched homoduplexes as well as mismatched
heteroduplexes. To identify heteroduplexes, a variety of techniques can be used,
including the following:

1. *Denaturing gradient gel electrophoresis* (DGGE), which uses a gel containing a gradient of a chemical denaturating agent such that heteroduplexes
   will denature at a lower concentration of the denaturant and migrate at a
   different position than a perfectly matched homoduplex
2. *Temperature gradient gel electrophoresis*, where a temperature gradient is
   used in the same manner
3. *Denaturing high-performance liquid chromatography* (dHPLC), in which
   gel electrophoresis is replaced by liquid chromatography such that heteroduplexes have a different retention time than homoduplexes

Recently, these methods have been largely replaced with *high-resolution melting
curve analysis* of PCR-amplified DNA [2]. In this method, *real-time PCR* is used
to generate a fluorescent signal proportional to the amount of amplicons generated. The fluorescence is typically generated by a dye, such as SYBR green, that
fluoresces only when intercalated into double-stranded DNA, and is continuously
measured with thermocyclers with built-in fluorometers. At the end of the amplification cycles, the thermocycler slowly increases the temperature while measuring
the resulting decrease in fluorescence as the DNA molecules transition from double
stranded to single stranded. The negative first derivative of the resulting melting
curve indicates the temperature point at which the majority of DNA molecules transition to single stranded, and is highly dependent on the primary DNA sequence.
Using high-resolution analysis of the DNA melting profile, single nucleotide changes
between two otherwise identical alleles can be easily distinguished. Because these
heteroduplex methods do not require any probes, the reagent costs are minimized.
In particular, the high-resolution melting curve analysis by PCR offers the flexibility
and efficiency of 98 or 384 well-plate analysis, which is a vast improvement over
older methods of heteroduplex analysis. Another major advantage of these heteroduplex methods is that they can identify unknown mutations and sequence deletions
or insertions that may be missed by some target probe or sequencing methods.

Several companies have commercialized instrumentation and reagent kits to perform high-resolution melting curve analysis, including

- 7500 or 7900HT Fast-Real Time PCR systems with HRM Software and
  MeltDoctor™ reagents by Applied Biosystems (Foster City, California)
- CFX96 or CFX384™ by Bio-Rad (Richmond, California)

- LightScanner system by Idaho Technology (Salt Lake City, Utah)
- Eco Real-Time PCR system by Illumina (San Diego, California)
- Rotor-Gene Q real-time cycler and Type-it HRM PCR Reagents by Qiagen (Valencia, California)
- LightCycler 480 by Roche Molecular Systems (Indianapolis, Indiana)
- eQ-PCR™ LC Warfarin genotyping kit, FDA-cleared for clinical laboratory testing in the Roche LightCycler, by TrimGen (Sparks, Maryland)

Currently, the gold standard method for sequence analysis is *Sanger sequencing*. The method involves primer extension using DNA polymerase and the four naturally occurring deoxynucleotides (dNTP), but with a proportion of the deoxynucleotides replaced by dideoxynucleotides (ddNTP), which are normally incorporated into DNA at the right position complementary to template strand, but are incapable of being extended by the polymerase due to a lack of 3′-hydroxyl group, therefore generating different size fragments that can be separated by gel or capillary electrophoresis. Because each ddNTP (A, C, T, or G) is labeled with a different label (or run in separate lanes), the sequence can be directly read from the order of the separated fragments. This method is accurate in detecting base substitutions in heterozygous alleles and in mutant DNA present at levels above 10% to 20%, but major drawbacks include lack of sensitivity to pick up smaller amounts of mutated DNA and gene copy number changes, such as duplication or deletions.

Another approach based on primer extension and dye-terminator nucleotide incorporation is the single-base primer extension method, exemplified by the SNaPshot™ system from Applied Biosystems [3,4]. In this assay, DNA is amplified by PCR, purified, and treated with alkaline phosphatase and exonucleotidase I to prevent template extension. Subsequently oligonucleotide primers and the SNaPshot mix containing DNA polymerase and labeled ddNTPs are added. Because there are no natural deoxynucleotides, only the appropriate labeled ddNTP is incorporated at each interrogated site. The resulting DNA fragments are then separated by capillary electrophoresis and the color of the incorporated ddNTP read directly on the electrophoretogram to reveal the single nucleotide polymorphism (SNP). Because both the distance of the primers to the SNP and the size of primers can be varied, various loci (up to 10) can be simultaneously assessed without overlap of electrophoretic peaks. This method has relatively lower costs because no labeled primers or probes are used, but a major disadvantage is the labor required to perform the various steps.

Pyrosequencing (Qiagen, Valencia, California) is an alternative method for obtaining relatively short sequence reads, with the major advantage of reliable quantification of the proportion of each base at each position [5]. The method was developed by Biotage AB (Uppsala, Sweden) and is based on the release of pyrophosphate during nucleotide extension catalyzed by DNA polymerase after serial addition of each deoxynucleotide, according to the following steps:

1. DNA is PCR-amplified with one of the primers containing biotin.
2. Single-stranded DNA templates are purified on streptavidin beads.
3. A sequencing primer corresponding to the sequence 5′ of the region of interest anneals to the template strand.

4. Serial addition of deoxynucleotide triphosphates begins (e.g., by adding dGTP to the mix).

5. In this example, if the base immediately following the primer in the template strand is a C, the polymerase present in the mix will incorporate the just added dGTP in the nascent strand, therefore releasing pyrophosphate.

6. If pyrophosphate is released, the second enzyme present in the mix, sulfurylase, will use adenosine phosphosulfate present in the mix and the just released pyrophosphate to form ATP.

7. If ATP is generated, the third enzyme, luciferase, will use the recently formed ATP and luciferin to produce a light pulse proportional to the amount of ATP, which is measured with a sensitive luminometer.

8. Remaining unincorporated deoxynucleotide triphosphates (in this example, dGTP) are converted to diphosphates (e.g., dGDP), which cannot be incorporated by the polymerase, by the addition of the fourth enzyme, apyrase.

9. In this example, if the base next to the primer is not a C, the first added dGTP will not be incorporated and no signal is generated. The instrument then adds the next base (e.g., a dCTP) and checks for generation of light if G is the base in the template strand. The process is repeated for dTTP and dATP and then new cycles of sequential addition of the four bases are performed until the reaction is terminated or no further light is generated.

10. By matching light peaks and the consecutive sequence of bases added, the sequence can be read. If the sequence has multiple nucleotides of the same type in a row (e.g., AAA), the light peak height will be roughly proportional to the number of homopolymer bases, in this example threefold of the single base peak. The method runs into difficulties with long homopolymer tracts beyond seven or eight bases, but proper primer selection in known regions can obviate this problem.

Major advantages of pyrosequencing include accurate sequence determination, sensitivity, automation, and the ability to quantify allele variants. The major disadvantages are the low multiplexing and the high cost of biotinylated primers and instrumentation. Variants of this technology have been adapted for high-throughput next-generation sequencing (see 454 Life Sciences and Ion Torrent products below).

The Invader™ assay (Third Wave Technologies/Hologic, Bedford, Massachusetts) [6] is based on the properties of a proprietary thermostable flap endonuclease named Cleavase™, which is able to recognize and cleave DNA hairpin and flap structures. The principles of the method are illustrated in Figure 10.1 and include the following steps:

1. First Cleavase reaction:
   a. PCR amplified DNA is denatured and hybridized with an oligonucleotide, called the Invader oligonucleotide, which is complementary to the sequence immediately 3′ of the SNP of interest.
   b. The allele-specific oligonucleotide is then added to the reaction. This oligonucleotide contains a 3′ region that is complementary to the specific SNP being assayed and to the bases immediately 5′ of the SNP. In addition, the 5′ region of the allele-specific oligonucleotide contains a

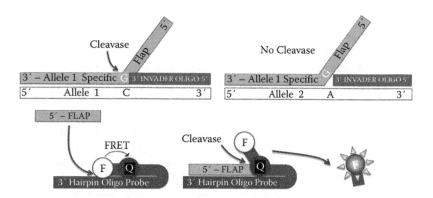

FIGURE 10.1 Invader™ first Cleavase reaction (on top) and second Cleavase reaction (on bottom). (F) represents the fluorescent label and (Q) the quencher.

defined sequence that does not hybridize to the target, therefore forming a single-stranded flap.

c. The Cleavase enzyme recognizes and cleaves the flap, which will then be released and serve as an Invader oligonucleotide for the second Cleavase reaction. Note that if the SNP is not complementary to the allele-specific probe a gap will be formed between the Invader nucleotide and the flap (Allele 2, Figure 10.1). Cleavase will not recognize and cleave this flap; therefore the reaction will not proceed for the second phase.

d. Because the hybridization temperature (around 63°C) is close to the annealing temperature of the allele-specific oligonucleotide, but lower than the annealing temperature of the Invader oligonucleotide, the allele-specific oligonucleotide is constantly being hybridized, cleaved, and released from the stable target DNA–Invader oligonucleotide molecules, therefore resulting in significant amplification of the signal.

2. Second Cleavase reaction:

a. The released 5′-flap oligonucleotide from the first Cleavase reaction hybridizes to a hairpin-containing oligonucleotide probe. This oligonucleotide contains at its 5′ end a fluorophore and a closely spaced fluorescence quencher.

b. When the 5′-flap oligonucleotide from the first reaction hybridizes to the hairpin-oligonucleotide, the short 5′-end of the probe becomes dislodged from the hairpin and will form a flap that is recognized and cleaved by Cleavase.

c. Upon release by cleavage, the labeled 5′-end of the oligonucleotide probe is no longer quenched by the quencher that remains attached to the hairpin, and the resulting fluorescence increase can be measured by a fluorometer.

d. This cycle of 5′-flap oligonucleotide annealing, cleavage of the fluorochrome, and release of the 5′-flap, which can then bind to another hairpin molecule, is repeated until all the hairpin probes have been cleaved, therefore amplifying the signal.

e. Commercially available duplex Invader assays have two different allele-specific oligonucleotides, each with a different flap region, and two different hairpin probes, one with FAM and one with Redmond Red fluorophores, so that the ratio of the two alleles can be measured by quantifying the fluorescence at the two specific wavelengths.

With the high level of amplification inherent in the method, a standard 4-hour reaction can yield a signal with amounts of target DNA as low as 10^{-21} mol. This corresponds to about 20 to 100 ng of unamplified genomic DNA. With initial PCR amplification of the DNA, the reaction time can be as short as 10 minutes to produce a robust signal. Major advantages of this method are that various inexpensive microtiter plate fluorometers can be used to measure the signal, and the ease of design and low cost of reagents, because the specificity is contained in the 5′-flap region and therefore many different allele-specific oligonucleotides can be designed for one common labeled hairpin probe. One disadvantage is that multiplexing is limited by how many nonoverlapping fluorophore probes can be included in one reaction (currently two with the duplex assays). Attempts to modify the Invader assay for a higher level of multiplexing using alternative methods for identification of the released 5′-flap include bead flow-cytometry [7], microarrays [8], or matrix-assisted laser desorption/ionization time-of-flight mass spectrometry (MALDI-TOF MS) [9], but these approaches are not widely used at present.

The Taqman assay (Applied Biosystems/Life Technologies, Carlsbad, California) uses the 5′-nucleotidase property of the Taq polymerase to remove nucleotides from a previously annealed probe, called the Taqman probe, as it synthesizes the new strand of DNA [10]. The process involves

1. PCR amplification of the DNA region containing the sequence of interest.
2. Denaturation of the PCR product and annealing to the allele-specific Taqman probe, which is designed to hybridize only to its specific targets and not the other alleles, even if they differ by only one nucleotide base. The Taqman probes, similar to the hairpin Invader probes described above, have a fluorophore bound to their 5′ nucleotide and a quencher molecule bound to an adjacent nucleotide. Upon binding of the Taqman probe to the target, the fluorophore remains quenched and no fluorescence is measured.
3. A primer corresponding to the region 5′ of the sequence of interest is added, together with Taq polymerase and all four deoxynucleotide triphosphates. The Taq polymerase extends the primer and when it reaches the previously bound Taqman probe, its 5′-nucleotidase activity releases the fluorophore-bound nucleotide, which is no longer quenched. The resulting increase in fluorescent is proportional to the amount of allele-specific Taqman probe bound and therefore allelic ratios can be measured by simultaneously adding allele-specific Taqman probes with different fluorophores.

The company developed Taqman reagents for a large number of human SNPs, and the assays can be easily automated in 96- or 384-well plates and read with widely available real-time PCR instruments. As with the Invader assay, the multiplexing is limited to the number of possible nonoverlapping fluorophores that can be read from a single well.

A related method developed by ParagonDx (Morrisville, North Carolina) uses *molecular-beacon probes* in the form of hairpins with a fluorescent dye on the 5'-end and a quencher on the 3'-end. These probes are designed to unfold the hairpin upon hybridization, such that the fluorophore is separated from the quencher, and yields fluorescence proportional to the amount of allele-specific beacon probe hybridized to its target. This provides a simple assay that can be completed in about an hour, from DNA extraction to results. Again, multiplexing is limited to the number of different fluorophores that can be measured in a single reaction. The assay has been cleared by FDA for warfarin PGx [11].

10.3 MID-MULTIPLEXING ASSAYS

In this section we describe methods with two orders of magnitude multiplexing, therefore providing a more comprehensive analysis of the various polymorphisms that may be involved in the pharmacogenetics of particular drugs. In contrast to the low multiplexing methods that interrogate only the most common variants and assume wild-type if these are not detected, higher multiplexing methods can assay for less common alleles of particular genes.

The first multiplexed pharmacogenetic method cleared by the FDA for clinical use was the AmpliChip® CYP450 (Roche Molecular Diagnostics, Pleasanton, California) [12], a microarray chip-based method that interrogates the most common polymorphisms in the cytochrome P450 genes *CYP2C19* and *CYP2D6*, which are involved in the metabolism of about 25% of all prescribed drugs. The method involves elaborate preparatory steps including DNA extraction and purification, multiplex PCR amplification of various regions of the *CYP2C19* and *CYP2D6* genes, fragmentation using DNAase I, and biotinylation of the DNA with terminal deoxynucleotidyl transferase (TdT). After hybridization to the array, the unbound DNA is washed, and the hybridized DNA is labeled on-chip with a streptavidin-coupled fluorescent dye (phycoerythrin). The microchip contains allele-specific probes for two variants of *CYP2C19* and 27 variants of *CYP2D6*, and is able to detect deletions and duplication of both genes. A major disadvantage of this assay is the high cost of microarray chips (≥$250) and of the instrumentation (>$150,000 for the GeneChip® platform).

Another microarray platform FDA-cleared for pharmacogenetics is the Infiniti™ platform from Autogenomics (Carlsbad, California). This method relies on multiplex amplification of the DNA of interest followed by primer-extension with SNP-specific primers and labeled dCTP. The SNP-specific primers contain a 3' region that hybridizes only to the allele containing the specific variant, and a 5' region that contains a unique artificial sequence called the "zip-code." The dCTP labeled DNA containing the SNP-specific primer and its unique zip-code at the 5' end is then denatured and hybridized to a microarray chip (BioFilmChip™) that contains probes complementary to each unique zip-code. This approach therefore provides the ability to quickly design specific primers for a variety of alleles that can be interrogated with the same microarray chip, because the chip hybridization specificity relies on the unique 5′ artificial zip-codes, and lower costs per chip (around $70). Another advantage is the high level of automation available, such that DNA extraction, PCR amplification, SNP-specific primer extension, microarray chip management, hybridization, washing, fluorescence reading, and interpretative report generation are all performed by a set of two instruments (the Infiniti Assist and the Infiniti Analyzer). The major disadvantages include the relatively high cost of the instrumentation (>$100,000), although it can perform a variety of genetic and proteomic assays; the slow turnaround time, with 6 to 8 hours typically needed to produce results; and the relatively low throughput (24 to 48 simultaneous microchips per run, with 1 to 4 samples per chip).

The *eSensor*® method (Osmetech/Genmark Dx, Carlsbad, California), uses microfluidic cartridges with 72 gold-plated electrode spots each containing a different capture probe [13]. The target region is amplified by PCR with a regular primer and a 5′ phosphorylated primer and single-stranded DNA are prepared by adding lambda exonucleotidase to digest the strand of the DNA extended from the phosphorylated primer. This requires opening the reaction tube post-PCR, which may introduce contamination in the reaction. Alternatively, asymmetric PCR can be performed with one primer in vast excess of the other to generate a predominance of single-stranded target DNA after exhaustion of the minority primer. After these offline preparation steps, the single-stranded DNA is mixed with allele-specific probes containing ferrocene labels and loaded into the eSensor cartridges. The instrument allows hybridization of the target DNA to the ferrocene probes, followed by hybridization of the target-ferrocene probe hybrids to the capture probes immobilized onto each gold-plated electrode. The proximity of the ferrocene ions to the electrodes generates an electrical signal that is proportional to the amount of target-DNA/ferrocene probes bound. The capture probes are directed against a conserved region of the gene, therefore allowing the capture of all allelic variants in one spot. The specificity is conferred by the ferrocene-labeled probes, because different ferrocene compounds generate alternating currents at different redox potentials. Currently, it is possible to simultaneously assay for four different ferrocene probes, therefore allowing up to four different alleles to be quantified on a single electrode. The entire assay can be completed in about 4 hours, with the online hybridization and reading taking about 30 minutes with the XT-8 system, which has a modular design of one to three towers, each capable of processing eight cartridges for a maximum of 24 independent, random-access simultaneous assays. With 30 minutes per cartridge, one three-tower system can process over 300 samples per 8-hour shift.

Another hybridization method with a unique detection approach is the Verigene® system (Nanosphere, Northbrook, Illinois) [14]. This method uses allele-specific probes labeled with gold nanoparticles, which have high stability, low background binding, and scatter light with an intensity equivalent to 500,000 fluorophore molecules, with resulting increased signal-to-noise ratio and high sensitivity, allowing about 0.25 to 2 µg of nonamplified DNA to be assayed. The DNA is sheared to a size of 300 to 500 bp by sonication, and hybridized to allele-specific capture probes immobilized on a slide array contained in a closed cartridge. At the same time, the target DNA is hybridized to gold-nanoparticle-labeled oligonucleotides complementary to a conserved region of the target, which is therefore sandwiched between the capture and the gold particles. After washing, detection of the gold particles is accomplished by deposition of elemental silver onto the gold particles and visualization of the strong light-scattering of silver-enhanced gold particles with the Verigene ID image analyzer. Major advantages of this method include its high sensitivity, allowing nonamplified DNA to be used, minimal hands-on labor with only two pipetting steps required, and a fast turnaround time of 1.5 hours for the first sample and 10 additional minutes for subsequent samples. The Verigene SP processor automates DNA extraction, reverse transcription (for RNA targets), PCR amplification (if needed), and target identification and analysis in up to four cartridges, making this method suitable for point-of-care and other environments without access to highly trained operators.

The xTAG® assay (Luminex Corporation, Austin, Texas), initially developed as Tag-It® at Tm Bioscience (Toronto, Canada), uses multicolor bead flow-cytometry for specific identification of hybridized probes [13,15]. The method employs a system of 24-mer oligonucleotides with isothermal melting temperatures that have been designed to minimize nonspecific hybridization to genomic DNA and between each other, allowing high-level multiplexing to occur in a single tube. These oligonucleotides are designed with their 3′ end matching the polymorphic site of interest such that primer extension can only occur if the corresponding base is present in the target, and a unique 5′ end artificial sequence tag similar to the zip-codes described for the Infinity system. The xTag procedure consists of

1. Multiplex PCR amplification of the target DNA.
2. Hybridization with the allele-specific primers.
3. If there is a perfect match with the polymorphic site, primer extension is catalyzed by DNA polymerase and various biotinylated deoxynucleotides (typically dCTP) are incorporated into the nascent strand. The biotin moieties allow the nascent DNA to be labeled with a streptavidin-conjugated dye, typically phycoerythrin.
4. The biotinylated strand just synthesized is denatured and hybridized to the corresponding antitag probes immobilized on 5.6 mm polystyrene beads. Each different antitag is immobilized on a bead uniquely identified by the ratio of two impregnated fluorophores. By changing the ratio of the two fluorophores during bead manufacturing, up to 100 different spectral addresses are available for bead identification.
5. The beads are read by the xMAP flow-cytometer, which uses laser-induced fluorescence to identify each bead's unique spectral address, as well as whether the bead contains hybridized phycoerythrin-labeled target DNA. Therefore, each allele, bound to a specific bead, can be identified and quantified, and up to 100 alleles or 50 biallelic variants can be analyzed simultaneously. With the most recent technology using three internal fluorophores (FlexMap-3D®), up to 500 different spectral addresses can be discriminated.

10.4   HIGH MULTIPLEXING AND NEXT-GENERATION SEQUENCING SYSTEMS

The advent of high-throughput methods for genetic analysis has made it possible to interrogate a large number of genetic variants that may play a role in predicting drug response and explaining unexpected drug effects. With the substantial technological advances and exponential drops in the cost of genome-wide DNA analyses together with the availability of user-friendly interpretative software it will soon become cost-effective to sequence or scan the entire genome of each individual at an appropriate time, such as neonatal screening or the first clinical visit as an adult. This will allow the health-care system to focus on preventive measures and screening approaches aimed at specific personal risks and diseases the individual might be predisposed to, and on the effective use of chemicals and therapeutic drugs to maximize their benefits and minimize their undesired effects based on a comprehensive understanding of the interplay between genetic variation, environmental factors, metabolism, and pharmacodynamics.

10.4.1   HIGH-PLEX SINGLE NUCLEOTIDE POLYMORPHISM (SNP) ASSAYS

One multiplexed system allowing high-throughput analysis is the *MassArray®* system (Sequenom, San Diego, California) that uses matrix-assisted laser desorption/ionization time-of-flight mass spectrometry (MALDI-TOF MS) to accurately measure the mass of DNA fragments [16]. Genotyping is performed by single-base primer extension with nucleotide terminators. Because each terminator has a different mass that can be accurately measured by MALDI-TOF MS, the base present at a particular SNP can be identified. Using different size primers allows multiplexing up to 40 SNPs per well. With a 384-well plate format, the system is capable of analyzing up to 138,000 genotypes per day.

Another high-throughput SNP analysis system is the *VeraCode system* (Illumina), which uses cylindrical glass micron microbeads embedded with high-density holographic bar codes that can be excited by a laser and imaged by the BeadXpress Reader system [17]. The system allows up to 384 different beads to be read on a single well of a 96-well plate, resulting in throughputs of about 300,000 genotypes in a 6-hour period. The company has a variety of systems allowing various degrees of multiplexing and a number of simultaneous samples to be processed, with up to 5-million SNPs per sample in the case of the HumanOmni5 bead microarray platform. This level of SNP testing allows the identification of any minor alleles with frequencies above 1%.

10.4.2   HIGH-THROUGHPUT LIBRARY PREPARATION AND POLYMERASE CHAIN REACTION (PCR) SYSTEMS

Currently available technologies for genome-wide sequencing require the preparation of libraries of DNA fragments that are distributed into massive parallel reaction cells where sequencing takes place. Ideally, each cell should contain DNA fragments amplified from a single molecule so that all signals derived from each base position will be uniform. The process of library preparation usually involves fragmentation of the genomic DNA into 150 to 500 bp fragments, followed by end repair and ligation of oligonucleotide adaptors that serve as template sequences for PCR amplification of each fragment. Several technologies have addressed the problem of separating the DNA fragments so that each PCR reaction results in amplification of one unique fragment. The most common technologies involve immobilization of each DNA fragment either on beads that are distributed into individual microwells, or on a flat surface in flow cells or arrays, where the signal can be visualized as labeled nucleotides are added, a process known as *sequencing by synthesis.*

In contrast to unbiased genome-wide sequencing, several enrichment strategies have been devised to reduce costs and increase the number of reads for each base (coverage) for those polymorphisms of interest. These strategies usually involve either parallelized PCR amplification of each target region of interest or selection by hybridization to capture probes immobilized in a solid array. In general, hybridization capture approaches are more scalable, as exemplified by the ability to capture the complete exome, but suffer from a lack of specificity resulting from capture of homologous regions and pseudogenes, which may consist of up to 50% to 70% of the irrelevant sequences. In contrast, single-locus PCR confers very good specificity but requires massive multiplexing capabilities to achieve large-scale coverage of the genome.

An example of the high-throughput PCR-based enrichment approach is the RainStorm™ system (RainDance Technologies, Lexington, Massachusetts) that uses picoliter-volume droplet manipulation in a disposable microfluidic chip with no moving parts or valves to aliquot the genomic DNA, mix it with PCR reagents, and perform discrete PCR reactions in each droplet at a rate of 10 million droplets per hour. The system currently supports droplet primer libraries of up to 20,000 unique primer pairs. Assuming an average amplicon size of 500 bp, the total coverage of amplified DNA can reach 10 Mb with current technology. The amplicons can then be prepared for high-throughput sequencing on the various next-generation sequencers.

Another PCR enrichment technology is available from the Fluidigm Corporation (San Francisco, California), which developed nanofluidic Integrated Fluidic Circuit (IFC) chips with the ability to individually assemble multiple PCR reactions inside the chip, with capacities up to 48 samples and 770 assays per chip in the case of the EP1™ Analyzer [18]. In a variation of the technology, the Digital Array™ IFC Chips partition one sample into hundreds of individual PCR reactions, allowing precise absolute quantification of target sequences, which is required for precise copy number variation assessment.

The *Selector* technology (Halo Genomics, Uppsala, Sweden) uses rolling circle PCR amplification, after digestion of the genomic DNA with restriction endonucleases, followed by denaturation and hybridization to a Selector probe library [19]. These probes are designed to hybridize to both ends of the restriction fragment, therefore facilitating circle formation and intramolecular reannealing. Because the probes are biotinylated, purification of the circles is performed with paramagnetic streptavidin beads before ligation and amplification. Amplification can be achieved by PCR or by rolling circle and multiple displacement amplification with Φ29 polymerase (e.g., the TempliPhi kit from Amersham, which obviates the need for a primer library or a thermocycler).

The SureSelect technology (Agilent, Santa Clara, California) uses capture hybrid-ization on a microarray to select DNA targets and prepare them for next-generation sequencing (NGS). Basically, the process entails fragmentation of the genomic DNA with ultrasound waves, ligation of NGS-specific adaptors to the DNA ends, size-purification, limited (six to eight cycles) PCR amplification with primers comple-mentary to the adaptors, and hybridization to 244,000 60-mer oligonucleotides immobilized on the SureSelect array. The capture oligonucleotides are designed to cover the regions of interest with considerable sequence overlap (30 to 57 bases) between consecutive 60-mer oligonucleotides. After washing in a stringent buffer, the selected DNA fragments are heat eluted and used for downstream NGS. For larger-scale enrichment, the DNA fragments are hybridized in solution with a library of biotinylated 120-mer RNA oligonucleotides. The longer RNA oligonucleotides offer stronger hybridization kinetics and tolerance to base changes in the target region, which will increase both the coverage and the number of unintended sequences cap-tured. After purification of the DNA-RNA hybrids on magnetic streptavidin beads, the capture RNA probes are digested and the purified DNA is ready for limited-PCR amplification and downstream NGS.

With the NimbleGen technology (Roche) the genomic DNA is similarly prepared by fragmentation and adaptor ligation. The single-stranded DNA is then hybridized to 2.1 million specific capture oligonucleotides immobilized on a microarray. After washing unbound DNA, the enriched fragment pool is eluted from the array and used for downstream NGS. NimbleGen also offers a solution capture protocol using biotinylated oligonucleotides and purification on streptavidin beads (SeqCap EZ).

10.4.3 Next-Generation Sequencing

Next-generation sequencing technology is constantly evolving so that only an overview of some current and evolving technologies will be provided in this section. The reader is encouraged to consult reviews (e.g., [20,21]) and Web sites of the various companies for details. In 2012, the most widely used NGS technologies include the Roche 454, various Illumina sequencers, Applied Biosystems SOLiD and Life Technologies Ion Torrent.

The SOLiD™ system (Applied Biosystems/Life Technologies) uses massive parallel oligonucleotide ligation sequencing of target DNA amplified on emulsion beads, achieving maximum read lengths of 35 to 50 bases, a throughput of about 20 to 30 gigabases (Gb) per day with about 30-fold coverage, and 99.99% base call accuracy. The sequencing process involves clonal amplification by emulsion PCR of the target DNA immobilized on beads, 3′ modification of the DNA to allow covalent attachment to a glass slide, deposition and attachment of the beads at high density on the array, denaturation of the DNA, hybridization of a primer to the adaptor sequence present at the 5′-end of all target DNA fragments, and ligation of the interrogation probes. These 8-mer oligonucleotides contain a unique dinucleotide (one of 16 pos-sible combinations) at the 3′-end, followed by a random sequence of six degenerate bases and one of four fluorescent labels attached to the 5′-end of the probe. If the 3′ dinucleotide matches the template sequence immediately following the primer, the interrogation probe is stabilized by ligation to the 3′-end of the primer and the fluo-rescent color corresponding to one of four dinucleotide combinations is registered on each array spot by an image snapshot. The last three bases at the 5' end of the interrogation probe plus the attached fluorescent label are then cleaved and washed away. The remaining ligated five bases of the interrogation probe now serve as primer for another cycle of ligation dependent on the following dinucleotide sequence in the template. This process is repeated six times, after which the ligated DNA strand is removed by denaturation to allow a second round of ligations. This time, the initiating primer is complementary to adaptor sequence but offset by 1 nucleotide (n-1) and the instrument performs another round of ligations and sequence imaging. A total of five rounds (using n, n-1, n-2, n-3, and n-4 primers) are performed so that each template base is covered twice for each bead. The sequence is inferred from combinatorial analysis of the sequential images based on the 16 possible combinations of dinucleotides.

The *Illumina* NGS sequencers are based on the Solexa Genome Analyzer and use random primer extension of immobilized DNA fragments with reversible fluorescent deoxynucleotide terminators to visualize the DNA sequence in high-density arrays. Similarly to other NGS methods, the genomic DNA is fragmented, adaptors are ligated, and the DNA is immobilized on a flow cell array. Amplification is performed by "bridge" PCR, where the template DNA bends and binds an adjacent oligonucleotide anchor that also serves as the PCR primer, therefore forming local clonal clusters. After amplification, only the forward strand of the template is retained, as the reverse strand is cleaved. Sequencing is performed by adding polymerase and the four dNTP terminators, each labeled with a different fluorescent dye. After incorporation of the first nucleotide terminator, an image is obtained, similarly to the SOLiD process, and the fluorescent dye is subsequently cleaved, unblocking the incorporated nucleotide for further extension. Multiple rounds of nucleotide terminator addition, imaging, and cleavage allow the sequencing to progress along the template for an average reading length of 36 to 100 bases. The current top-of-the-line sequencer, the HiSeq 2000 system with a dual flow cell, has a throughput of about 25 Gb per day, with more than 30-fold coverage of 80% to 90% of the bases and a raw error rate of 1% to 2%, allowing full sequencing of two human genomes in about 8 days.

The pyrosequencing technology has been adapted for large-scale genomic sequencing by the *454 Life Sciences* company, currently owned by Roche Diagnostics. This technology uses single-molecule emulsion PCR to amplify the template DNA and massive parallel pyrosequencing reactions for deriving the sequence. The process involves the usual DNA library preparation steps of fragmentation and adaptor ligation, followed by limiting dilution of the library to single-molecule concentration, annealing to a bead-immobilized oligonucleotide complementary to the adaptor, and isolation of each bead into an oil-water emulsion microvesicle containing PCR reagents and primers. After clonal amplification of the target DNA to several million molecules per bead, the instrument delivers each bead to one of 3.4 million picoliter flow-cells in a plate, where pyrosequencing reactions similar to those described in Section 10.2 are performed to obtain the sequence of the DNA attached to each bead. The flagship Genome Sequencer FLX Titanium System can obtain read lengths of 400 bases with about 1 million individual reads per 10 hour run and a base call accuracy of >99%.

A variant of pyrosequencing has been developed by Ion Torrent/Life Technologies, using electrochemical detection of the proton released when a deoxynucleotide is incorporated into the nascent DNA during the polymerase extension reaction. As with the 454 approach, DNA is amplified by clonal emulsion PCR, and beads are deposited in individual nanowells in a semiconductor microchip. Each of the four dNTPs is sequentially added, and the sequence can be read from the electrochemical signal generated, similarly to pyrosequencing, but with the advantage of using only natural dNTPs and polymerase, because apyrase, luciferase, or sulfurylase enzymes are not required, which improves speed and lowers costs. In addition to lower reagent costs, the initial investment in instrumentation can be less than one tenth of other NGS sequencers, as the Ion Torrent's Personal Genome Machine costs about $50,000. With the 318 chip containing 12 million sequencing wells and read lengths of >200 bp (predicted >400 bp in 2012), a 1 Gb region can be sequenced in less than 2 hours with base call accuracy of >99.99%.

Other companies are working to reduce reagent use and costs while increasing read lengths and speed of NGS, by focusing on technologies able to asynchronously sequence unamplified single molecules of DNA without stopping between read steps (sometimes called *third-generation sequencing*, [22]). For example, Helicos Biosciences uses poly-A tailing of single-stranded DNA targets to hybridize them to flow-cells containing 600 million to one billion immobilized polyT oligonucleotides, allowing massive parallel sequencing of unamplified DNA by cycles of sequential addition of each fluorescent reversible terminator dNTPs, imaging, and label cleavage. Read lengths of 25 to 55 bases with >99.995% accuracy at >20x coverage are achieved at a throughput of greater than 1 Gb per hour and 21 to 35 Gb per run.

Another example of third-generation sequencing is the Pacific Biosciences SMRT™ technology, which uses nanometer-size holes in a microchip, each containing a single active f29 polymerase molecule. Incorporation of fluorescent-labeled deoxynucleotides is detected by the longer duration of fluorescence of polymerase-bound nucleotides being incorporated compared to free unincorporated nucleotides. Cleavage of the fluorescent label, which is attached to the triphosphate backbone, occurs naturally as the polymerase releases the pyrophosphate moiety. This technology avoids the costly and time-consuming steps of sequential reagent addition, washing, and scanning common to second-generation sequencing platforms, and given the high processivity of the f29 polymerase read lengths of over 1,000 (sometimes up to 10,000) are possible with unamplified genomic DNA. Using hairpin adapters and rolling circle DNA synthesis, multiple reads for the same molecule are possible, increasing accuracy. While the system currently uses ~75,000 sequencing "holes," much higher throughput will be possible with significant increases in the number of parallel reactions.

Ultimately, large-scale sequencing technologies may be able to produce sequence information at high speed without using reagents—for example, by simple high-throughput scanning of DNA molecules by physical methods, such as visualization by transmission electron microscopes (in development, e.g., by Halcyon Molecular and ZS Genetics), scanning tunneling microscopy [23], or physical measurement of the nucleotide bases as DNA flows in single file through a nanopore (e.g., Oxford

Nanopore Technologies, NobleGen Biotechnologies, IBM/Roche) [24–27]. These approaches will lead to truly inexpensive (<$100) and reliable genome-wide sequencing that will eventually enable routine genetic risk assessment for every individual.

10.5 CONCLUSIONS

Current technologies for pharmacogenetic analysis are rapidly evolving. On one hand, the development of technologies for low-multiplexing, specific analysis of a restricted number of genetic variants of high significance for predicting drug response (such as *CYP2C9, 2C19, 2D6*, and *VKORC1* polymorphisms) is focused on user-friendly, rapid, low complexity methodologies and instrumentation appropriate for point-of-care use and for obtaining fast results of timely relevance to drug prescribing. On the other hand, the development of inexpensive, fast, and reliable whole-genome scanning or sequencing technologies will lead to universal, one-time testing of the population resulting in a paradigm shift away from individualized specific genetic assays and toward interpretation of genetic information for rational and personalized use of preventive and therapeutic measures to improve health.

REFERENCES

1. Southern, E.M., Detection of specific sequences among DNA fragments separated by gel electrophoresis. *J Mol Biol*, 1975, 98(3): 503–517.
2. Temesvari, M., et al., High-resolution melting curve analysis to establish CYP2C19 *2 single nucleotide polymorphism: comparison with hydrolysis SNP analysis. *Mol Cell Probes*, 2011, 25(2–3): 130–133.
3. Bender, K., SNaPshot for pharmacogenetics by minisequencing. *Methods Mol Biol*, 2005, 297: 243–252.
4. Yuferov, V., et al., Search for genetic markers and functional variants involved in the development of opiate and cocaine addiction and treatment. *Ann NY Acad Sci*, 2010, 1187: 184–207.
5. Marsh, S., Pyrosequencing applications. *Methods Mol Biol*, 2007, 373: 15–24.
6. Mein, C.A., et al., Evaluation of single nucleotide polymorphism typing with invader on PCR amplicons and its automation. *Genome Res*, 2000, 10(3): 330–343.
7. Rao, K.V., et al., Genotyping single nucleotide polymorphisms directly from genomic DNA by invasive cleavage reaction on microspheres. *Nucleic Acids Res*, 2003, 31(11): e66.
8. Lu, M., et al., A surface invasive cleavage assay for highly parallel SNP analysis. *Hum Mutat*, 2002, 19(4): 416–422.
9. Berggren, W.T., et al., Multiplexed gene expression analysis using the invader RNA assay with MALDI-TOF mass spectrometry detection. *Anal Chem*, 2002, 74(8): 1745–1750.
10. Schleinitz, D., J.K. Distefano, and P. Kovacs, Targeted SNP genotyping using the TaqMan(R) assay. *Methods Mol Biol*, 2011, 700: 77–87.
11. Babic, N., et al., Comparison of performance of three commercial platforms for warfarin sensitivity genotyping. *Clin Chim Acta*, 2009, 406(1–2): 143–147.
12. Heller, T., et al., AmpliChip CYP450 GeneChip: a new gene chip that allows rapid and accurate CYP2D6 genotyping. *Ther Drug Monit*, 2006, 28(5): 673–677.
13. Lee, C.C., et al., Evaluation of a CYP2C19 genotype panel on the GenMark eSensor(R) platform and the comparison to the Autogenomics Infiniti and Luminex CYP2C19 panels. *Clin Chim Acta*, 2011, 412(11–12): 1133–1137.

14. Lefferts, J.A., P. Jannetto, and G.J. Tsongalis, Evaluation of the Nanosphere Verigene System and the Verigene F5/F2/MTHFR nucleic acid tests. *Exp Mol Pathol*, 2009, 87(2): 105–108.
15. Maurice, C.B., et al., Comparison of assay systems for warfarin-related CYP2C9 and VKORC1 genotyping. *Clin Chim Acta*, 2010, 411(13–14): 947–954.
16. Gabriel, S., L. Ziaugra, and D. Tabbaa, SNP genotyping using the Sequenom MassARRAY iPLEX platform. *Curr Protoc Hum Genet*, 2009, Chapter 2: Unit 2, 12.
17. Lin, C.H., et al., Medium- to high-throughput SNP genotyping using VeraCode microbeads. *Methods Mol Biol*, 2009, 496: 129–142.
18. Wang, J., et al., High-throughput single nucleotide polymorphism genotyping using nanofluidic dynamic arrays. *BMC Genomics*, 2009, 10: 561.
19. Johansson, H., et al., Targeted resequencing of candidate genes using selector probes. *Nucleic Acids Res*, 2011, 39(2): e8.
20. Voelkerding, K.V., S.A. Dames, and J.D. Durtschi, Next-generation sequencing: from basic research to diagnostics. *Clin Chem*, 2009, 55(4): 641–658.
21. Su, Z., et al., Next-generation sequencing and its applications in molecular diagnostics. *Expert Rev Mol Diagn*, 2011, 11(3): 333–343.
22. Schadt, E.E., S. Turner, and A. Kasarskis, A window into third-generation sequencing. *Hum Mol Genet*, 2010, 19(R2): R227–R240.
23. Ivanov, A.P., et al., DNA tunneling detector embedded in a nanopore. *Nano Lett*, 2011, 11(1): 279–285.
24. Luan, B., et al., Base-by-base ratcheting of single stranded DNA through a solid-state nanopore. *Phys Rev Lett*, 2010, 104(23): 238103.
25. Lieberman, K.R., et al., Processive replication of single DNA molecules in a nanopore catalyzed by phi29 DNA polymerase. *J Am Chem Soc*, 2010, 132(50): 17961–17972.
26. Stoddart, D., et al., Single-nucleotide discrimination in immobilized DNA oligonucleotides with a biological nanopore. *Proc Natl Acad Sci USA*, 2009, 106(19): 7702–7707.
27. ten Bosch, J.R., and W.W. Grody, Keeping up with the next generation: massively parallel sequencing in clinical diagnostics. *J Mol Diagn*, 2008, 10(6): 484–492.

Index

A

ABCB1, 5, 167, 179
Acamprosate, 199
Acetaldehyde, 5, 17, 71, *See also* Aldehyde
    dehydrogenase (ALDH) and
    associated genes
  chronic alcohol abuse markers, 59–60
  liver toxicity, 32–33
  serotonin metabolism and, 55
Acetone, 117
N-Acetyl-β-hexosaminidase, 56
Acetylcholine, 67–68
ACN9, 70
Acquired immunodeficiency syndrome (AIDS), 155
Adderall, 103
Addiction, *See* Dependence and addiction
Addiction treatment, pharmacogenomics aspects,
    195–206
  buprenorphine, 167, 197
  disulfiram and acamprosate, 199–200
  drug biochemical effects, 203–205
  genome-wide association studies, 195, 204
  medication metabolism, 196–197
  medication site of action, 197–203
  methadone, 111, 176, 196–197
  opioid receptor antagonist (naltrexone),
    70–71, 182, 198
  smoking cessation, 203–205
  SSIs, 200–202
ADH genes, 60–63, 71
Adolescent neural development, 27, 33
Adolescent suicide, 30
Adulterants, urine drug testing, 119
Adult respiratory distress syndrome (ARDS), 37
Afghanistan, 166
Age and alcohol consumption, 50
Alanine aminotransferase, 56
Alcohol abuse (chronic use or dependency)
  binge drinking definition, 19
  diagnostic criteria, 50
  hazardous drinking definition, 18–19
  metabolic enzymes and, 5
  naloxone therapy, 9
  societal cost, 17, 29
  U.S. statistics, 17–18
Alcohol abuse, biochemical markers of chronic
    use (slate markers), 47, 48, 49, 52–54
  long-term use markers, 56
    acetaldehyde adducts, 59

N-acetyl-β-hexosaminidase, 56
    carbohydrate deficient transferrin (CDT),
      58–59
    γ-CDT, 59
    γ-glutamyltransferase (GGT), 57–58
    lipid markers, 57
    liver damage, 56
    mean corpuscular volume, 57
    nutritional status, 56
    sialic acid, 60
  short-term direct markers, 54
    direct ethanol measurements, 53–55
    ethylglucuronide, 55–56
    fatty acid ethyl esters, 55
    5-hydroxytryptophol, 55
Alcohol abuse, health hazards of, 29, 50
  bone, 37
  brain damage, 33–36
  cancer, 37–38, 50
  cirrhosis and other liver diseases, 18, 21, 29,
    31–33, 38
  endocrine system, 37
  fetal alcohol syndrome, 38–39, 77
  heart disease, 30, 36
  immune system, 36–37
  life span reduction, 29–30
  mortality statistics, 18
  stroke, 36
  traffic accidents, 48–49
  violent behavior, 30–31
Alcohol abuse or dependence treatment, 50,
    See also Addiction treatment,
    pharmacogenomics aspects
  disulfiram and acamprosate, 199–200
  naloxone, 9
  naltrexone, 70–71, 182, 198–199
  pharmacogenetics of, 70–71
Alcohol-cocaine interactions, 113, 144
Alcohol consumption, 11
  blood alcohol levels, 16, 19–22, 54–55, *See*
    also Blood alcohol levels
  cocaine use and, 113, 144
  demographic associations, 50
  historical perspective, 13–14
  markers of, 53–55
  moderate, heavy, and binge drinking
    definitions, 18–19
  prevalence, 48
  U.S. statistics, 17–18

Alcohol consumption, health benefits of
      moderate drinking, 12, 22
  arthritis, 28
  cancer, 27
  common cold, 28–29
  dementia and Alzheimer's disease, 26
  heart disease, 22–25
  life extension, 27–28
  stroke, 25
  type 2 diabetes, 25–26
Alcohol dehydrogenase (ADH) and associated
      genes, 5, 16, 54, 60–63
Alcohol Dependence Scale (ADS), 68
Alcohol dependence susceptibility, genetic markers
      of (trait markers), 48, 49, 51–52, 60
  acetylcholine, 67–68
  ACN9, 70
  alcohol and aldehyde dehydrogenases, 60–63
  associated polymorphisms (table), 61–62
  catechol-*O*-methyltransferase, 66
  dopamine, 64
  dopamine metabolizing enzymes, 65–66
  dopamine receptors, 64–65
  dopamine transporters, 65
  ethnic variations, 51–52
  GABA, 66–67
  glutamate, 68–69
  molecular profiling technologies, 71–78
    epigenetics, 72–75
    genomics, 72–75
    metabolomics, 75–76
    proteomics, 76
    transcriptomics, 71–72
  monoamine oxidase, 66
  neuropeptide Y, 69–70
  neurotransmitter systems, 63–64
  opioids, 70
  serotonin, 69
  specific markers, 60
Alcoholic content of drinks, 12
  proof, 14
  standard drink definition, 14, 21
Alcoholic dementia, 34
Alcoholic drink calories, 15
Alcoholic hepatitis, 31, 32
Alcoholic odor in breath, 21
Alcohol metabolism, 15–17, 60
  dependency susceptibility markers, 60–63,
      See also Alcohol dependence
      susceptibility, genetic markers of
  endogenous ethanol production, 21
  ethanol measurement issues, 55
  liver toxicity and, 32–33
  markers of chronic abuse, 55–60, *See also*
      Alcohol abuse, biochemical markers
      of chronic use (slate markers)
  metabolomics, 75–76

Alcohol oxygenase, 158
Alcohol screening and intervention, 51
Alcohol tax, 14
Alcohol Use Disorders Identification Test
      (AUDIT), 68
Alcohol withdrawal effects, 35, 51, 65, 68–69,
      70, 162
Aldehyde dehydrogenase (ALDH) and associated
      genes, 5, 17, 51, 60, 63, 71
Alfentanil, 112
Alfentanyl, 172, 178
Alprazolam, 107, 108
Alzheimer's disease, 26–27, 35, 71
Aminotransferases, 56
Amobarbital, 106, 107
d-Amphetamine, 202
l-Amphetamine, 129
Amphetamines and related compounds, 93, 100,
      103–106, 129–130, *See also* MDMA;
      Methamphetamine; *specific drugs*
  confirmation methods, 101
  designer drugs, 104
  dopamine and, 6
  interference studies, 106
  isomer resolution analysis, 105–106
  medical uses, 129
  pharmaceutical sources, 103–104
  pharmacogenetics of abuse, 132–135
  screening and confirmation methods,
      105–106
  tolerance effects, 129
  volatility, 106
  workplace drug testing, 105
AmpliChip®, 219
Amyl nitrates, 117
Anabolic steroids, 96
Analgesic properties, opiates and opioids, 7, 109,
      111–112, 180
Anandamide, 162, 185
Anhydroecgonine methyl ester (AEME), 144
Antabuse, 199
Antidepressant therapy, 202
Antioxidants, in red wine, 25
Antisocial behavior, 31
Antisocial personality disorder, 51
Anxiety and alcohol withdrawal, 51
Apnea, 145
Appetite stimulation, 155, 163
Appetite suppression, 103, 105, 129
2-Arachidonoyl glycerol, 185
β-Arrestin, 183
Arthritis, 28
Asians and Pacific Islanders, 51–52
Aspartate aminotransferase, 56
Ativan, 107
Attention deficit-hyperactivity disorder (ADHD)
      treatment, 129

Australia, 19
Austria, 19

B

Babylonians, 13
Bagging, 117
Barbiturates, 96, 101, 106–107
Beer, 13
Behavioral genetics, 2, 8
Benzodiazepines, 70, 96, 101, 107–109
Benzoylecgonine (BE), 5, 143–144
N-Benzylpiperazine (BZP), 116
Bilirubin, 56
Binge drinking, 19, 26, 30, 35
    biochemical markers, 57
Bladder cancer, 27
Blood alcohol levels, 16, 19–22
    direct ethanol measurements, 54–55
    legal limits for driving, 19–20
    postmortem levels, 21–22
    Widmark formula, 20–21
Blood-brain barrier (BBB), 5, 131, 159, 163, 167
Blood testing for drugs of abuse, 97
Body mass index (BMI), 162
Bone and alcohol abuse, 37
Brain damage, alcohol-related effects, 33–36
Breast cancer, 37, 38
Breast-feeding mothers, 174
Breath analyzers, 54
Bufotenin, 115
Buprenex, 109
Buprenorphine, 109, 110, 167, 173, 179, 197
Bupropion, 201, 202, 204
Butabarbital, 106, 107
Butalbital, 106, 107
Butane, 117
1,4-Butanediol (1,4-BD), 115
Butyl nitrates, 117
Butyrylcholinesterase (BChE), 5, 145–146
B vitamins, 34

C

Calcitonin gene-related peptide (CGRP), 183
Calories of alcoholic drinks, 15
Campral, 199
Cancer
    alcohol abuse associations, 37–38, 50
    moderate alcohol consumption benefits, 27
Cannabidiol, 155
Cannabinoid receptors, 159–161
    polymorphisms, 161–162
Cannabinoids, 93, 155–156, *See also* Marijuana
    absorption, 156–157
    confirmatory methods, 102
    metabolism, 157–158

    opioid dependency and, 185
    pharmacogenetics, 158–159
    tolerance effects, 155
Cannabis sativa, 114–115
Carbohydrate deficient transferrin (CDT), 58–59
    γ-CDT, 59
Carboxylesterase 1 (CES1), 145
Cardiac arrhythmia, 118, 177
Cardiovascular disease, 22–23
Catalase, 16
Catecholamines, 100
Catechol-*O*-methyltransferase (COMT), 66, 132, 184, 202–203
Catha edulis, 129
Cathinone, 129
γ-CDT, 59
Centrax, 107
China, 13, 19, 165–166
Chlordiazepoxide, 107
Chloroform, 117
Cholinesterase, 5, 145–147
Chronic atrophic gastritis (CAG), 27
Cirrhosis of the liver, 18, 21, 29, 31, 56
Citalopram, 201
Clarithromycin, 175
Cleavase™ reactions, 216–218
Clockwise hysteresis, 138
Clonazepam, 107, 108
Club drugs, 102–104, 115–117, *See also specific drugs*
Cocaethylene, 113, 144
Cocaine, 93, 113–114, 137–138
    confirmatory methods, 102
    CYP2D6 inhibition, 4
    dopamine and, 6, 138, 202–203
    ethanol use and, 113, 144
    immunoassays and confirmation methods, 114
    metabolic enzymes
        butyrylcholinesterase, 145–147
        carboxylesterase 1, 145
    metabolism and metabolites, 5, 143–144
    paranoia and, 141
    pharmacogenetics of abuse, 139
        dopamine-β-hydroxylase, 142
        dopamine receptors, 139–140
        dopamine transporters, 140–141
        norepinephrine transporter, 142
        opioid receptors, 142–143
        serotonin transporter, 141–142
    pharmacology, 138–139
    street use, 138
    tolerance effects, 138
    toxicity, 113–114, 139
    treatment approaches, 147, 200–202
Coca teas, 113
Codeine, 96, 109, 110, 166, 172, 174

Colon cancer, 37, 38
Common cold, 28–29
Congestive heart failure, 23, 36
Control group issues, 2, 8
Coronary heart disease, 22–23
Corticomesolimbic dopamine system (CMDS), 67
Cortisol, 37
Crack cocaine, 138, 144
Creatinine, 119
Cross-reactivity, drug immunoassays, 98, 99
Cushing's syndrome, 37
Cyclic adenosine monophosphate (cAMP)
          activity, 160
CYP2A1, 176
CYP2A4, 171
CYP2A6, 4, 106
CYP2B6, 4–5, 176–178, 196–197
CYP2C, 106
CYP2C19, 176, 219
CYP2C8, 106, 174, 179
CYP2C9, 106, 158–159
CYP2D6, 4, 132, 134, 171, 173, 174, 175–177,
          179–180, 185–186, 196, 219
CYP2E1, 16, 32
CYP3A, 106
CYP3A3, 108
CYP3A4, 5, 108, 115, 158–159, 171–179, 196
CYP3A5, 4, 172, 176, 178, 185, 186
CYP3A7, 172
Cytochrome P450 enzymes, 16, 134, 171, 195,
          See also specific CYP enzymes
AmpliChip® system, 219

D

Dalmane, 107
Darvon, 109
Date rape drugs, 109, 115, 116, 117
Delavirdine, 179
Delta (δ) opioid receptors, 7, 142, See also Opioid
          receptors
Dementia, 26–27, 34
Demerol, 109
Denaturing gradient gel electrophoresis (DGGE),
          214
Denaturing high-performance liquid
          chromatography (dHPLC), 214
Denmark, 19
Dependence and addiction, 1, See also Alcohol
          dependence susceptibility, genetic
          markers of; specific drugs
     dopamine receptors and, 139–140, 183–184
     genetic gap issues, 7–8
     opiates, 109
     pharmacogenomics aspects, See Addiction
          treatment, pharmacogenomics aspects
     serotonin associations, 184

Depression, 71, 161–162
     antidepressants, 202
De Ritis ratio, 56
Designer drugs, 104, 113, See also specific drugs
Dexedrine, 103
Dextroamphetamine, 202
Dextromethorphan, 96, 115
Diabetes, 23, 25–26
Diagnostic and Statistical Manual of Mental
          Disorders (DSM-IV), 50
Diazepam, 107, 108
Dibucaine number, 145
Didrex, 103
Dihydrocodeine, 109, 174–175
Dilaudid, 109
Diltiazem, 178
Dimethyltryptamine (DMT), 115, 179
Diphenhydramine, 115
N,N-Dipropyltryptamine (DPT), 115
Distilled alcohol, 12
Disulfiram, 71, 199–200, 202
DNA fragment libraries, 222–223
DNA methylation, 8, 52, 76–77
DNA sequencing, See also Pharmacogenetics
          testing methodologies
     Cleavase™ reactions, 216–218
     high-resolution melting curve analysis,
          214–215
     hybridization capture approaches, 223–224
     microarray systems, 71, 219–221
     molecular beacon probes, 219
     molecular profiling technologies of alcohol
          dependency markers, 71–78
     next-generation, 71, 76, 77, 224–227
     PCR, 212, See also Polymerase chain reaction
     pyrosequencing, 215–216, 225–226
     Sanger sequencing, 215
     single-base primer extension method, 215
     Southern blot, 212
     Taqman assay, 218–219
     third-generation, 226
DOI, 116
Dopamine, 6, 64, 100, 139
     cocaine effects and, 6, 138, 202–203
     metabolizing enzymes, 65–66
Dopamine-β-hydroxylase (DBH), 65–66, 142,
          204
Dopamine receptors, 64
     alcohol dependency susceptibility markers,
          65
     cocaine pharmacogenetics, 139–140, 203
     nicotine addiction treatment
          pharmacogenetics, 203–204
     opioid pharmacogenetics, 183–184
Dopamine transporters, 65, 132, 140–141, 202
Doral, 107
Driving, legal blood alcohol level limits, 19–20

Dronabinol, 155
Drugs of abuse (DOA), 93–97, *See also specific substances*
    commonly abused drugs list, 93–96
    defining, 93
    pharmacogenomics challenges, 1–3
    testing for, 97–100, *See also* Testing for drugs of abuse
Drug transporters, 4, 5
Drunken Monkey Hypothesis, 12
Dudley, Robert, 12
Duragesic, 109
DWI ("driving with impairment"), 19
Dynorphin, 143, *See also specific receptors*

E

Ecstasy, *See* MDMA
Efavirenz, 179
Egypt, ancient, 13, 165
EKG effects, 177
Eldepryl, 103, 105
Emagrece Sim, 105
Endocannabinoids and cannabinoid receptors, 159–162, 185
Endocrine system and alcohol abuse, 37
Endogenous alcohol production, 21
Endogenous opioids, 7, 182, 197, *See also* Opioid receptors
Endorphins, 7, 197
Enkephalins, 7
Ephedrine, 103, 106, 129
Epigenetics, 8, 52, 72–75
Erasistratus, 166
Erythromycin breath test (ERMBT), 158
Erythroxylum coca, 113, 137
eSensor® system, 220
Esophageal cancer, 38
Estazolam, 107
Estrogen levels, 37
Ethanol measurement, 54–55
Ethnicity
    alcohol dependency susceptibility, 51–52
    alcohol toxicity and, 32
    genetic variation, 1–2
Ethyl alcohol, as commonly abused drug, 93
Ethylglucuronide, 55–56
2-Ethylidene-1,5-dimethyl-3,3-diphenylpyrrolidine (EDDP), 111, 112, 176
Ethylmorphine, 172, 175
Extreme sports, 8

F

False positive drug tests, 99–100, 115
Fatty acid amide hydrolase, 162

Fatty acid ethyl esters (FAEEs), 55
Fatty acid metabolism, 33
Fatty liver, 31, 32, 58
Fentanyl, 96, 109, 110, 112–113, 167, 178
Fermentation, 12
Fetal alcohol spectrum disorder, 38–39, 77
Fetal alcohol syndrome, 38–39
Flunitrazepam, 96, 107, 109, 115, 179
Fluoride number, 145
Fluoxetine, 200–201
Flurazepam, 107, 108
Food consumption and alcohol metabolism, 15–16
Free-basing, 138
Freon replacements, 117
Freud, Sigmund, 137

G

GABA, 7, 66–67, 116, 183, 199
Gallstones, 38
Gamma-aminobutyric acid (GABA), 7, 66–67, 116, 183, 199
Gamma-butyrolactone (GBL), 115
Gamma-CDT, 59
Gamma globulins, 56
Gamma-hydroxybutyrate (GHB), 96, 102, 115, 116
Gas chromatography-mass spectroscopy (GCMS), 99, 108, 110, 114
    amphetamine confirmation, 105, 106
    cannabinoids, 157–158
Gastric cancer, 27, 38
GBL, 115
Gelsolin, 76
Gender differences
    alcohol consumption, 50
    alcohol metabolism and blood levels, 16, 21
    alcohol toxicity, 32, 33, 35
Genetic gap, 2, 7–8
Genetic screening for alcoholism susceptibility, 51, *See also* Alcohol dependence susceptibility, genetic markers of
Genome Sequencer FLX Titanium System, 225
Genome-wide association studies (GWAS), 3, 52
    addiction treatment medications, 195, 204
    alcohol dependency, 72–75
    cost-effective systems, 222
    limitations, 3, 8
Genomics, 72–75
Genotyping methods, 60
Georgia, 20
GHB (gamma-hydroxybutyrate), 96, 102, 115, 116–117
GIRK genes, 183
Glucocorticoid actions, 35, 37
Glucuronyl transferase, 4, 108, 171, 173, 174

Glue sniffing, 117
Glutamate, 68–69, 183, 185
Glutamate receptors, 27, 68, 185, 197
γ-Glutamyltransferase (GGT), 57–58
Glutaraldehyde, 119
"Golden Triangle" of opium production, 166
GRM6, 185, 197
GRM8, 185, 197

H

Hair testing, 55, 97
Halcion, 107
Hallucinogens, *See* Club drugs; *specific drugs*
Hazardous drinking, defined, 18–19, *See also*
	Alcohol abuse
Health effects of alcohol consumption, *See*
	Alcohol abuse, health hazards of;
	Alcohol consumption, health benefits
	of moderate drinking
Heart disease risk
	chronic/excessive alcohol consumption and,
		30, 36
	healthy alcohol consumption and, 22–24
	red wine effects, 25
Heart failure, 23, 36
Helicobacter pylori, 27
Hemopexin, 76
Hepatitis, 31, 32, 56
Herbal teas, 113
Herbathin, 105
Heritable variation, genetic gap issues, 7–8
Heroin, 93, 109, 110, 166–167
	metabolism, 5
	naltrexone therapy, 198
	pharmacogenetics of metabolism, 173–174
	sigma receptors and, 179
High-density lipoprotein (HDL) cholesterol, 24, 57
High-performance liquid chromatography
	(HPLC), 214
High-resolution melting curve analysis, 214–215
Hippocrates, 13, 166
Histidine-rich glycoprotein, 76
Histone modifications, 52, 76
Historical perspective of alcohol use, 13–14
HIV infection, 36–37
5-HTTLPR, 141–142, 201–202
Huffing, 117
Human carboxylesterase 1 (hCES1), 145
Hungary, 20
Hyaluronic acid, 56
Hybridization capture approaches, 223–224
Hydrocodone, 96, 109, 110, 111, 175
Hydromorphone, 96, 109, 110, 175
m-Hydroxybenzoylecgonine (m-HOBE), 144
5-Hydroxyindole-3-acetic acid (HIAA), 55
α-Hydroxylprazolam, 108

4-Hydroxymethamphetamine, 132
Hydroxynorephedrine, 106
5-Hydroxytryptophol, 55
Hypothalamus, 37

I

IgA, 56
IgG, 56
IgM, 56
Illumina next-generation sequencing system, 225
Immune system effects, 36–37
Immunoassays, 97–99
	barbiturates, 107
	benzodiazepines, 108
	club drugs, 116
	cocaine, 114
	confirmatory methods, 99–103
	marijuana, 114–115
	opiates and opioids, 110–113
	phencyclidene, 115
	specificity and cross-reactivity issues, 98, 99
India, 13
Infection risk and alcohol abuse, 36–37
Infiniti™ system, 219–220
Inhalants, 96, 103, 117–118
Inhibitory G-protein, 183
Insomnia, 51
Integrated fluidic circuit (IFC) chips, 223
Interference studies, 106, 110–111
Invader™ Cleavase™ reactions, 216–218
Invalid test results, 119–120
4-Iodo-2,5-dimethoxyamphetamine (DOI), 116
Isomer resolution analysis, 105–106
Italy, 19

J

Japan, 19

K

Kallikrein, 76
Kappa (κ) opioid receptors, 7, 70, 142–143, 179,
		See also Opioid receptors
Ketamine, 96, 103, 115, 117
Ketoconazole, 179
Khat, 129
Klonopin, 107
Korsakoff's syndrome, 34
Kupffer cells, 33
Kwashkiorkor, 56

L

Laboratory accreditation, 100
Legal alcohol limits for driving, 19–20

Legal E, 116
Legal X, 116
Levodopa, 202
Librium, 107
Life extension benefits of moderate alcohol
        consumption, 27–28
Life span reduction, alcohol abuse and, 29–30
Lipid markers, chronic alcohol use, 57
Liquid chromatography-tandem mass
        spectroscopy (LCMSMS), 99, 108,
        110, 114
Liver diseases, alcohol-related, 18, 21, 29, 31–33,
        38
    markers, 56, 59
Lofentanil, 112
Loperamide, 5
Lorazepam, 107, 108
LSD (lysergic acid diethylamide), 96, 102, 115,
        116
Lung cancer, 27, 37
Lysergic acid diethylamide (LSD), 96, 102, 115,
        116
Lysophosphatidic acid receptors, 161

M

Major depression, 71
Marijuana, 114–115, 155, 163, *See also*
        Cannabinoids
    absorption, 156–157
    cannabinoid receptors, 159–161
    immunoassays and confirmation methods,
        114–115
    mechanism of action, 159–161
    medical uses, 155
    metabolism and pharmacogenetics, 157–159
    passive inhalation, 114–115
    polymorphisms of cannabinoid receptors,
        161–162
    treatment possibilities, 162
Marinol, 155
MassArray® system, 222
Mast cells, 36–37
Matrix-assisted laser desorption/ionization time-
        of-flight mass spectrometry (MALDI-
        TOF MS), 222
MDA (methylenedioxyamphetamine), 104, 105,
        130, 131
MDMA (methylenedioxymethamphetamine,
        ecstasy), 93, 100, 101, 104–106, 115,
        129–131
Mean corpuscular volume (MCV), 57
Meconium, 55, 114, 144
Medical marijuana, 155
Melanocortin-1-receptor (*MC1R*) gene, 182
Memory loss, 27
Meperidine, 96, 109, 113, 177–178

Mescaline, 96
Mesopotamia, 13
Metabolic enzymes, *See specific enzymes*
Metabolism of drugs, *See* Pharmacokinetics;
        specific processes and substances
Metabolomics, 75–76
Methadone, 109, 111, 176–177
    ABCB1 polymorphisms and, 167
    addiction treatment (maintenance therapy),
        111, 176, 196–197
    immunoassays and confirmation methods,
        110, 112
    metabolism, 111, 176
    metabolism, pharmacogenetics, 4–5, 176–177
    pain management applications, 111–112
Methamphetamine, 93, 100, 103–106, 129, 131,
        See also Amphetamines and related
        compounds
    confirmation methods, 101
    heritability of use, 132–133
    isomer resolution analysis, 105–106
    pharmaceutical sources, 103–104
    pharmacogenetics of abuse, 132–135
    pharmacokinetics, 132, 202
    screening and confirmation methods,
        105–106
    sigma receptors and, 179
    toxicity, 131–132
5-Methoxy-*N,N*-α-methyltryptamine (5-MeO-
        AMT), 115–116
Methylenedioxyamphetamine (MDA), 104, 105,
        130
Methylenedioxymethamphetamine, *See* MDMA
α-Methylfentanyl, 113
3-Methylfenyl, 113
Methylphenidate, 96, 129
1-Methyl-4-phenyl-4-propionoxypiperidine
        (MPPP), 113
N,N-α-Methyltryptamine (AMT), 115
m-hydroxybenzoylecgonine (m-HOBE), 144
Microarray sequencing systems, 71, 219–221, *See
        also* Genome-wide association studies
MicroRNAs, 8, 77, 155
Microsomal alcohol oxygenase, 158
Microsomal ethanol oxidizing system (MEOS),
        16
Midazolam, 107
Middle Eastern countries, 20, 165
Moderate alcohol consumption, defined, 18
Molecular-beacon probes, 219
Molecular genetics testing methodologies, *See*
        DNA sequencing; Pharmacogenetics
        testing methodologies
Molecular profiling technologies, 52, 71–78
    epigenetics, 72–75
    genomics, 72–75
    metabolomics, 75–76

proteomics, 76
transcriptomics, 71–72
6-Monoacetylmorphine, 5
Monoamine oxidase (MAO), 66, 100
Monoaminergic signalling, 6–7, *See also specific*
        neurotransmitters
Morphine, 96, 109, 110, 166, 167
    endogenous, 174
    pharmacogenetics of metabolism, 173–174
Mortality and alcohol use, 18, 27–28, 29–30
Mu (μ) opioid receptors, 7, 70, 142–143, 175,
        179, 180, 197–198, *See also* Opioid
        receptors
Multidrug resistance protein (mdr), 167
Muscarinic receptors, 67
Myocardial infarction, 23, 113–114

N

Naloxone, 9, 110, 183
Naltrexone, 70, 71, 182, 198–199
Narcan, 110
Narcolepsy treatment, 129
National Institute on Drug Abuse (NIDA),
        commonly abused drugs list, 93–96
National Laboratory Certification Program
        (NLCP), 100
Netherlands, 19
Neurodegenerative disorders, alcohol
        consumption effects, 26–27, 35
Neuropeptide substance P, 183
Neuropeptide Y, 69–70
Neurotoxic effects of alcohol, 33–36
Neurotransmitters, 139, *See also specific*
        neurotransmitters
    alcohol effects, 63–64
    opioid pharmacogenetics, 183–184
    stimulant effects, 100
Next-generation sequencing, 71, 76, 77, 224–227
Nicotinamide adenine dinucleotide (NAD), 16
Nicotine, 93
    CYP2A6 and metabolism, 4
    treatment pharmacogenetics, 203–205
Nicotinic acetylcholine receptors, 67–68
NimbleGen technology, 224
Nitrates, 115, 117
NMDA (*N*-methyl-aspartate) receptors, 27, 35,
        68–69, 183
Noncoding RNAs (ncRNAs), 8, 72, 77
Norcocaine, 144
Norcodeine, 111, 174
Nordiazepam, 108
Norephedrine, 106, 132
Norepinephrine, 100, 138, 139, 183
Norepinephrine transporter, 132, 142
Normeperidine, 177–178

Norpseudoephedrine, 106
Nutritional status markers, relevance for chronic
        alcohol abuse, 56

O

Ondasetron, 202
Opana, 109
Opiate addiction treatment
    buprenorphine, 167, 197
    methadone maintenance, 111, 176, 196–197
    opioid antagonist, *See* Naltrexone
Opiates, 109–113, 165, *See also* Opioids; *specific*
        types
    addictive properties, 166
    analgesic properties, 109
    confirmatory methods, 101–102
    definition and structure, 166
    history of use, 165–166
    immunoassay specificity issues, 99
    interference studies, 110–111
    pharmacogenetics, *See* Opioids,
            pharmacokinetics and
            pharmacogenetics
    testing for, 110–113
    toxicity, 110
Opioid receptors, 7, 109, *See also specific*
        receptors
    alcohol dependence susceptibility markers, 70
    buprenorphine activity, 179
    cocaine pharmacogenetics, 142–143
    naloxone and antagonism, 9, 110, 183
    nociceptive pathways and, 7, 183
    pharmacodynamic effects, 180–183
    pharmacogenomics of addiction treatment,
            195
    types (table), 180
Opioids, 96, 109, *See also* Opiates; *specific types*
    analgesic properties, 7, 109, 180
    classification (table), 168–169
    definition and structure, 166–167
    dopamine receptors and addiction, 139–140,
            183–184
    metabolism, 167
    nociception, 7, 183
    opiate assays and, 110
    pharmacodynamics, 179–185, *See also*
            Opioid receptors
        neurotransmitter systems, 183–185
        opioid agonists and antagonists, 183
        opioid receptors, 179–182
Opioids, pharmacokinetics and
            pharmacogenetics, 167–179,
            See also Addiction treatment,
            pharmacogenomics aspects; Opioid
            receptors

ABCB1 single nucleotide polymorphisms, 167
metabolic enzymes, 171–173, 196–197, *See also* CYP2D6; CYP3A4
specific opioids, 173–179
urine detectability (table), 170–171
Opium, 93, 166
OPRD1, 7, 142, 182
OPRK1, 7, 70, 142, 143, 182
OPRM1, 7, 9, 70, 71, 142–143, 180, 182, 197, 198–199
Osmolar gap, 54
Oxazepam, 107, 108
Oxycodone, 96, 109, 110, 111, 175–176
Oxycontin, 109
Oxymorphone, 109, 111

P

Pacific Islanders, 51–52
Pain management contexts for drug testing, 97–98
Pain treatment, opiates and opioids, 7, 109, 111–112
Pancreatic cancer, 38
Papaver somniferum, 109, 165
Parkinson's disease, 65, 113, 161–162
Paroxetine, 201
PCP (phencyclidene), 96, 102, 115, 179
Pemoline, 129
Pentazocine, 109, 179
Pentobarbital, 106, 107
Pentozifylline, 201
Percodan, 109
Personal Genome Machine, 226
Personality disorder, 51
P-glycoprotein (P-gp), 4, 5, 167
Pharmacodynamics, 1, 6–7
relevant genes (table), 6
Pharmacogenetics of drugs of abuse, *See specific substances*
Pharmacogenetics testing methodologies, 211–212, 225–226
FDA approved assays (table), 212
high-multiplexing systems, 222–224
high-plex SNP assays, 222
library preparation and PCR systems, 222–224
low-multiplexing assays, 212–219
high-resolution melting curve analysis, 214–215
Invader™ Cleavase™ reactions, 216–218
molecular beacon probes, 219
polymerase chain reaction, 212
Sanger sequencing, 215
screening methods, 212, 214
Southern blot, 212
Taqman assay, 218–219

mid-multiplexing assays, 219–222
next-generation sequencing, 71, 76, 77, 224–227
pyrosequencing, 215–216, 225–226
third-generation sequencing, 226
Pharmacogenomics, 1, 211
aspects of addiction treatment, 195–206, *See also* Addiction treatment, pharmacogenomics aspects
drugs of abuse challenges, 1–3, 9
genetic gap issues, 7–8
Pharmacokinetics, 1, 3–5, 211–212, *See also specific substances*
alcohol metabolism, 15–17
metabolic enzymes, 4–5, *See also specific enzymes*
monoaminergic signalling, 6–7, *See also* Dopamine; Serotonin; other relevant neurotransmitters
relevant genes (table), 4
Phencyclidine (PCP), 96, 102, 115, 179
Phenmetrazine, 103
Phenobarbital, 106–107
Phentermine, 103, 106
Phenylephrine, 103, 106
Phenylpropanolamine, 106
pH measurement, urine specimen validity testing, 119
Phosphatidylethanol (PEth), 57
Plasma kallikrein, 76
Polymerase chain reaction (PCR), 212
DNA fragment libraries, 222–223
high-resolution melting curve analysis, 214–215
high-throughput enrichment approach, 223–224
multiplex approaches, 219–221
next-generation sequencing methods, 224–227
Polyphenolic compounds, 25
Poppy seeds, 110
Positron emission tomography (PET), 203
Postmortem alcohol levels, 21–22
Prazepam, 107
Pregnane X receptor (PXR) ligands, 171
Prescription drugs of abuse, 96, *See also specific drugs*
Primate diets, 12
Prodynorphin, 70
Proenkephalin, 182
Proof, 14–15
Propane, 117
Propiomelanocortin (POMC), 182–183
Propofol, 113
Propoxyphene, 96, 109, 110
Propylhexedrine, 103, 106

ProSom, 107
Protein kinase A (PKA), 160
Proteomics, 76
Pseudocholinesterase, 5, 145
Pseudo-Cushing's syndrome, 37
Pseudoephedrine, 103
Psilocin, 115
Psilocybin, 96, 115
Psychiatric disorders, 65
Pure food and Drug Act, 14
Pyridinium chlorochromate, 119
Pyrosequencing, 215–216, 225–226

Q

Quazepam, 107

R

RainStorm™ system, 223
Real-time polymerase chain reaction, 214–215
Red wine, 15, 21, 25, 27, 29
Remifentanyl, 113, 178
Renal cell carcinoma, 27
Reproductive system effects, 37
Restoril, 107
Restriction fragment length polymorphisms
        (RFLPs), 140, 212
Resveratrol, 25
Rhabdomyolysis, 131, 139
Rheumatoid arthritis, 28
Rimonabant, 162
RNA sequencing, 72
Rohypnol, 107, 109
Rolling circle PCR amplification, 223, 226
Romania, 20
Romans, 13

S

Salvia divinorum, 96
Salvinorin A, 166
Sanger sequencing, 215
Sativex, 155
Schizophrenia, 65, 66, 71, 161, 184
Secobarbital, 106, 107
Selective serotonin reuptake inhibitors (SSIs),
        200–202, See also Serotonin
Selector technology, 223
Selegiline, 105
Selenoprotein P, 76
Serax, 107
Serotonin, 6, 139
    addiction associations, 184
    alcohol effects, 31, 55, 69
    cocaine effects and, 138

receptors, 184
reuptake inhibitors, 183, 200–201
Serotonin transporter, 69, 132, 141–142, 179, 184,
        201–202
Sertraline, 201, 202
Serum gamma globulins, 56
Sexual function and alcohol abuse, 37
Sialic acid, 60
Sigma receptors, 179
Single-base primer extension method, 215
Single nucleotide polymorphisms (SNPs), See
        also specific genes
    alcohol consumption associated gene
            identification, 67–70
    cocaine pharmacogenomics, 141
    DNA melting profile, 214
    genotyping methods, 60
    high-plex assays, 222
    limitations of genome-wide association
            studies, 3
    marijuana pharmacogenomics, 158, 162
    nicotine treatment pharmacogenomics,
            203–204
    opioid pharmacogenomics, 167
    SNapShot™ approach, 215
    SSRI pharmacokinetics, 201–202
Single stranded conformation polymorphisms
        (SSCPs), 214
Slate markers, 48–49, See also Alcohol abuse,
        biochemical markers of chronic use
SLC6A2, 142
SLC6A4, 141–142, 179, 184, 201–202
SLCA3, 65
Small RNAs, 52
SMART analysis, 76
Smoking and heart disease, 22
SMRT™ third-generation sequencing technology,
        226
SNapShot™ system, 215
Snorting inhalants, 117
SOLID™ next-generation sequencing system,
        224–225
Southern blot, 212
Specific gravity, 118–120
Specimen validity testing for drug testing,
        118–120
Spiking, 112
SR141716A, 162
Standard drink of alcohol, 21
Steatohepatitis, 56
Steatosis, 31
Stimulants, See Amphetamines and related
        compounds; Cocaine
Stroke risk and alcohol consumption, 25, 30, 36
Sublimaze, 109
Succinylcholine, 145

Sudden sniffing death syndrome, 118
Sufentanil, 112
Suicide, 30
Sulfentanyl, 178
Sumerians, 165
SureSelect technology, 224
Sweden, 20
Switzerland, 19
Sympathetic nervous system, 100
Sympathomimetic amines, 100–106,
          See also Amphetamines and
          related compounds; MDMA;
          Methamphetamine

T

Talwin, 109
TaqIA, *See also* Dopamine receptors
Taqman assay, 218–219
Temazepam, 107, 108
Temperature gradient gel electrophoresis, 214
Testing for drugs of abuse, 97–100, *See also*
          Immunoassays; Workplace drug
          testing; *specific substances*
    confirmatory methods, 99–103
    false positives and negatives, 99–100, 115
    isomer resolution analysis, 105–106
    laboratory accreditation, 100
    pain management contexts, 97–98
    specimen validity testing, 118–120
    spiking, 112
    windows of detection for urine, 98
Testosterone, 31, 37
Δ9-Tetrahydrocannabinol (THC), 114–115, 155,
          See also Cannabinoids
    absorption and bioavailability, 156–157
    cannabinoid receptors, 159–162
    metabolism, 157–158
    pharmacogenetics, 158–159
Tetranectin, 76
Thailand, 19
Thebaine, 166
Thiamine deficiency, 33, 34
Thiopental, 106
Thioridazine, 115
Third-generation sequencing, 226
Tiagabine, 201
Tobacco, nicotine pharmacogenetics, *See*
          Nicotine
Tobacco use, alcohol abuse correlation, 50
Tolerance effects
    amphetamines, 129
    benzodiazepines, 108
    cannabinoids, 155
    cocaine, 138
Toluene, 117

Topiramate, 67
Traffic accidents, alcohol related, 48–49
Trait markers, 48, 49, *See also* Alcohol
          dependence susceptibility, genetic
          markers of
Tramadol, 109, 178–179
Transcriptomics, 71–72
Transferrin, 58–59, 76
Treatment for addiction, *See* Addiction treatment,
          pharmacogenomics aspects
Triazolam, 107, 108
Triglycerides, 57
TRPV1, 183
Turkey, 19
Type 2 diabetes, 25–26

U

UGT1A enzymes, 179
UGT2B7, 171, 173, 174
Ultram, 109
Uridine diphosphate glucuronyl transferases
          (UGTs), 4, 106
Urine testing, 97
    methadone, 111
    opioid detectability (table), 170–171
    specimen validity testing, 118–120
    windows of detection, 98

V

Valium, 107
Vanilloid receptors, 161
VeraCode system, 222
Verigene® system, 220–221
Versed, 107
Vicodin, 109
Violent behavior, alcohol abuse and, 30–31
Vitamin D, 37
Vitamin E, 25
Vitronectin, 76
Volatile organic compounds (VOCs), 117, *See*
          also Inhalants

W

Wernicke-Korsakoff syndrome, 34
Whiskey tax, 14
Whole-genome linkage studies, 72, *See also*
          Genome-wide association studies
Widmark formula, 20–21
Windows of detection, drugs in urine, 98
Wine
    health benefits, 25, 27, 28–29
    historical perspective, 13
    red wine, 15, 21, 25, 27, 29

Withdrawal effects
  alcohol, 35, 51, 65, 68–69, 70, 162
  amphetamines, 131
  benzodiazepines, 108
  MDMA, 131
  methadone and opiates, 111, 176, 184
Women's reproductive system effects, 37
Workplace drug testing, 97, *See also* Testing for
        drugs of abuse
  amphetamines, 105
  cutoff concentrations, 98, 99
  laboratory accreditation, 100

X

Xanax, 107
xTAG® assay, 221

Y

Yeast, 12, 13

Z

Zip-code, 219
Zolpidem, 115